Cereal-Based Food Products

Manzoor Ahmad Shah
Kappat Valiyapeediyekkal Sunooj
Shabir Ahmad Mir
Editors

Cereal-Based Food Products

Springer

Editors
Manzoor Ahmad Shah
Department of Food Science and Technology
Government Degree College for Women
Anantnag, Jammu & Kashmir, India

Kappat Valiyapeediyekkal Sunooj
Department of Food Science and Technology
Pondicherry University
Puducherry, India

Shabir Ahmad Mir
Department of Food Science and Technology
Government College For Women
Srinagar, Jammu & Kashmir, India

ISBN 978-3-031-40307-1 ISBN 978-3-031-40308-8 (eBook)
https://doi.org/10.1007/978-3-031-40308-8

This Springer imprint is published by the registered company Springer Nature Switzerland AG
The registered company address is: Gewerbestrasse 11, 6330 Cham, Switzerland

Paper in this product is recyclable.

Preface

Cereal grains and their products are staples in the diet of almost every culture of the world and have made an important contribution to daily nutrient requirements. Cereal grains are high in carbohydrates, are good sources of protein, and provide varying amounts of fiber, vitamins, and minerals. The nutritional composition of grains may vary depending on the variety and environmental growing conditions. A number of cereal products are prepared from all the regions of the world. Cereals are processed into many products such as bread, cookies, cakes, pasta, etc., which are consumed daily by the majority of the population and play an important role in human nutrition.

This book is the first of its kind, focusing on the preparation methodology and quality evaluation of cereal products. The unique features of this book will be the focus on different types of cereal products including bread, biscuit, cakes, tortillas, pasta, noodles, and popped products. The contributing authors addressed the cereal-based functional products, 3D printing, and packaging of cereal-based products. This book is suitable for students, scientists, and professionals in food science, food processing, product development, food marketing, nutrition, and other health sciences. This book will be typical in its form, and we hope that it will represent an example for future works in this and related areas. We aimed to summarize and review the outcomes and knowledge gained so far in the said area. We hope that readers will find this book a useful resource for their research or studies and that it will help develop cereal-based products.

Anantnag, Jammu & Kashmir, India — Manzoor Ahmad Shah
Puducherry, India — Kappat Valiyapeediyekkal Sunooj
Srinagar, Jammu & Kashmir, India — Shabir Ahmad Mir

Contents

Chapter 1
Introduction to Cereals

Elismara das Graças Costa and Poliana Mendes de Souza

1 Introduction

Food, health and disease are interconnected, currently the search for functional foods is growing and this concept encompasses foods or ingredients that in addition to providing nutrients to the body, provide additional health benefits. In recent years, cereals have entered the group of foods with potential for the development of functional products, as they have different characteristics. Consumers have given great importance to a balanced diet, where macro and micronutrients are present in a balanced way (Masisi et al., 2016).

There are several scientific studies that demonstrate the importance of consuming whole grains, and associate the intake of cereals with the reduction of chronic diseases such as cancer, cardiovascular diseases, obesity and diabetes (Mir et al., 2016). Grains are a source of starch, dietary fiber, nutritious proteins and lipids that contain essential fatty acids, these foods are included by the nutritional guidelines at the base of the nutritional food pyramids. Cereals such as wheat, barley, corn, rice, the pseudocereal amaranth are a source of bioactive peptides, which contribute to increasing the value of the protein quality of the grains. (Malaguti et al., 2014).

E. das Graças Costa
Universidade Federal dos Vales do Jequitinhonha e Mucuri. Instituto de Ciência e Tecnologia, Curso de Engenharia de Alimentos, Diamantina, Minas Gerais, Brazil

P. M. de Souza (✉)
Universidade Federal dos Vales do Jequitinhonha e Mucuri. Instituto de Ciência e Tecnologia, Curso de Engenharia de Alimentos, Diamantina, Minas Gerais, Brazil

Universidade Federal dos Vales do Jequitinhonha e Mucuri. Instituto de Ciência e Tecnologia, Diamantina, Minas Gerais, Brazil
e-mail: poliana.souza@ict.ufvjm.edu.br

M. A. Shah et al. (eds.), *Cereal-Based Food Products*,
https://doi.org/10.1007/978-3-031-40308-8_1

In addition to the nutritional importance, cereals have economic importance, as there are numerous applications that grains within the industry, generating various products, such as bread (wheat), cookies (rice), cornmeal (corn), beer (barley), bioethanol (sorghum), skin care products (oats), alcoholic beverages (rye), among others. (Almeida, 2019; Farinhas, 2022). Knowing the importance of cereals, the main objective of this work was to make a bibliographic review of the main cereals consumed in the world and some products and by-products obtained from them.

Cereals are defined as the seeds or edible grains of grasses, such as rice, wheat, rye, oats, and corn. Cereal grain presents great structural similarity, being composed of three tissues: husk, endosperm and germ, despite the parity, they can be distinguished in relation to chemical composition and nutritional value. In grains, compounds such as carbohydrates, water, protein, lipids, fiber, B vitamins, vitamin E, iron and minerals are found, providing 10–20 times more energy than most fruits and vegetables. (Mir et al., 2015; Takeiti, 2022) (Fig. 1.1).

Many studies have been investigating the benefits of consuming whole grains, that is, those that have not undergone refinement within the industry, as nutrients are lost in this process. Fibers are highlighted on this subject, studies reveal that the consumption of fibers present in cereals reduces cholesterol; blood pressure control; decreased incidence of cancer; greater control of diabetes, among others.

Cereals are considered the most important food group in the world, when compared to other food crops produced around the globe, being consumed all over the planet and providing about 50% of the protein and energy for the human diet. There are numerous cereal-based foods, such as bread, especially those made with wheat and rye; can be used in the production of alcoholic beverages such as beers, whiskey, vodka, Japanese sake, among others. Some cereals are also used in the pharmaceutical, health and beauty industries (Takeiti, 2022).

Fig. 1.1 Cereals. (Source: Canva, 2022)

Wheat

Wheat (*Triticum aestivum* L.) is a cereal belonging to the Poaceae family, considered one of the main food crops in the world. Currently, wheat has a prominent economic and nutritional role, and wheat flour is used for various purposes within the industry (Rustgi et al., 2019).

Its cultivation would have started in antiquity, the domestication and use of wheat is a result of human efforts to protect themselves from hunger, in the beginning, consumption was only of grains, which were transformed into porridge or mixed with fish and fruits (Fig. 1.2). It was the Egyptians, around 4000 BC, who discovered the fermentation of wheat and with that, the production of bread began. The grain became popular around the world, the Chinese around 2000 BC used it to make flour, noodles and pastries. It is currently grown worldwide, providing about a quarter of the annual global demand for vegetable proteins, carbohydrates and dietary fiber, being the main ingredient in the manufacture of breads, cakes, cookies, among others (Rustgi et al., 2019).

A quality wheat grain is the result of several factors, including the interaction of growing conditions (soil and climate), incidence of (pests and diseases), harvesting, drying and storage operations. The wheat grain (Fig. 1.3) is a single seeded grain, where the seed and the fruit cannot be separated, the shape is normally oval and composed of several different tissues: the germ, endosperm, which is packed with starch grains to provide energy for germination; the cell wall has a thick layer of aleurone, which surrounds the endosperm; and the pericarp. Aleurone contains B vitamins and minerals, the endosperm is composed of approx.

The wheat grain is composed by 88% of carbohydrates, formed basically by starch, the germ is one of the richest sources of vitamins B and E, proteins, lipids and minerals are also present in this structure. The whole grain and its derivatives are rich in fermentable carbohydrates (dietary fibers), resistant starch and

Fig. 1.2 Wheat and wheat flour. (Source: Canva, 2022)

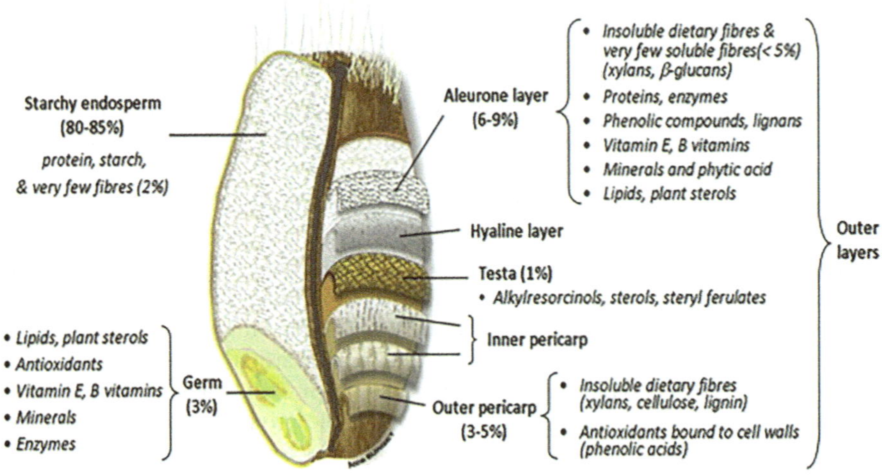

Fig. 1.3 Wheat grain. (Source: Brouns et al., 2013)

oligosaccharides, compounds that bring benefits to the intestinal microbial (Gwirtz et al., 2014).

Wheat grains can be classified according to the degree of hardness (Scheuer et al., 2011):

- Soft wheat: low gluten content, the dough obtained from this wheat is characterized by being weak, with low water absorption, the flour is very fine, normally used in the production of cookies, crackers, cakes and pies.
- Durum and durum wheat (hard wheat): high in protein and high in water absorption, it is ideal for the production of bread and fermented products. Durum wheat is used in the manufacture of pasta, spaghetti and other pasta, it is ideal for the production of these foods due to its unique color and quality for cooking.

To obtain the flour, it is necessary that the wheat goes through the milling process, at this stage the endosperm is reduced to semolina, then the bran is separated from the germ, these procedures are important because they allow the elaboration of delicious products (Atwell & Finnie, 2016). Milling and the choice of equipment involved in the process influence the industrial quality of the flour, which can be classified according to the degree of endosperm extraction and the ash content, into whole, white, common and special flour (Costa et al., 2008).

The main protein found in wheat is gluten, this substance is elastic, insoluble in water and is the result of the complex mixture between hundreds of proteins, mainly gliadin and glutenin. Gluten directly influences the quality and structure of food dough. There is a group of people who have the so-called celiac disease, a permanent intolerance to gluten, the only treatment for this pathology is the exclusion of protein from the diet (Biesiekierski, 2017).

Wheat flour is the main raw material for baking, its rheological properties, such as viscosity, plasticity, elasticity and extensibility, influence the quality of the product, so knowledge about them is essential within the industry (Gutkoski et al., 2008).

The carbohydrates present in the wheat grain influence the texture, and when subjected to heat treatment, they undergo gelatinization and retrogradation. Gelatinization is a process that causes changes in viscosity, absorption and solubility, there is formation of a paste, the result of this process is softness and moisture in the manufactured food. Retrogradation is the rearrangement of gelatinized starch chains after cooling the paste that compromises the desirable technological characteristics that were obtained through gelatinization, it is undesirable in breads. Starch is the carbohydrate that will be broken down by enzymes during fermentation, so the greater the amount of this substance, the greater the amount of sugars for fermentation, resulting in greater gas production (Farinhas, 2014).

Gluten is related to the extensibility, elasticity and strength of the dough, in addition, the characteristics of glutenin allow the expansion of the dough, allowing the entrapment of carbon dioxide bubbles generated in the fermentation (Farinhas, 2022).

Use of Wheat

Wheat flour is the main ingredient in the preparation of breads, cakes, cookies, savory pies, sweet pies, pasta and, in addition, it can be used to bread vegetables and meats. Bread is the main product of the culinary use of flour, the type of flour influences the final properties of the product, French bread is the most consumed in Brazil and white flour is used for its production (Brasil, 2014).

The separation of wheat flour into starch and gluten, allows the starch to be used by the confectionery, syrups and sweeteners industry. Gluten has been used as an additive in bread formulations, a fortifier in breakfast cereals and pet foods, and as an extender in meat, fish and poultry products. Starch has been used for the production of biodegradable adhesives, paper coatings and packaging (Takeiti & Trigo, 2015).

Currently, there are procedures being done to modify wheat flour for a specific use. The treatment with chlorine gas is a modification process, which seeks to obtain a flour that will be used in the manufacture of cakes with high amounts of sugar and liquids (high ratio cake, type of sponge cake), this procedure improves the properties of breadmaking and sensory characteristics, this type of flour is widely used in the United States of America (Ferrari et al., 2022).

Heat treatment has been used to selectively modify the qualitative properties of wheat (or other) flour. The flour that goes through this process can be called "inactivated flour", the treatment can be severe, leading to the denaturation of all the protein or gently modifying the properties. The processed flour has a wide application, being used in breading, batter, toppings, soups, sauces, baby food, biscuits, wafer and cake flours, fresh doughs, frozen products, "binders" (e.g. for sausages and cereal bars), "vehicle" for pre-mixes, tablets, among others (Farinhas, 2022).

Rice

Rice (*Oryza sativa* L.) is one of the most important cereals in the world, being a source of protein in several regions of the world. Rice consumption is ancient and would have started in Southeast Asia, currently this continent is responsible for a consumption of up to 90% of all world production (Mir & Bosco, 2014). The cultivation of this cereal requires a place with plenty of water, so the ambient temperature will remain within suitable ranges and, in traditional systems. The main form of consumption is in grain and a small portion is consumed from ingredients in processed products (Farinhas, 2022; Walter et al., 2008).

The rice grain (Fig. 1.4) is a caryopsis, being formed by husk, germ skin and endosperm. This cereal is rich in proteins, mineral salts (phosphate, iron and calcium) and B vitamins. Another compound present in the grain is starch, which is formed by amylose and amylopectin, the amylose content is one of the most important parameters for determine the quality and agglutination capacity of the grains (Bassinello & Castro, 2004).

During processing, the rice goes through the polishing process, this step causes a reduction in the nutrients present in the outer layers of the grain, other factors such as the conditions involved in the cultivation (temperature, humidity, fertilization) and the forms of preparation of the cereal for consumption also influences the nutritional value of rice (Farinhas, 2022).

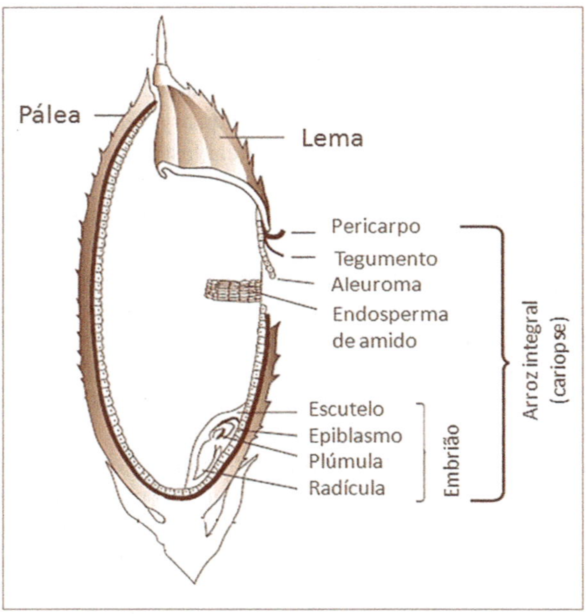

Fig. 1.4 Rice grain. (Source: Mclean et al., 2002)

Use of Rice

There are different types of rice consumed in the world, it is the cereal that has the largest number of varieties, with the commercialization of more than 1000 different types and this large number is the result of climatological differences, nature of soils, regional culture.

In Brazil, the Ministry of Agriculture, Livestock and Supply (MAPA), through Ordinance 06/2009, classifies rice as shown in Table 1.1 below:

Rice provides various types of pre-gelatinized flour, which are obtained by cooking broken rice, either by extrusion, steam or pressure. First the rice is cooked, then dried and ground. The flour obtained is fine and silky, with an amount of approximately 6 and 7% of protein, and does not contain gluten, which makes it a great choice for people with celiac disease. In the United States and Asian countries, rice flour is used in the production of breads, pasta, sauces, extruded breakfast products, cereal bars, as it is easy to digest, it is recommended for the production of foods for children, the elderly and people with disabilities (Mir et al., 2020).

Rice bran has been used as an additive in animal feed, as fertilizer, or discarded as unusable waste. It can be used in the human diet as a low-cost source of nutrients. It has fibers that retain water and oil, this property makes it an option for use in the development of industrialized products, such as in the production of laxatives by the pharmaceutical industry, as a depollutant in the adsorption of heavy metals from water for industrial purposes or domestic consumption (Bassinello & Castro, 2004).

The waste generated by the rice husk has a slow biodegradation, if not disposed of correctly, they end up causing great environmental impacts, due to this, alternatives have emerged for the use of the husk, which can be used in the manufacture of various products, such as adhesives; toxic material absorbents; animal food; fertilizers for vegetable gardens and orchards; metal polishing; building material in the manufacture of bricks, among others (Nunes et al., 2017).

Rice bran oil is not popularly consumed in the world, it is considered healthy and has its highest consumption rate in Asian countries, the benefits associated with this

Table 1.1 Rice classification

Rice	Grains from the species *Oryza sativa L.*
Processed rice	Mature product that has undergone a processing process and is deprived, at least, of its shell.
Peeled or brown rice	Product from which only the skin was removed.
Parboiled rice	Rice that undergoes a hydrothermal process (parboiling) in which the paddy rice is immersed in drinking water at a temperature above 58 °C, followed by total or partial gelatinization of the starch and subsequent drying).
Polished rice	Product that, during processing, has the germ, the pericarp and most of the inner layer removed.
Sailor	Grain that preserves the husk after processing.
Quirera	Fragment of rice that passes through circular holes of 1.6 mm in diameter.

Source: MAPA, 2009

oil include a reduction in bad cholesterol or low density protein (LDL) and tends to increase the good cholesterol level, or high density protein (HDL), this is due to the substance oryzanol (Bassinello & Castro, 2004).

The broken grains and broken grains can be used in animal rations, as adjuvants in breweries, during the fermentation stage. Broken rice is a source of starch and has a low glycemic index, which makes it an option for balancing meals, it can be used in the production of frozen foods, breakfast cereals and for the use of fats in snacks (Bassinello & Castro, 2004).

Corn

Corn is part of the three most produced cereals in the world, being cultivated on almost all continents, this cereal has great economic importance due to several applications within the industry, its use ranges from human and animal food; production of by-products by large industries in different areas, such as: chemical, pharmaceutical, beverage and fuel; production of biodegradable films and packaging. About 70% of world corn production is used for animal feed, in developed countries this percentage may be higher. For human consumption, only about 15% of world production has this purpose, either directly or indirectly (Paes, 2006; Regitano-Darce et al., 2015).

According to the characteristics of the grain (Fig. 1.5) corn can be classified into five types: toothed, hard, floury, popcorn and sweet. Colors vary from black to red, but the most common color of the grain is yellow, this cereal is formed by four structural tissues: endosperm, germ, pericarp (husk) and tip. In Brazil, the predominant production is the hard type, and in temperate countries, the toothed type predominates. The types of corn differ in the shape and size of the grains, defined by the structure of the endosperm and the size of the germ (PAES, 2006).

Corn (Fig. 1.6) is a source of energy, being composed of approximately 72% of starch, 9.5% of proteins, 9% of fiber and 4% of oil, despite this composition, the protein quality is lower than that of rice and of wheat, due to the limitation in essential amino acids, especially in lysine and tryptophan (PAES, 2006).

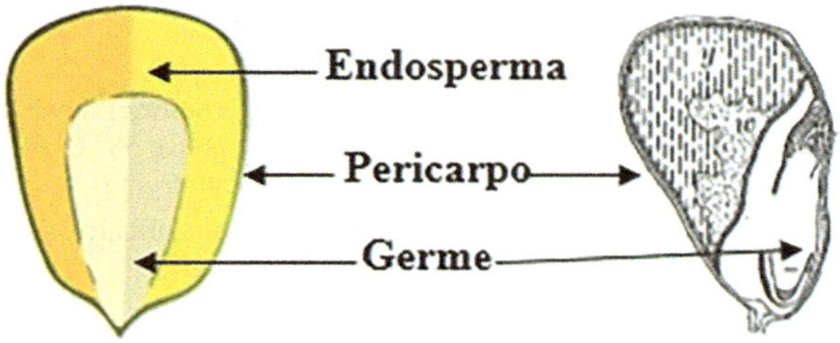

Fig. 1.5 Corn grain. (Source: Adapted from Canva, 2022)

Fig. 1.6 Corn. (Source: Canva, 2022)

The oil found in corn, but specifically extracted from the germ, which has a relatively large size, is considered of great importance for human nutrition, as it prevents cardiovascular diseases and fights high serum cholesterol. Corn is also a source of insoluble fiber (hemicellulose, cellulose and lignin), this component helps control cholesterol, improves constipation, among others. The predominant mineral in the grain is phosphorus, present in the form of potassium and magnesium phytates, other minerals are also present, but in smaller amounts such as chlorine, calcium, sodium, iron, zinc, among others (Paes, 2006).

Use of Corn

The dry grains are transformed into several products within the industry, after harvesting the corn goes through some steps in order to obtain derivatives that are obtained through two different processes: dry milling and wet milling (Regitano-Darce et al., 2015; Strazzi, 2015).

Dry milling: this method seeks to separate the germ, pericarp and endosperm, a physical break is made in the grain, the germ is removed and the oil is obtained from it, the remaining endosperm, called hominy, can be used as a product or sent for additional processes, providing new products. The result of the milling are products such as flour (bread), cornmeal (pasta, bakery and cooking), bran, extruded, hominy, grits and refined oil. In Table 1.2 we can see the products obtained with dry grinding and the applications (Regitano-Darce et al., 2015; Strazzi, 2015).

Wet milling: the corn goes through a process called maceration, this step is carried out in large tanks with water heated between 50 and 54 °C and a quantity of sulfurous acid solution for a period of 36–48 h, which results in the separation of starch and proteins. Soluble components are extracted from the grain; the water is

Table 1.2 Dry grinding: products and applications

Products	Main applications
Hominy	Sweet popcorn, corn flakes, biju flour, alcoholic beverages
Canjiquinhas	Snacks (snacks)
Screams	Breweries, snacks, mining, extruded to replace Styrofoam
Flours and cornmeal	Baby food, chlorine, bakery, prepared mixes for cake, biscuits and pasta, weaving, explosives, animal feed industries
Pregelatinized corn flour	Instant food, fertilizer, oil well drilling, smelting, feed industry.
Corn flakes and pre-cooked cornmeal	Soups, cookies, feed industries
Corn bran and defatted germ	Animal food

Source: Brazilian Association of Corn Industries (Abimilho) (Strazzi, 2015)

Table 1.3 Wet grinding – products and applications

Products	Main applications
Maize starch	Powders for desserts, bakery, prepared mixes for cakes and others, baking powder, infant foods, meat products, mustards, soups, pasta, pharmaceuticals, fermentation processes, paper, corrugated board, weaving, mining, explosives, adhesives, chalk, cosmetics, cleaning products, industrial water recovery
Modified starches	Gummies, beverages, instant foods, baby foods, frozen pre-cooked foods, flavor and coloring vehicles, sauces, paper, corrugated cardboard, gummed tape, weaving, oil well drilling
Dextrin	Adhesives, sandpaper, abrasive papers, fabric printing, multiwall paper bags, cardboard, mining
Glucose and maltose syrups	Hard candies, chewable candies, chewing gum, candies, fruit cream and juice, jellies and jams, toppings, meat and sausage products, prepared mixes (cakes, baby food, pudding powders), syrups, ice cream, beers, soft drinks, sauces, alcoholic beverages, breakfast foods, bakery, sauces, pharmaceuticals
Corn liquor	Antibiotics, pharmaceuticals, enzymes, fermentation products, animal feed
Germ, gluten and fiber	Feed formulation.

Source: Brazilian Association of Corn Industries (Abimilho) (Strazzi, 2015)

drained and concentrated in evaporators, the concentrated liquor is obtained. Subsequently, the corn kernels are coarsely milled and hydrocyclones separate the germ from the corn, the resulting material is fibrous and separated through sieving. Several products can be obtained from this method, such as starch (modified starch), dextrins, glucose syrup, refined oil, fiber and gluten. In Table 1.3 we have more products and applications (Strazzi, 2015).

If the corn does not pass through any of the mills, it is destined for the segment that uses corn in natura (green corn), which is sold in bulk or in trays, and can be consumed roasted, cooked, or used in the preparation of pamonha, corn cream and cakes. There is great potential for growth in human consumption of corn, the industry has been seeking to provide diverse and quality by-products, this expansion should accompany the increase in the consumption of functional, prebiotic, ready, frozen and quick-prepared foods. (Strazzi, 2015).

Fig. 1.7 Barley. (Source: Canva, 2022)

Barley

Barley is considered the fourth most important cereal in the world, behind only corn, rice and wheat in terms of production (Fig. 1.7). Its cultivation would have started around 10,000 years ago. Its most significant application is as a raw material for the brewing industries, including beer and whiskey, a small scale of production is used in the composition of flours or flakes for baking, in the production of medicines and in the formulation of dietary products, it is also used for feed animals, mainly cattle and pigs (Zhu, 2017).

The barley grain is caryopsis, being coated with a husk that fulfills the protective function from harvest to the process within the industry. Caryopsis is composed of bran, a layer of aleurone, endosperm and embryo, with the grain composed of approximately 3.8% germ, 18.1% bran (pericarp and testa) and 78.1% endosperm (including aleurone). Barley contains a high level of proteins and minerals when compared to other cereals, such as maize and sorghum, and low in lipids.

The bark has in its composition Beta-glucans, cellulose, lignin and pentosans, phenolic compounds and mineral substances, the pericarp has a chemical composition similar to the bark. The forehead is composed of cellulose, phenolic compounds and pigments of lipid origin. Aleurone is rich in phytic acid, lipids and proteins, it is the main enzyme-producing tissue. The embryo during malting produces gibberellic acid which will induce the production of enzymes by the aleurone layer, such as alpha-amylases, limit dextrinases, endo-glucanases, endo-proteases and xylanases. The endosperm is composed predominantly of starch and proteins (Martins & Rodrigues, 2015).

Use of Barley Within the Industry

Barley is commonly used in the production of beer, this is because this cereal has an ideal amount of starch and this compound undergoes a process of structural modification during the malting stage being converted into sugars, such as maltose and glucose. This cereal is also rich in proteins, which provide essential amino acids for the growth of yeasts involved in the brewing process, in addition, nitrogenous substances are present in the composition, which are important in foam formation (Junior et al., 2009; Oliveira, 2011).

Currently, there is interest in the use of barley for food production, as this cereal has functional components such as y-glucans, tocols (such as tocotrienols and tocopherols), polyphenols (phenolic acids, proanthocyanidins and catechins) and others. The compounds present are responsible for health effects, such as the reduction of blood cholesterol, regulation of the glycemic index and antioxidant activity, so barley has been used in the composition of some foods such as pasta and bread (Zhu, 2017).

Barley can be eaten raw or cooked and added to salads, has been used to replace rice and in dishes such as soups, risottos and soufflés. Currently, the use of barley for the production of "tea" has been gaining prominence, being called "barley coffee", but this cereal does not contain caffeine. Bran can be used in the production of breads, pies and cakes, as a substitute for wheat (Azevedo, 2018).

In countries like Russia barley is used to make pancakes, in Poland it is consumed ground with butter, in Japan and India in the form of fermented paste, in Korea it is used to replace rice and in the production of paste and soy sauce, in Morocco is part of the ingredients in the production of soups, porridge and bread. In Western countries, it is usually consumed for breakfast, used in the manufacture of soup, porridge, baby food, among others (Castro & Paredes, 2017; Sayd, 2014). There is a desire to use the by-product of barley malt for the development of products, due to its nutritional value, but studies on the subject are scarce (de Souza et al., 2020).

Sorghum

Sorghum (Sorghum bicolor L. Moench) is a cereal that belongs to the grass family. The cultivation of this cereal would have started in Africa, but India is also mentioned as a region of possible origin. The plant we have today is the result of domestication made by man over generations. Sorghum (Fig. 1.8) has the advantage of being able to be cultivated in semi-arid regions, it adapts well to tropical and subtropical ecosystems, unlike other cereals (Hwang et al., 2002; Ribas, 2003).

In continents such as Asia and Africa, sorghum is grown for human consumption, and in countries such as Brazil, Australia and the United States it is produced mainly for animal feed (Taleon et al., 2012).

Fig. 1.8 Sorghum. (Source: Canva, 2022)

This cereal is considered one of the most versatile and efficient food species, both from a photosynthetic point of view and in terms of maturation speed. The grain is composed of the following components: bran (forehead pericarp, 7%), germ (embryo, 9%), and endosperm (storage tissue, 84%). In 2003, sorghum was the staple food of more than 500 million people in different countries (Ribas, 2003; Queiroz et al., 2021).

There are five groups of sorghum, namely, grain, forage, sugar, biomass and broom. Those aimed at human and animal food are grain and forage. Saccharin and biomass are used in the production of second generation bioenergy and ethanol. Sorghum broom due to its characteristics is used in the manufacture of brooms (Landau et al., 2020; Silva, 2019).

The chemical composition, nutritional value and productivity vary according to the type of sorghum, as this cereal has different stems, leaves and panicles. The grain is small and supplies grain to the industry, it has little green mass. The forage is tall, reaching up to 3.0 m, providing an excellent yield of green mass, being used in the production of silage. There are still hybrids, which provide a good production of green mass and produce grains satisfactorily.

Use of Sorghum Within the Industry

With an increase in the search for sorghum-based foods, this has been happening because the grains of this cereal are gluten-free, making it an excellent option for the development of "gluten-free" foods. In addition, the compounds present in sorghum place it in the group of functional foods, but the concentration of substances varies according to the genotype (Asif et al., 2010).

Table 1.4 Types of sorghum and use

type of sorghum	Use
Grain	Human food, animal feed, industry
Saccharine	In the production of ethanol, sucrose, fructose, animal feed
Forager	Cutting, silage and hay
Biomass	Second generation ethanol production, bioenergy production
Broom	Broom production

Source: Barcelos, 2012

Sorghum is a cereal considered "smart crop", its consumption occurs from fresh or floury grain, it can be used as raw material for starch, wax, beverages, oils and bioenergetic production as well as biofuel. We have four groups of sorghum, the grain; forage for silage and/or sugar; forage for grazing/green cutting/hay/mulching; broom. In Table 1.4 we can see the types of sorghum and some uses (Silva, 2019).

Sweet sorghum has been the subject of research as an alternative source of raw material for the production of second generation bioenergy (ethanol). This cereal is considered an "energy plant", for its high photosynthetic efficiency and adaptation to adverse conditions. The raw material used in the manufacture of alcohol is plant biomass, so far the ethanol produced from sweet sorghum has shown high quality (Dos Santos et al., 2015).

Photosynthetic biomass sorghum has been an excellent option in the energy matrix for steam production, an essential process in the sugar and ethanol production processes or can be used in the generation of second generation ethanol, because the genotypes that are present in this cereal after passing through the acid/base treatment, they become a source of fermentable sugars for the production of ethanol (Almeida, 2019).

Another application of biomass sorghum is energy production, this cereal has characteristics related to the combustion process such as calorific value, fixed and volatile carbon contents and alkali metal contents (de Oliveira, 2015).

Pellets are solid biofuels obtained from sawdust and sawmill by-products, and can also be obtained from biomass sorghum, this biofuel is widely used in industries, pizzerias, bakeries, hotels, farms and laundry facilities, with the European Union being the largest market consumer (Simeone et al., 2018).

Grain sorghum is the most economically important sorghum, being the fifth most cultivated cereal in the world, it is used in human and animal food. Production is low-cost and grain sorghum provides food with high nutritional value and a source of bioactive compounds. For human consumption, there are flours, breads, fermented and non-fermented porridges, alcoholic beverages, couscous, use of whole grain and infant transition foods Sorghum provides a flour that can be used for the production of pasta such as instant noodles, chocolate, extruded products such as snacks, breakfast cereals, among other foods (Martino et al., 2014).

Forage sorghum is widely used in silages for cattle feed, has dry matter availability and nutritional value, this plant produces silages with good fermentative characteristics and stands out for having the appropriate concentration of soluble carbohydrates, essential for lactic fermentation, being after maize, the most important crop for silage production (Ribas, 2010).

Europeans introduced sorghum broom in Brazil, the cultivation and use spread throughout the country, obtaining broom from this cereal, ended up becoming an alternative activity for domestic artisans and broom manufacturers, generating an alternative source of income. The seeds are supplied to domestic animals, the stem and leaves that are left are also reused, being used as mulch or in the composting of plant residues (Foltran et al., 2016).

Oat

Oat is an annual grass belonging to the family Poaceae, tribe Aveneae and genus Avena. In Brazil, the main cultivated species are white oat, yellow oat and black oat. This cereal is characterized by its good taste, dietary properties and activity that stimulate metabolic changes in the body. Despite the benefits, only a small part of the production is destined for human consumption and only approximately 5% of the harvest enters world trade.

The oat grain is composed of husk (25%), pericarp, testa, aleurone (9%), starchy endosperm (63%) and embryo (3%). When compared to other cereals, oats (Fig. 1.9) have a greater amount of total protein, the amino acids present are of high nutritional value. Crude fat is a source of essential unsaturated fatty acids, contains several vitamins, minerals and fiber. The consumption of oat flour or bran is favorable to health, the dietary fiber present is composed of pectins, ß-glucans, mucilages, some hemicelluloses and resistant starch, components that provide different health benefits (Biel et al., 2009).

Fig. 1.9 Oat. (Source: Canva, 2022)

Oats are a source of antioxidant, but phenolic compounds can be lost during processing, so it is important to control the variables used in hydrothermal treatment. This cereal is rich in soluble dietary fiber, and B-glucan is a compound that provides benefits to human health and gives oats the status of a functional food (Crestani et al., 2010; Klajn et al., 2012).

Use of Oats

Within the industry, oats provide grains for human consumption, can be used as an ingredient for the production of infant foods, breakfast cereals (hot or cold), granola, muslins, cereal bars, an additional component to thicken soups, sauces and to increase the volume of meat products. Oats have excellent moisture absorption, and can be used in bread making for the production of breads, cakes and cookies, this property guarantees greater durability to these products, as it delays aging. Another interesting property is the ability to stabilize lipid components due to its antioxidant properties (De Mori & Minella, 2012; Webster, 2016).

The search for ingredients of natural and organic origin for the health and beauty industry has grown, oats due to their composition rich in starches, antioxidants, anti-inflammatory, beta-glucans and exclusive components such as avecins and avenatramides have emerged as excellent option for the cosmetics and pharmaceutical industry. Oat (Avena sativa) goes through the milling process in a standardized process that results in extremely small and patterned particles, it is through this method that colloidal oatmeal is obtained, which can be used as a skin protector, moisturizer, soothing, anti-aging, irritant and antioxidant (Kurtz & Wallo, 2007).

Rye

Rye (Secale cereale L.), is a grain that would have originated in Southwest Asia, is the eighth most harvested and produced cereal in the world, and the second most used for bread making, behind only wheat. Rye (Fig. 1.10) presents low economic risk, this is because this cereal has the ability to grow in environments with low temperatures, in low fertility soil, in addition, this cereal has a deep and abundant root system that allows it to absorb water and nutrients under drought conditions. Rye can be grown organically, as it requires little use of fertilizers, pesticides and herbicides.

Rye grain is composed of approximately protein-13.4%, lipids-1.8%, carbohydrates-80.1%, soluble pentosan-2.2%, insoluble pentosan-7%. The flour that is obtained by grinding the grain is rich in soluble dietary fiber that are mainly composed of pentosans, this compound increases the viscosity and are responsible for

Fig. 1.10 rye. (Source: Canva, 2022)

the structure in rye breads, in addition to this technological characteristic, pentosans produce physiological effects, such as: decrease in triglycerides; decrease in blood cholesterol levels; reduction in glycemia, among others already studied, due to the benefits it is recommended the inclusion of rye in the diets for diabetics, hypertensive (Aman, 2006).

Use of rye

The use of rye is very versatile, its production can be destined for human consumption in the form of flour, which will later be an essential ingredient in the production of bread, cookies, cakes and other baked goods; in breakfast cereals; in the production of clear distilled beverages such as whiskey and vodka. This cereal can be used for the production of adhesive mixtures and glues, pharmaceuticals, cosmetics, alcohol and other products within the chemical, health and beauty industries. Animal feed is another that benefits from rye, the grain is mixed with feed, another functionality is in the form of forage, grazing, hay and silage (De Mori et al., 2013).

Rye bread is normally dark, sour and commonly consumed in countries such as Finland, Poland, Belarus, Russia, Germany, Sweden, Denmark, among other countries, where this food is biased with cultural food traditions.

Rye is used in the pharmaceutical industry, this grain when infected by the fungus *Claviceps purpurea* ends up originating important alkaloids such as ergotamine, ergometrine, among others, the functionalities found in these compounds allow them to be used as a therapeutic form for postpartum hemorrhages; migraines; orthostatic hypotension, Parkinson's disease; pathologies associated with hyperprolactinemia and senility (De Mori et al., 2013).

For animal feed, one of the options is direct grazing, but if the grains are contaminated by the Claviceps fungus, the animal can be intoxicated. The best way to feed animals with this cereal is by mixing it with other cereals, a proportion smaller than a third, due to high palatability and stickiness when chewing, research using enzymes seeks precisely to improve the palatability to feed animals and this has increased the percentage of grains in animal feed.

Rye grains can also be used for the production of alcoholic beverages, being a prominent ingredient in the production of Canadian whiskey or rye whiskey; in gin, a drink of Dutch origin; and in beers, such as roggenbier, in Germany, rye-PA, in the United States, sahti, in Finland and kvass, in Russia and in Eastern European countries.

The use of rye is not limited to the food industry, the properties and composition of this food allow it to be used in the production of adhesive mixtures; rye straw is used to manufacture paper and straw agglomerates (Nascimento Jr & Baier, 2006).

2 Other Cereals

Triticale

Triticale (X. triticosecale Wittmack) is the result of the hybridization of two distinct species, wheat (Triticum aestivum L.) and rye (Secale cereale L.), the functionality and high yield of wheat and the durability of rye, which made the triticale grain (Fig. 1.11) bread-making and rustic. In Brazil, this cereal has been used in animal feed, as plant cover for soil protection and green manure. However, the southern

Fig. 1.11 Triticale. (Source: Canva, 2022)

region of the State of São Paulo uses triticale as an alternative to wheat, due to the characteristics that this grain presents (Mcgoverin et al., 2011).

Triticale has a chemical composition more similar to wheat than to rye. This grain contains proteins, starch, polysaccharides, proanthocyanidins, lignans, phenolic acids. Triticale straw has significant antioxidant activity. Despite the benefits that the consumption of this cereal provides to human health, most of its production is intended for animal feed and only a small portion is used for baking, triticale products supplied were found in countries such as Canada and the United States in 1980s, however, these foods lost marketing strength due to the cultivation of wheat and barley in these countries (Mcgoverin et al., 2011).

In animal feed, triticale can be used for pigs, poultry, ruminants (goats, sheep and cows), being used in the form of grains, forage, silage, hay, as a diet formulation (Salmon & Gomex-Macpherson, 2004).

Modern triticale cultivars have emerged as a competitive raw material for the production of ethanol, this cereal has an autoamylolytic enzyme system that contributes to the conversion of large amounts of starch into fermentable sugars, but the production of bioethanol from grains has generated debate, as it may imply limitations on the supply of triticale for food (Solomon, 2010).

Amaranth

Amaranth is considered a pseudocereal, its leaves can be consumed as a vegetable and the grains as a cereal, this food is practically unknown in Brazil, but consumed in several parts of the world, being included in the list of the 36 most promising crops to feed humanity. The first civilizations to cultivate amaranth would have been the Aztec, Maya and Inca, this pseudocereal was considered a sacred food, being part of religious ceremonies, a practice that was prohibited with the arrival of colonizers, and the cultivation of amaranth almost disappeared (Pinto, 2007).

The leaves of some species are rich in calcium, protein, beta-carotene, phosphorus and iron, but there are anti-nutritional substances such as phytic acid, polyphenols and oxalic acid that limit commercialization. Grains are composed of protein, lipid, starch and dietary fiber. This pseudocereal has been cited as a functional food, as it has the ability to reduce serum cholesterol levels in animal tests (Mendonça, 2006).

This grain can be used in human and animal food, it can also be used in soil protection as fodder during the off-season season. Among the foods that can be obtained from amaranth are flours, breakfast cereals, pasta, gluten-free cookies, in China, several products are made from amaranth, such as spaghetti, wines, vinegar, among others. In Australia, in Africa and Asia, amaranth grains (Fig. 1.12) and leaves are consumed in the diet.

Fig. 1.12 Amaranth. (Source: Canvas, 2022)

When subjected to heating, the flour tends to become gelatinous and viscous due to the starch content present, this characteristic allows it to be used as a thickener in jellies and preserves that do not require the addition of sugar (Farinhas, 2022).

Quinoa

Like amaranth, quinoa is a pseudocereal. This grain originates from the Andes, being cultivated all over the world, was introduced in Brazil in the 90 s, where it underwent processes of genetic modification resulting in the BRS Piabiru variety, adapted for cultivation in Brazilian soils. This food has excellent protein quality and high content of vitamins and minerals, being intended for human and animal food (Spehar & Santos, 2002).

Quinoa stands out when compared to other cereals, such as wheat and barley, this grain has high protein quality and an excellent balance of essential amino acids, such as lysine, which is usually limited in other cereals. It is reported that the consumption of a portion of approximately 40 g brings health benefits, due to the essential nutrients present, but it is necessary that further studies are carried out on the subject, as there is a lack of research (Graf et al., 2015).

Its grain is consumed in various ways, such as soups, in the composition of breakfast cereals, or transformed into flour that will later be used in the production of cookies, bread, biscuits, pasta, chips, pies and pancakes. The grains can be fermented and used in the production of beer. In animal feed, quinoa (Fig. 1.13) is used as feed.

Fig. 1.13 Quinoa. (Source: Canva, 2022)

3 Conclusion

The importance of cereals is notorious, despite this most of their production is destined for animal feed, and only a small portion is intended for human consumption, providing the most diverse foods. The researchers believe that the consumption of cereals should increase, due to the functional properties.

References

Almeida, L. G. F. (2019). Etanol de segunda geração utilizando sorgo biomassa (Sorghum bicolor). Trabalho apresentado ao programa de Pós-Graduação em Biocombustíveis – Universidade Federal dos Vales Jequitinhonha e Mucuri.

Aman, P. (2006). Health aspects of rye. In: *EUCARPIA: International Symposium on rye breeding and genetics*. Germane. p. 22.

Asif, M., et al. (2010). Uses of sorghum grain in gluten-free products. *Cereal Foods World, 55*(6), 285–291.

Atwell, W., & Finnie, S. (2016). *Wheat flour*. Elsevier.

Azevedo, K. (2018). *Cevada: um cereal nutritivo e cheio de benefícios*. Dicas de mulher.

Barcelos, C. A. (2012). Aproveitamento das Frações Sacarínea, Amilácea e Lignocelulósica do Sorgo Sacarino [Sorghum bicolor (L.) Moench] para a Produção de Bioetanol. Tese de Doutorado. Escola de Química, Universidade Federal do Rio de Janeiro.

Bassinello, P. Z., & Castro, E. D. M. (2004). *Arroz como alimento. Embrapa Arroz e Feijão-Artigo em periódico indexado*. ALICE.

Biel, W.; Bobko, K.; Maciorowski, R.(2009). Chemical composition and nutritive value of husked and naked oats grain. Journal of Cereal Science.v. 49, p 413–418.

Biesiekierski, J. R. (2017). What is gluten? *Journal of Gastroenterology and Hepatology, 32*, 78–81.

Brasil. (2014). Ministério da Saúde. Secretaria de Atenção à Saúde. Guia alimentar para a população brasileira. 2. ed. Departamento de Atenção Básica.

Brouns, F. J., van Buul, V. J., & Shewry, P. R. (2013). Does wheat make us fat and sick? *Journal of Cereal Science, 58*(2), 209–215.

Canva. Imagens. (2022). Disponível em: https://www.canva.com/. Acesso em 17 de março de 2022.
Castro, J. D. Q., & Paredes, S. L. S. (2017). Promoción de la cebada variedad Cañicapa com propuesta culinaria a madres de familia de laCdla. Banco de la Vivienda Del Cantón Milagro. 2017. 163 f. Monografia (Graduação em Licenciatura em Gastronomia) – Universidade de Guayaquil.
Costa, M. D. G. D., et al. (2008). Qualidade tecnológica de grãos e farinhas de trigo nacionais e importados. *Food Science and Technology, 28*(1), 220–225.
Crestani, M., et al. (2010). Conteúdo de β-glucana em cultivares de aveia-branca cultivadas em diferentes ambientes. *Pesquisa Agropecuária Brasileira, 45*, 261–268.
De Mori, C., & Minella, E. (2012). Aspectos econômicos e conjunturais da cultura da cevada. Documentos online n° 139. Empresa Brasileira de Pesquisa Agropecuária – EMBRAPA trigo.
De Mori, C., Nascimento Jr, A., & Miranda, M. Z. (2013). Aspectos econômicos e conjunturais da cultura do centeio no mundo e no Brasil. Embrapa Trigo, Passo Fundo.
de Oliveira, N. S. S. (2015). Características forrageiras de genótipos de sorgo submetidos a diferentes densidades de plantas e épocas de corte. 2015. Trabalho de Conclusão de Curso (Bacharelado em Agronomia) – Universidade Federal de São João Del Rei.
de Souza, I. S., et al. (2020). Malte de cevada obtido da produção de cerveja artesanal: análise físico-química e destino deste subproduto. *Fag journal of health (FJH), 2*(3), 370–376.
dos Santos, R. F., et al. (2015). Sorgo sacarino na produção de agroenergia. *RevistaBrasileira de EnergiasRenováveis, 4*(1).
Farinha, é tudo igual? Nem um pouco, e nem de longe! Pizzas & Massas, São Paulo, n. 13, 2014. Disponível em: http://pizzasmassas.com.br/upload_arquivos/201606/2016060066250001466793518.pdf. Acesso em 26 de março de 2022.
Farinhas: De trigo, de outros cereais e de outras origens. Aditivos & Ingredientes, São Paulo, 2022. Disponível em: http://insumos.com.br/aditivos_e_ingredientes/materias/98. pdf. Acesso em 27 de março de 2022.
Ferrari, G. T., Proserpio, C., Stragliotto, L. K., Boff, J. M., Pagliarini, E., & Oliveira, V. R. D. (2022). Salt reduction in bakery products: A critical review on the worldwide scenario, its impacts and different strategies. *Trends in Food Science & Technology, 129*, 440–448.
Foltran, D. E., Sawazaki, E., & Freitas, R. S. (2016). Novos cultivares de sorgo vassourapara agricultura regional. *Pesquisa e Tecnologia, 13*(1).
Graf, B. L., et al. (2015). Innovations in health value and functional food development of quinoa (Chenopodium quinoa Willd.). *Comprehensive Reviews in Food Science and Food Safety, 14*(4), 431–445.
Gutkoski, L. C., et al. (2008). Efeito do período de maturação de grãos nas propriedades físicas e reológicas de trigo. *Food Science and Technology, 28*(4), 888–894.
Gwirtz, J. A., Willyard, M. R., & Mcfall, K. L. W. (2014). Wheat: More than just a plant. MÜHLENCHEMIE. Future of flur: a compendium of flur improvement.
Hwang, K. T., et al. (2002). Properties, composition, and analysis of grain sorghum wax. *Journal of the American Oil Chemists' Society, 79*(6), 521–527.
Junior, A. A. D., Vieira, A. G., & Ferreira, T. P. (2009). Processo de produção de cerveja. *RevistaProcessosQuímicos, 3*(6), 61–71.
Klajn, V. M., et al. (2012). Compostos antioxidantes em aveia. *Revista brasileira de agrociencia, 18*(4), 292–303.
Kurtz, E. S., & Wallo, W. (2007). Colloidal oatmeal: History, chemistry and clinical properties. *Journal of drugs in dermatology: JDD, 6*(2), 167–170.
Landau, E. C., et al. (2020). Dinâmica da produção agropecuária e da paisagem natural no Brasil nas últimas décadas: sistemas agrícolas, paisagem natural e análise integrada do espaço rural. *Embrapa Milho e Sorgo-Livro científico (ALICE).*
Malaguti, M., et al. (2014). Bioactive peptides in cereals and legumes: agronomical, biochemical and clinical aspects. *International Journal of Molecular Sciences, 15*(11), 21120–21135.
Martino, H. S. D., et al. (2014). *Por que utilizar o sorgo na alimentação humana?* Embrapa Milho e Sorgo-Capítulo em livro científico (ALICE).

Martins, V. M. R., & Rodrigues, M. A. (2015). Produção e tecnologia de cereais: processo de maltagem da cevada. Livro de atas das Jornadas do Lúpulo e da Cerveja: novas oportunidades de negócio, pp. 37–51.
Masisi, K., Beta, T., & Moghadasian, M. H. (2016). Antioxidant properties of diverse cereal grains: A review on in vitro and in vivo studies. *Food Chemistry, 196*, 90–97.
Mcgoverin, C. M., et al. (2011). A review of triticale uses and the effect of growth environment on grain quality. *Journal of the Science of Food and Agriculture, 91*(7), 1155–1165.
Mclean, J., Dawe, D., Hardy, B., & Hettel, G. (2002). *Rice almanac: Source book for the most important economic activity on earth* (p. 298). IRRI.
Mendonça, S. (2006). *Efeito hipocolesterolemizante da proteína de amaranto (Amaranthus cruentus BRS-Alegria) em hamsters*. Embrapa Agroenergia-Tese/dissertação (ALICE).
Mergoun, M., Gomez-Macpherson, H. (Ed.). (2004). Triticale improvement and production. Rome: FAO.
Salmon, D. F., Mergoum, M., & Gomez-Macpherson, H. (2004). Triticale production and management. *Triticale improvement and production, 179*, 27–32.
Mir, S. A., & Bosco, S. J. D. (2014). Cultivar difference in physicochemical properties of starches and flours from temperate rice of Indian Himalayas. *Food Chemistry, 157*, 448–456.
Mir, S. A., Bosco, S. J. D., Shah, M. A., Mir, M. M., & Ganai, S. A. (2015). Rice: Parboiling and milling properties. *International Journal of Food Engineering, 11*(6), 777–787.
Mir, S. A., Bosco, S. J. D., Shah, M. A., Mir, M. M., & Sunooj, K. V. (2016). Variety difference in quality characteristics, antioxidant properties and mineral composition of brown rice. *Journal of Food Measurement and Characterization, 10*, 177–184.
Mir, S. A., Shah, M. A., Bosco, S. J. D., Sunooj, K. V., & Farooq, S. (2020). A review on nutritional properties, shelf life, health aspects, and consumption of brown rice in comparison with white rice. *Cereal Chemistry, 97*(5), 895–903.
Nascimento Jr, A., & Baier, A. C. (2006). Introdução. In: CULTIVO de centeio. 2. ed. Passo Fundo: Embrapa Trigo. (Embrapa Trigo. Sistemas de produção, 1).
Nunes, O. M., Borges, G. D. R., & Wohlenberg, J. (2017). o uso da casca de arroz como alternativa energética: um estudo de caso no município de dom pedrito – RS. 2017. 21 f. Curso de Tecnologia em Agronegócio, Unipampa, Dom Pedrito.
Oliveira, N. A. M. (2011). Leveduras utilizadas no processo de fabricação da.
Paes, M. C. D. (2006). *Aspectos físicos, químicos e tecnológicos do grão de milho*. Embrapa Milho e Sorgo-Circular Técnica (INFOTECA-E).
Pinto, W. C. (2007). Las especies del género Amaranthus (Amaranthaceae) en Venezuela. *Revista de la Facultad de Agronomía, 1*, 190–195.
Queiroz, V. A. V. et al. (2021). Sorgo para alimentação humana.
Regitano-Darce, M. A. B., Spoto, M. H. F., & Castellucci, A. C. L. (2015). Processamento e industrialização do milho para alimentação humana. *Visão Agrícola, Piracicaba, 13*, 138–140.
Ribas, P. M. (2003). Sorgo: introdução e importância econômica. Embrapa Milho e Sorgo-Documentos (Infoteca-E).
Ribas, M. N. (2010). Avaliação agronômica e nutricional de híbridos de sorgo com capimsudão, normais e mutantes bmr – portadores de nervura marrom. 140 f. Tese(Doutorado em zootecnia). Universidade Federal de Minas Gerais, Escola de Veterinária, BeloHorizonte.
Rustgi, S., et al. (2019). Wheat seed proteins: Factors influencing their content, composition, and technological properties, and strategies to reduce adverse reactions. *Comprehensive Reviews in Food Science and Food Safety, 18*(6), 1751–1769.
Sayd, R. M. (2014). Variabilidade, parâmetros genéticos e caracterização agronômica e molecular de genótipos de cevada nua (*Hordeum vulgare L.* var. *nudum Hook f.*) sob irrigação no cerrado. 2014. 80 f. Dissertação (Mestrado em Agronomia) – Universidade de Brasília.
Scheuer, P. M., et al. (2011). Trigo: características e utilização na panificação. *Revista Brasileira de Produtos Agroindustriais, 13*(2), 211–222.

Silva, L. C. D. M. (2019). Cinética de secagem dos grãos e caracterização física e química durante o armazenamento de farinha de sorgo granífero. Tese (Doutorado agronomia) Programa de Pós-Graduação em Ciências Agrárias – Agronomia do Instituto Federal de Educação, Ciência e Tecnologia Goiano – Campus.

Simeone, M. L. F., et al. (2018). Genótipos de sorgo com potencial para sacarificação. *Embrapa Milho e Sorgo-Boletim de Pesquisa e Desenvolvimento (INFOTECA-E).*

Solomon, B. D. (2010). Biofuels and sustainability. *Annals of the New York Academy of Sciences, 1185*(1), 119–134.

Spehar, C. R., & Santos, R. L. D. B. (2002). Quinoa BRS Piabiru: alternativa para diversificar os sistemas de produção de grãos. *Pesquisa Agropecuária Brasileira, 37*(6), 809–893.

Strazzi, S. (2015). Derivados do milho são usados em mais de 150 diferentes produtos industriais. *Revista Visão Agrícola–USP/ESALQ, 13*, 146–150.

Takeiti, C. Y. (2022). Cereais e Grãos. Agência Embrapa de Informação Tecnológica, (AGEITEC). Disponível em: https://www.agencia.cnptia.embrapa.br/gestor/tecnologia_de_alimentos/arvore/CONT000fid57plx02wyiv80z4s47384pdxjo.html. Acesso em 29 de março de 2022.

Takeiti, C., & Trigo, Y. (2015). Brasília: Agência Embrapa de Informação Tecnológica. Disponível em:http://www.agencia.cnptia.embrapa.br/gestor/tecnologia_de_alimentos/arvore/CONT000girlwnqt02wx5ok05vadr1qrnof0m.html. Acessoem 29 de março de 2022.

Taleon, V., et al. (2012). Effect of genotype and environment on flavonoid concentration and profile of black sorghum grains. *Journal of Cereal Science, 56*(2), 470–475.

Walter, M., Marchezan, E., & De Avila, L. A. (2008). Arroz: composição e características nutricionais. *Ciência Rural, 38*, 1184–1192.

Webster, F. (2016). *Oats: Chemistry and technology*. Academic.

Zhu, F. (2017). Barley starch: Composition, structure, properties, and modifications. *Comprehensive Reviews in Food Science and Food Safety, 16*(4), 558–579.

Chapter 2
Breads

Nicola Gasparre, Cristina M. Rosell, and Maneka Malalgoda

1 Introduction

Probably, 15,000 years ago, hunter-gatherers that produced flour from wild grasses in southwest Asia (Arranz-Otaegui et al., 2018) could not imagine that they were laying the foundations for the birth of a food product that would become one of the most consumed foods for all mankind: the bread. The product in question was far from the "contemporary" bread concept. In fact, it could look like a common flatbread, especially those prevalent in the Middle East. As the years roll on, the discovery of the dough volume increase, due to the yeast action, and the exploration of different cooking techniques, were crucial for developing a bread increasingly like the modern-day one. Subsequently, bread variations in shape, texture, color, and taste, linked to the available raw materials and the bread-maker's know-how extended the concept of bread. Currently, the bakers' expertise, the explanations provided by the food scientists and an increased and more accurate knowledge of the raw materials and processing techniques have allowed the production of a very affordable, nutritious, and sensory appealing food. In addition, new optimized processes and formulations combined with phenomena of immigration and more and more demanding consumers have prompted the consumption of autochthonous breads at thousands of kilometers away from the original geographical region.

N. Gasparre · M. Malalgoda
Department of Food and Human Nutritional Sciences, University of Manitoba, Winnipeg, Canada

C. M. Rosell (✉)
Department of Food and Human Nutritional Sciences, University of Manitoba, Winnipeg, Canada

Institute of Agrochemistry and Food Technology (IATA-CSIC), Valencia, Spain
e-mail: Cristina.Rosell@umanitoba.ca

M. A. Shah et al. (eds.), *Cereal-Based Food Products*,
https://doi.org/10.1007/978-3-031-40308-8_2

To better understand the magnitude of the bread as staple food, around 180 million tons were produced all over the world in 2020 with an average volume per person of 24 kilograms (Statista, 2022). The global bakery industry, especially the bread sector, has been growing steadily over the years. In 2021, the revenue from the bread sector was recorded at approximately 0.44 trillion U.S. dollars and is expected to grow at a compound annual growth rate (CAGR) of 5.15% from 2022–2027 (Statista, 2022). Globally, the innovation of the bakery industry is a huge driver for its continued growth on the different types of bread. China, India, United States, Japan, and Germany represent the top five countries in which most of the revenue is created (Statista, 2022).

Despite the size of these numbers, bread represents a food product in continuous updated, always trying to embody a handy choice for the consumers. The following chapter focuses on the main ingredients used in breadmaking showing the versatility in the production methods depending on the type of bread consumed around the world. A description of the quality aspects related with the bread and its acceptance by the consumers will complete the overview.

2 Ingredients Using in Bread Production

There are several key ingredients that are needed in bread production, while other ingredients enhance the texture, flavor or nutritional quality of bread. The essential ingredients in breadmaking comprise flour, water, yeast, and salt. These ingredients allow developing a dough with visco-elastic properties that can be shaped/molded into the desired form, and then baked.

Wheat Flour

Flour is the predominant ingredient in bread and in most cases for a 0.57–0.75 kg loaf of bread, approximately 0.45 kg of flour is needed (Bevier, 1913). Starch accounts for majority of the functional component and its techno-functional role in breadmaking is related to flour hydration and crumb fixing structure after baking. Starch hydration is related to the degree of starch damage that occurs during the milling process (Greer & Stewart, 1959). Damage starch has high ability to absorb water compared to undamaged/native starch (Barrera et al., 2013). The extent of starch damage can vary between approximately 2–15% in most commercially available wheat flours (Tara & Bains, 1972). The other important physical property of starch is the gelatinization that occurs during the baking process. When heating, starch absorbs water, with a simultaneous swelling till the granule structure disrupts; those are the changes associated to gelatinization. The leaching out of starch polymeric chains into the aqueous intergranular phase causes a sharp viscosity that lasts until the physical rupture of the granules, in which viscosity starts dropping down (Rosell & Collar, 2009). After baking, starch recrystallization that is mainly influenced by the amylose leads to the formation of a jelly structure in which, its components (amylopectin and amylose) try to rearrange in an ordered configuration. The retrogradation of the other starch polymer, amylopectin, occurs very slowly due to its branched nature and it has been associated with staling over time and the crumb firming (MacRitchie, 1984). It must be mentioned that this chain reorganization provokes also a progressive water migration to the crust making it softer and causing a firmness increase of the crumb; these two main changes are closely related to the bread staling during storage (Ringsted et al., 2017). Therefore, the physicochemical properties of wheat starch play a significant role in breadmaking (Kusunose et al., 1999).

Gluten forming proteins, which can be categorized as glutenin and gliadin is the next functional component in flour. These proteins account for 75–80% of total protein in wheat flour (Shewry, 2019). Gluten forming proteins play a critical role in the visco-elastic properties of dough, which are vital for bread and bakery product development. The gliadins are alcohol soluble and are monomeric in structure, whereas the glutenins are soluble in dilute acid and alkali and are mainly polymeric (Cook & Rose, 1934). The gluten network, which develops during the dough mixing process is primarily associated with gluten proteins, hence these proteins provide structure to bread (Delcour & Hoseney, 2010). This network is a result of different non-covalent interactions, which as hydrogen bonds, ionic bonds and hydrophobic interactions (Wieser, 2007). Therefore, gluten proteins are important in the techno-functionality of wheat flour when used in breadmaking. Bread end product quality is associated with dough quality, which is characterized by the visco-elastic properties of dough. These properties are impacted by both protein content and quality, of which the latter depends on the physicochemical properties of gliadin and glutenin proteins (Singh & Singh, 2013). Gluten is made up of polymeric glutenin and monomeric gliadin that are responsible for the functionalities of the final macromolecule. In the dough system, the gliadin protein performs as a plasticizer and are responsible for the viscous and extensible nature of dough, whereas the glutenins impart elasticity and cohesiveness (Ooms & Delcour, 2019). The qualities of the gluten network are regulated by the quaternary structures, in which the major forces involved are mainly intermolecular disulfide bonds (glutenin-glutenin) and hydrogen bonding interactions (glutenin-gliadin) (Aussenac et al., 2001). Together, these proteins are the primary determinants of wheat dough visco-elasticity, hence dough and protein qualities.

The lipid content of wheat flour is approximately 2–2.5% (Pareyt et al., 2011). Although significantly low in amount, flour lipids have been associated with dough characteristics, baking quality as well as bread staling. Lipids in wheat flour can be either polar or non-polar lipids. In wheat, polar lipids are mostly located in the endosperm, and accounts for 70–75% of total lipids in the kernel (Finnie et al., 2009). Non-polar lipids are mainly found in the germ and the aleurone layer. In terms of end-product bread quality, wheat flour lipids have been associated with loaf volume, and lipid content, composition and extractability have varying effects on this characteristic (Macritchie, 1979). In fact, during the dough mixing process, approximately 70% of flour lipids become bound or incorporated in the gluten network, of which the majority are free lipids (Pareyt et al., 2011). With reference to gliadin and glutenin, these two protein types interact with different types of free lipids. The binding of lipids to gluten protein allows them to align at the interface of gas cells, hence increasing gas cell stability throughout the breadmaking process. The role of lipids with reference to loaf volume is associated with the ability of free polar lipids stabilize the liquid lamellae separating gas cells in the dough (Sroan & MacRitchie, 2009). Moreover, hydrogen bonds and hydrophobic forces seem to be at the base of the interaction between glutenin and glycolipids while gliadin probably links to phospholipids. However, non-polar lipids can destabilize gas cells, and can therefore reduce loaf volume. During the baking process, lipids can interact with amylose and form complexes, which has a softening effect of bread crumb (Pareyt et al., 2011).

Pentosans are the major group of non-starch polysaccharides in wheat flour and are composed of arabinoxylans and arabinogalactans (Gan et al., 1995). Pentosans have high water binding capacity and can also undergo oxidative gelation, thus can influence water absorption during dough mixing, thereby impact dough rheological properties (Baker et al., 1943). Arabinoxylans can also form gels in the dough matrix through covalent linkages (Nishitsuji et al., 2020). Studies have shown that water extractable arabinoxylans can stabilize gas cells and thereby contribute to gas retention and has a positive impact on loaf volume (Hoseney, 1984). Most importantly, the water extractable arabinoxylans stabilize protein foams and protect against thermal disruption. These pentosans also increase gas retention by way of facilitating slow diffusion of carbon dioxide produced during the fermentation process (Kiszonas et al., 2015). Water unextractable arabinoxylans have shown to disrupt dough structure and interfere in the gluten network, which could result in reduced dough quality. The unextractable fraction can block gluten protein access to sufficient water during dough development resulting in detrimental impacts on dough rheological behavior. In addition to starch, proteins, lipids and pentosans, white flour also contains some moisture and minerals, which could range between 12–14% and 0.45–0.65%, respectively.

Water

Water is the second most important ingredient in bread after wheat flour. Without adequate levels of water, dough features are greatly affected during processing (Schiraldi & Fessas, 2012), but also the water quality is important, that is the levels of salts in the water. The primary purpose of adding water to dough is for hydration and for ingredient functionality to be unleashed. Water is also needed to disperse ingredients in the dough and is a medium for enzymes action and microbial fermentation. Dough temperature can also be controlled through water addition. In general, the total level of water in bread dough can range between 55–65%, but it depends on wheat flour type and the flour extraction rate. If dough is too dry or tough, water could be deficient, whereas water could be in excess if dough is too sticky. Water is also associated with crumb softness. In order to maintain product quality, water pH and hardness are monitored. Even in the quality of the final bread, water content is important since the migration of water is linked with the staling process.

Yeast

Baker's yeast is primarily composed of *Saccharomyces cerevisiae* and is the primary leavening agent used in bread (Ali et al., 2012). Yeast also contributes to bread flavor and aroma. Yeast is available in a number of different forms, including compressed, granular, cream, dried pellet, instant and frozen, that differed in their physical structure and the ratio of yeast solids to moisture. During the fermentation process yeast consumes sugars in the dough and produces carbon dioxide and ethanol (Struyf et al., 2017). Other metabolites such as organic acids, glycerol and aroma compounds are also produced, all of which are important for end-product bread quality. The functionality of yeast in bread dough can be impacted by dough ingredients, conditions during fermentation, pre-growth condition of yeast and the strain of the yeast itself (Randez-Gil et al., 1999). In terms of the amount of fermentable carbohydrates available, this can be impacted by factors such as, type of flour, types of carbohydrates in the flour used and enzymes in yeast that degrade carbohydrates to fermentable sugars (Cauvain & Young, 2007). Other factors such as, salt level, vitamin and mineral composition and presence of bran can also impact the activity of yeast in bread dough (Struyf et al., 2017).

Salt

The role of salt in breadmaking is three folds, namely, contribution to gluten network development during dough mixing, controlling yeast activity – salt can inhibit yeast activity during fermentation and controlling water activity in final baked products (Belz et al., 2012), apart from its contribution to taste. Salt has also been associated with the formation of elastic crumb during the baking process (Sluimer & Sluimer, 2005). The ionic characteristic of salt is important in the mixing process as it impacts the gluten network development (Larsson, 2002). Salt can also cause changes in flour water absorption. Inadequate levels of salt can cause excessive fermentation and can reduce bread end-product quality, especially cause poor texture and open grain structure (Matz, 1992).

Technological Aids

Although flour, water, salt, and yeast are the essential ingredients for bread, other ingredients such as fats, surfactants, emulsifiers, dough conditioners, hydrocolloids and enzymes are used to improve the quality and shelf life of products. In terms of fats, lard has been traditionally used in baking for a number of years (Cauvain & Young, 2007). Shortening is another form of fat frequently used (Pareyt et al., 2011). Typically, 2–5% of shortening is added on flour basis during bread production. During the mixing process, shortening plasticizes and lubricates the dough (Mehta et al., 2009) and during baking shortening prevents the coalescence of air bubbles which results in fine crumb structure (Brooker, 1993).

Surfactants, used at 0.3–1% in bread formulations, can be either dough strengtheners that interact with the gluten matrix or crumb softeners that complex with gelatinized starch (Delcour & Hoseney, 2010). Surfactants can cause increased air incorporation during the mixing process resulting in improved crumb structure and this is mainly caused by surfactants' role in reducing surface tension (Junge & Hoseney, 1981). An emulsifier is a type of surfactant that has hydrophilic and hydrophobic functional groups, and can therefore, interact with both the aqueous and lipid phase of an emulsion stabilizing it (Gómez et al., 2004), although the functionality is greatly dependent on the type of emulsifier (Garzón et al., 2018). Emulsifiers such as, diacetyl tartaric acid ester of monoglycerides (DATEM) and sodium stearoyl lactylate (SSL) serve as dough strengtheners and are active during proofing, handling and transportation (Armero & Collar, 1996).

Hydrocolloids or gums are also used in breadmaking as dough conditioning reagents. When hydrocolloids are added to wheat flour, the water absorption of the flour is impacted as these compounds are hydrophilic, therefore interact with water (Ferrero, 2017). Interestingly, when hydrocolloids interact with gluten proteins, there is no impact on the development of the gluten matrix (León et al., 2000). In addition, specific hydrocolloids such as hydroxypropylmethylcellulose (HPMC) have anti-staling effects, such as reducing crumb hardening (Guarda et al., 2004). Hydrocolloids such as xanthan gum and alginate strengthened the dough, where dough stability during fermentation can be achieved through the addition of hydrocolloids, and gums such as kappa-carrageenan, xanthan gum and HPMC have shown to have a positive impact of bread specific volume (Rosell et al., 2001).

Enzymes are also used in breadmaking for purposes such as improving dough rheological properties, improving loaf volume and other properties such as crumb texture and crust color and also increasing shelf life (Dura & Rosell, 2016; Kornbrust et al., 2012). Enzymes capable of cross-linking proteins, such as transglutaminase can contribute to the formation of the gluten network (Kuraishi et al., 2001). Transglutaminase can also form large insoluble polymers when using in wheat dough (Larré et al., 2000). Oxidative enzymes such as glucose oxidase can facilitate the formation of disulfide bonds in the gluten network and cause gelation of water extractable pentosans, which ultimately leads to improved dough rheological properties (Hoseney & Faubion, 1981).

Overall, the type of additive used during breadmaking depends on the raw ingredients used (mainly flour properties), process conditions and desired end-product quality. Therefore, there can be a wide range of choices for bakers in terms of what kind of additives can be used, although bakers need to consider synergistic or antagonistic effects of the additives used and choose the optimum amount of additives for excellent end-product quality.

Novel Ingredients

With increasing consumer interest in bakery products with innovative textural and flavor properties as well as increased nutritional value, there is interest in using pulse flours in bread. In fact, addition of pulse flour increases bread protein -related nutritional quality. Fenn et al. (2010) reported a 8–32% increase in proteins when adding pea flour in bread formulations and Simons et al. (2014) reported an increase up to 17.4% upon adding pinto bean protein. The impact of pulse flour addition on dough rheological properties and baking performance can be varied and determined by the type and amount of flour used and pre-treatment of the pulse ingredients (Hall, 2021). The particle size of the flour can also impact its ability to interact with the dough network (Borsuk et al., 2012). From a nutritional perspective, the addition of pulse flours leads to better amino acid profile of bread, therefore is advantageous in many fronts (Boukid et al., 2019). With reference to gluten-free bread, rice flour is a commonly used flour (Wu et al., 2019). Although rice flour lacks the techno-functional properties of wheat gluten, with additives such as hydrocolloids, a palatable product can be developed. Pseudocereals are also be used for this purpose, especially to improve the nutritional value of gluten-free products and previous work has shown that amaranth and buckwheat flour together with emulsifiers and hydrocolloids can produce gluten-free bread with improved end-product quality (Alvarez-Jubete et al., 2009).

3 Preparation Methods

As pointed out before, breads that arrive on our tables, are the result of many different variables such as raw materials and processing operations based on the baker's gained knowledge. Thus, it would not be surprising being in front of two breads, belonging to the same type and manufactured in the same area but with distinct sensorial features. There is no "standard breadmaking" and the variations applied by the bakeries are intended to satisfy their customers' expectations and to make their work more agile. Therefore, the common goal to be pursued by the insiders is the transformation of the flour in a pleasant food. To achieve this objective, bakers resort to a process that, in general, entails the following major stages in which simultaneous physico-chemical changes are taking place:

- kneading (mixing),
- proofing (fermentation),
- heating (baking or steaming) (Rosell, 2011b).

Given that, this chapter aims to provide a wide view of the bread category, the inclusion of those belonging to the gluten-free type has been necessary. For this reason and because of their different rheological behavior, the term dough will refer to the wheat-based breads while batter describes their gluten-free counterparts.

Kneading

The development of a final product with the desired technological and sensorial attributes primarily depends on kneading stage. This step represents the first approach to the breadmaking, in which all the ingredients, traditionally flour, water, yeast, and salt interact. Kneading aims to increase the bulk homogeneity modifying also its functional properties (Fellows, 2009). In fact, the distribution and the hydration of the different flour components, the application of the mechanical energy to the dough and the air bubbles incorporation are key events that occur during kneading. In this stage, wheat dough is subjected to large uni-and biaxial deformations, along with the stretching and alignment of the proteins; all this leads to the creation of a three-dimensional viscoelastic structure with gas-retaining properties. An optimum kneading time is required to obtain a totally developed dough that shows maximum resistance to extension and intermediate rheological features between a viscous liquid and an elastic solid. This rheological behavior is mainly controlled by starch, protein, and water. Small un-mixed particles, deriving from under-mixing conditions, could create some problems during the fermentation and, on the contrary, over-mixing leads to weakening of the protein network with consequent stickiness and poor dough consistency (Rosell, 2011a). In a fully developed dough, the structural element is the protein film that results from the successions of the different reactions of depolymerization and/or conformational arrangements (Delcour et al., 2012). The interaction between protein and lipid seems to have a certain effect on the dough structuring process and bread volume. In fact, during mixing most of the free lipids in flour are combined with gluten proteins. According to McCann et al. (2009), non-polar lipids are trapped in the gluten network interacting with it through hydrophobic forces. Hence, the co-occurrence of physicochemical processes is fundamental for the dough development. These processes mainly depend on a set of factors such as quality and quantity of ingredients, water temperature and resulting dough temperature, mixing intensity, design and movement path of the working element of the mixer (Nabiev, 2019). Among the kneading machines, twin-arm mixers, continuous mixers, and planetary mixers, are the most employed ones in breadmaking (Berk, 2018). Another goal to achieve during kneading is the dough aeration that is positively correlated with a better product structure, in fact, the capacity of entrapping air bubbles allows to obtain breads with superior texture and appearance qualities. According to Sadot et al. (2017), kneading with a spiral agitator at higher rotation speed offered doughs with greater amount of incorporated air; thus, the higher the agitator revolutions, the higher was the air volume incorporated in dough. However, as pointed out by Meerts et al. (2017), gluten protein resulted more vulnerable in an oxygen-lean environment when over-mixing was applied even if it shows recuperation signs during resting period.

After kneading, but before dividing, there is a period in which the newly formed dough is left to rest with the main objective of improving its softness, extensibility, and machinability (Cauvain & Young, 2009). Resting period also has a positive effect on the fermentation process, in fact, during this step a significant increase of the yeast activity takes place with the carbon dioxide bubbles that start growing (Jha et al., 2017). Once this phase is completed, relaxed dough is subjected to the dividing and shaping steps, respectively before entering the proofing stage.

Proofing

The leitmotif that interconnects between the main stages of breadmaking is still about the air bubbles. Those originated during dough kneading represent the nucleation sites for their subsequent growth through proofing and cooking (Martin et al., 2008). Throughout this phase, a sequence of complex biochemical reactions, that go by the name of fermentation, bring about structural and mechanical changes of the dough with the formation of some compounds involved in the final product flavor. During this stage, enzymatic starch degradation led to sugar release that is used as substrate by yeast to start the alcoholic fermentation with consequent production of carbon dioxide and alcohol. Usually, to create ideal condition for the yeast activity, proofer temperature is set between 25–32 °C, while humid atmosphere is needed to reduce the dough weight loss. In response to the carbon dioxide production, cells start to grow, while gluten network surrounding them begins to stretch and, consequentially, dough volume increases by a factor of three or four. As temperature increases, gas bubbles come into contact and if the gluten-starch matrix is not extensible enough, cell walls break down giving rise to single larger gas bubble. These coalescence phenomena take principally place when weak flours (low protein content) are used or in over-proofing conditions (Cauvain, 2015). The appearance of a few large gas cells may result dramatic when looking for a regular alveolation in the final products. However, the presence of liquid lamellae surrounding growing gas cells could represent a valid protection mechanism offered by the gluten-starch matrix, especially during the end of proofing, when gas bubbles get closer to each other. Hence, their stability is pivotal in keeping the gas cells presence along the whole breadmaking process (Sroan & MacRitchie, 2009). After fermentation, when the dough volume is doubled, cooking usually represents the last phase in breadmaking, nevertheless, to fit in with consumers' demand, traditional breadmaking flow sheet had to undergo changes. Freezing and thawing have been applied to stop the process before or after fermentation in order to obtain frozen, partial-baked and in store freshly baked breads (Rosell, 2009b).

Heating

At the beginning of this stage, as the temperature increases, a residual proofing still takes place in the central section of the loaf and continues until the yeast inactivation is achieved at around 45 °C. After that, protein begins to denature and the extensible gluten encircling the gas cells turns into semi-rigid structure and with that starts the crumb formation. Heating step is also responsible for the inactivation of the different endogenous enzymes. The starch performances are clearly appreciable when the temperature rises and the role of protein takes less significance, in this scenario, this polymer passes through various stages such as swelling, deformation, fragmentation, disintegration, solubilization, and reaggregation. On the other hand, sugars released can participate with some amino acids in non-enzymatic browning reactions that are responsible for the color and aroma of the crust (Rosell, 2011a). The gradual formation of this latter provides a physical barrier to an additional expansion, moreover, when the dough foam changes to the sponge state, internal gas pressure becomes equalized bringing an end to the expansion (Cauvain, 2015). Thus, it seems clear that gas is retained if cells remain closed, for this reason, the time of pore opening during heating stage takes on great importance to guarantee an optimal expansion level of the loaf as well as, the final crumb texture (Grenier et al., 2021). It has been described that an early pore opening led to a breakdown of the cell walls resulting in less inflated breads and low-density crumb; on the opposite side, breads characterized by high inflation and homogeneity of gas fraction were obtained when the pore opening was delayed (Nicolas et al., 2016). Definitively, breadmaking can be viewed as a not-static process in which biochemical, rheological, and microbiological changes synergically occur over the course of the main steps above-described.

4 Types of Bread

Many different types of breads have been developed through the millennia, which very often reflect the adaptation of the breadmaking process to the raw materials existing in a determined area, or even several strategic processing changes implemented to optimize resources. Then, by combining different ingredients and by adapting process following their requirements, have resulted in thousands of different breads spread all over the world. In order to make reading easier, a classification of the main world's bread types has been proposed (Fig. 2.1). Yeast breads, flatbreads (leavened and unleavened), steamed breads, and gluten-free breads represents the main categories considered.

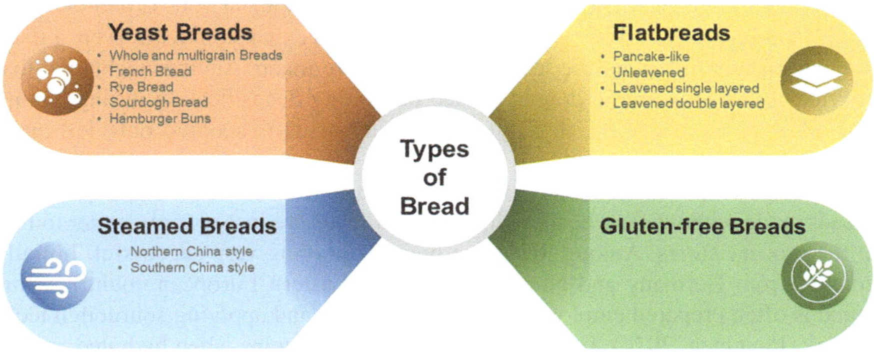

Fig. 2.1 Main categories containing different types of bread consumed around the world

Fig. 2.2 Typical yeast bread

Yeast Breads

This category gathers up the largest number of breads that are mostly produced following the unit operations described above. They are consumed especially in Europe, North America and Australia with white bread that represents the most consumed variety (Fig. 2.2). Belonging to this category, the French bread "baguette" is one of those that has made itself known all over the world for its curious shape. This thin long loaf seems to have been introduced in France by one of the Austrian queens that habitually consumed it in Vienna (Edwards, 2007). Generally, wheat flours with a protein content ranging from 9.5 to 11.0% (14% moisture) are the raw materials intended for the baguette production; they are left to proof longer and are characterized by medium density as well as a hard and crisp crust (Cauvain, 2015). To better understand the crust formation, Della Valle et al. (2012) analyzing the behavior of 64 French breads during baking, reported that the complete crust development (average final thickness 1.7 mm) was reached at 160 °C and moisture of 5%. Traditionally, to improve the gas retention and expansion in the oven, a fixed number of cuts following the length of the loaf are made. Another technique that provides cross cuts on the dough surface promoting degasification during baking and allowing to reduce the density of the final breads is adopted outside France (Cauvain, 2015).

Especially nowadays and because of more consumers' concerns about diet, yeast breads made with whole wheat, multi grain and seeds are increasingly easier to find at the grocery stores. Among them, rye bread is proving very successful. Initially coming from Germany and Nordic, Baltic, and Eastern European countries, this bread is often prepared employing whole-grain flour and applying sourdough technology (Poutanen, 2014). Unlike wheat gluten, rye proteins when hydrated are not able to form a tridimensional structure, but pentosans and starch assume greater significance within the production process owing to their water-binding capacity that improve viscosity, which is crucial for the crumb structure formation. To obtain the best results in terms of final products features, rye flour containing a pentosan:starch ratio of 1:16 to 1:18 is considered optimal. Given that rye starch gelatinizes at lower temperatures (50–70 °C) at which the α-amylases activity is very high, a normal level of salt addition and the lactic acid fermentation (sourdough) lower the pH, limiting the enzymatic damage (Rosentrater & Evers, 2018). By the acidification step, pentosans solubilizes and their water absorption capacity increases; during baking, the bound water can be released and absorbed by the starch that would contribute to obtain softer crumbs (Decock & Cappelle, 2005).

Sourdough in breadmaking is one of the oldest techniques to create leavened baking goods. Technically, sourdough is defined as a mixture of flour and water, spontaneously fermented by lactic acid bacteria and yeasts, and having acidification and leavening capacities (Gobbetti, 1998). Breads with different technological and nutritional profiles have been obtained; in fact, they regularly present more intense flavor, an acidic taste, higher crumb firmness, and an increased shelf life compared to yeasted ones. Depending on the tradition and the intended purpose, three types of sourdough are scientifically recognized. Type I involves a daily feeding and refreshments to keep the metabolic state of the microorganisms active; type II involves higher temperatures (>30 °C) and longer fermentation period (up to 5 days) while type 3 (dried) is used principally as flavoring agent (De Vuyst & Neysens, 2005). Taking into account the sourdough type I, Arora et al. (2021) analyzed 246 research articles about salty products (breads and substitutes) from 47 countries all over the world. Authors reported that before the final fermentation, sourdough is subjected to a variable number of refreshments (2–10) while the median fermentation time is 4 h and the most common temperature is set at 30 °C to accomplish an appropriate acidification, leavening power and cell densities of lactic acid bacteria and yeasts. Sourdough can be added in solid or liquid form and the median value of its inoculum percentage is 20%, ranging from 10 to 25%. Compared to yeast bread, those made from sourdough fermentation present better qualities in terms of texture, specific volume, crust color, crumb structure, flavor and shelf life (Siepmann et al., 2018). From a sensorial point of view, breads obtained through the sourdough fermentation usually achieve better score during test, because of their more complex aroma profile compared to the yeast-based ones. In fact, sourdough breads principally present superior quantities of acids, alcohols, aldehydes, esters, and ketones (Pico et al., 2015); additionally, higher concentrations of taste-active compounds, such as glutamate and glutathione notably improve the umami taste (Zhao et al., 2015). Nutritionally speaking, De Vuyst et al. (2017) described the phytase activity of the lactobacilli that could help reducing the antinutritional effect of the phytic acid while Garzon et al. (2021) pointed out that using sourdough fermentation boosted the total phenolic content and the antioxidant activity in microalgae enriched breads. Lastly, due to the longer time required for sourdough fermentation with respect to the conventional one, more starch is metabolized in lactic acid, ethanol, carbon dioxide, and organic acids with a consequent reduction of its availability for the digestion so as the final products show lower glycemic index (Siepmann et al., 2018).

Within yeast bread category also buns are particularly relevant, especially in North America where they are largely consumed filled with hamburger and hot dog. By comparing the formulations of typical breads and buns, four main differences come up, buns require higher amount of fat and sugar but less water while the recommended flour must be high in functional protein content. Regarding the process, some pre-fermentation usually precedes the doughmaking. One of the options, that allows to generate the essential amounts of acid to produce good buns, is the water brew. This fermenting liquid contains most of the yeast and some of the salt and sugar are mixed and fermented with a brew buffer for 60–90 min. Another alternative is the fermentation of part of the flour to produce a sponge; the rest of the ingredient is added after the fermentation. To achieve a softer and more elastic crumb, the addition of more water could help. After a suitable proof step (38 °C, 78% relative humidity, for 50–55 min), buns are covered with seed and then typically baked at 220 °C for 7.5–9.5 min depending on the weight of the product (Cauvain, 2015).

Flatbreads

Discoveries at ancient sites in Mesopotamia, ancient Egypt, and the Indus civilization point to the flatbread as the earliest form of processed food precursor of the most common bread. From there, a vast and heterogeneous assortment of bread types with a thickness ranging from a few millimeters to a few centimeters, has begun to populate larger areas across the world. Behind their spread among the first nomadic populations there seems to be an increased shelf-life, since they can be twice cooked (dehydrated) reducing the mold growth and improve their transport performance as they can be carried by stacking them (Pasqualone, 2018). Currently, food market is experiencing a steep rise demand for these products moving from a rural community to the modern society. In fact, the global market size of flatbreads was valued at USD 38 billion in 2018 and is estimated to reach almost USD 63 billion by 2026 with a compound annual growth rate more than 6%

(AlliedMarketResearch, 2022). One of the strong points that characterize their increasing appreciation in the Western countries is that flatbreads can contain food inside them and can be consumed as a complete dish (Pasqualone, 2017).

Flatbreads processing is very similar to that employed for the yeast breads. It involves ingredients kneading (flour, water, salt, sometimes slight amounts of fatty ingredients, with or without yeast), leavening in some of them, shaping (by pouring the batter on a griddle or by sheeting and eventually punching the dough) and baking. A huge variety of flatbreads can be produced from dense and elastic gluten containing doughs, as well as batters obtained from gluten-free raw materials. The wheat-based ones are typically spread into sheets with a rolling pin while the batters are directly poured onto the cooking side (Pasqualone, 2018). Flatbreads can be categorized based on the dough, the presence or absence of a leavening phase, the baking system used, and the thickness and structure of the finished product, in the following types (Pasqualone, 2018):

(i) Pancake-like breads, obtained from batters.
(ii) Unleavened flatbreads with consistent dough.
(iii) Leavened flatbreads with consistent dough, shaped as a thick sheet and eventually punched (single layered).
(iv) Leavened flatbreads with consistent dough and shaped as a thin sheet (double layered).

The first group of flatbreads requires a batter preparation (wheat or pulses flour) generally followed by a fermentation. Ethiopian *injera* made from teff flour (Neela & Fanta, 2020), the Somali *anjero* obtained from sorghum or maize flour (Kamal-Eldin, 2012), the Yemeni *lahoh* and *kisra* from Sudan both prepared form sorghum flour (Lyons, 2016; Zaroug et al., 2014) are examples of pancake-like breads. In the Arabian Peninsula, in Egypt, and in northern Iraq, *ruquag* is obtained from wheat flour while in the Maghreb region, wheat durum semolina is employed in the production of *ghrayef.* From India, two varieties obtained from local rice and black beans can join this category of flatbreads, *pitha* and *dosa*. In Italy wheat flour allows to obtain *testarolopontremolese*, and *panigaccio*, while chesnut flour is required to produce *neccio*(Pasqualone, 2018).

The unleavened flatbreads with consistent dough are characterized by a thin layer and higher chewiness. One of the most consumed is tortilla from south America than originally was obtained by maize and nowadays wheat flour is also used. Maize tortilla production is basically composed by three main steps: lime-cooking, stone grinding and baking. Because of its effect on the dough machinability and tortilla features, lime-cooking (calcium oxide 1%) is pivotal for the whole process, while stone grinding assures the correct distribution of the hydrated starch and protein as well as controlling the particle size of the final *masa*(Serna-Saldivar & Chuck-Hernandez, 2019). This is typically cooked in an electric or gas griddle at 177–232 °C for 30–60 s and the freshly baked product appears with two distinct layers and dark spots (Waniska, 2015). *Yufka* from Turkey (Basman & Köksel, 2001), Iranian *taftoon*(Karizaki, 2017), the *lavash* produced in Armenia, Iran, Turkey, and India (Pahwa et al., 2016), the *chapati* and *roti* from India and Pakistan

(Gocmen et al., 2009), the *marquq* from Lebanon (Zurayk & Rahman, 2008), the Italian *piadina* and *kesra* from Algeria and Tunisia (Kezih et al., 2014) possess the attributes to be classified as unleavened flatbreads.

Single-layered flatbreads are traditionally leavened using sourdough fermentation but currently, brewer's yeast is used to produce a softer product with 3–4 cm of thickness. To avoid excessive expansion during baking, the flat dough is pressed with fingertips or punched (Karizaki, 2017). Among the flatbreads belonging to this category, *matlouh* from Algeria, Tunisia and Morocco (Kezih et al., 2014), the Turkish *pide* and *bazlama*(Levent & Bilgiçli, 2012), *barbari* and *khoshk*from Iran (Karizaki, 2017), the Indian and Pakistani *naan* (Fig. 2.3) and *khameeriroti* (Farooq et al., 2014; Sakandar et al., 2020), as well as the Italian *crescentina*(Pasqualone, 2018) are the most known products. Owing to the high amount of oil and because they are usually garnished with other ingredients, pizza and focaccia cannot be included in the flatbread category (Pasqualone et al., 2011).

The double layers that characterize the following group of flatbreads are obtained when the dough disc subjected to baking is not previously punched. In the oven, the higher viscoelastic properties contribute to keep the inflation and dough takes the form of a balloon. During cooling a deflation occurs, and the final products results soft and elastic with the characteristic inside pocket that can be filled with vegetables or meat. Currently, the double-layered flatbreads are the most consumed in Western countries. Lebanese and Syrian *khobz*, the Moraccan*batbout*, *samoon* from Iraq, *pita* from Greece (Fig. 2.4) (Pasqualone, 2018), the Egyptian *baladi*(Abdel-Kader, 2000) and *pucciasalentina* from Italy (De Bellis et al., 2019) can be considered as double-layers flatbreads.

Focusing on the cooking method, flatbreads basically differ from the spongy and low-density ones for the oven types and processing time. Due to their lower thickness and mass, all flatbreads require less time (few minutes) and usually do not have crust. Habitually, fire plates and vertical ovens are used to bake the majority of the flatbreads; nevertheless, all the double-layered flatbreads are cooked into domed ovens that ensure high expansion (Pasqualone, 2018).

Fig. 2.3 Indian and Pakistani *naan*, single layered flatbread

Fig. 2.4 *Pita* from Greece, double layered flatbread

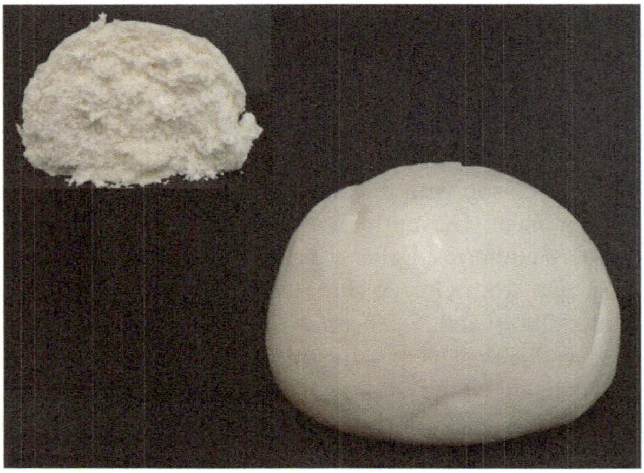

Fig. 2.5 Northern style steamed bread

Steamed Breads

Breads belonging to this group basically differs from other in the heating step. In fact, this type of fermented doughs undergoes a steaming step before consumption. Sometimes called *mantou*, these steamed buns, have been originally consumed in some parts of China as a staple food over two millennia, nowadays are well known around the globe. In China, owing to the cultural, geographic differences and depending on the composition of the basic ingredients and the adopted processing method, two types of steam breads can be roughly identified, northern and southern style (Zhang et al., 2007; Zhu et al., 2001). In north China, a semi-arid region, wheat is widely cultivated and steamed breads with high firmness, elasticity and chewiness are greatly appreciated (Fig. 2.5). Due to the warm weather that characterize South China, rice is the most cultivated cereal grain and consequently its flour is used to produce a softer and less chewy steamed bread (Huang & Miskelly, 2016). These oriental breads are generally considered healthier because of the low steaming temperature (100 °C) at which Maillard reaction does not occur, besides there are no acrylamide and furan releases as well as better nutrients retention (Zhu, 2014). On the other hand, it has been reported that these steamed goods are characterized by a high glycemic index (Zhu, 2019). According to the method published by the former Ministry of Commerce of China (SBT 10139–93), refined wheat flour (100 g) is mixed with 48 mL water (38 °C) containing dried yeast (1 g). Then, the ingredients are mixed by glass stick/chopstick and hand-kneaded for 3 min to form dough before proofed at 38 °C for 1 h. The proofed dough is further hand-kneaded for another 3 min and mounded to the shape of a loaf and left at room temperature for 15 min. The dough is then steamed for 20 min before wrapped with gauze and cooled at room temperature for 40–60 min (Zhu, 2014). Sometimes, sourdough fermentation is preferred due to its action in reducing dough water absorption, development time and stability that allow the production of steamed bread with higher specific volume (Wu et al., 2012). The Chinese homemade fermentation with sourdough provides another step before steaming, that consists in neutralizing the excess sourness produced by lactic acid bacteria with alkali (soda or sodium carbonate solution).

Gluten-free Breads

Emulating the viscoelastic properties and the gas retention capacity observed in wheat-based doughs in the absence of gluten is still an open challenge for food scientists. Usually, the basic raw materials utilized in gluten-free breadmaking are flours coming from, rice, maize, pseudocereals and pulses; all of these are united by the gluten absence. It is therefore essential to mimic gluten behavior to obtain gluten-free breads with close characteristics to those conventional. During the last two decades, extensive research have been carried out to investigate new raw materials functionalities and the impact of different physicochemical and biochemical treatments on semi-finished and final products. Nevertheless, mixing different flours, starches, and different additives, is still the most adopted way to obtain gluten-free breads. Among the technological strategies adopted in gluten-free breadmaking, annealing and heat moisture treatment have been proposed. Due to their expensive cost, flours pre-gelatinization fits as a more affordable alternative. Bourekoua et al. (2016) observed that hydrothermally treating rice and corn flours allowed to obtain gluten free breads with higher specific volume and reduced hardness and chewiness. Appreciated for their water holding capacity, anti-staling effect, textural improvement capability and gelling properties, hydrocolloids are widely used for the gluten-free product manufacturing. One of the most utilized in gluten-free breadmaking is hydroxypropylmethylcellulose that helps to improve batter consistency and promotes final products with desirable crumb hardness, cohesiveness and resilience; nevertheless, its impact greatly depends on dosage and hydration level (Morreale et al., 2018). Some studiespointed out the quality enhancement following the addition of emulsifiers such as distilled monoglycerides, lecithin, and di-acetyl tartaric ester of monoglycerides (Demirkesen et al., 2010; López-Tenorio et al., 2015). Particularly, lecithin reduced the elasticity of the batter, distilled monoglycerides decreased the crumb staling rate, while di-acetyl tartaric ester of monoglycerides deeply affected the crumb structure (Nunes et al., 2009). To fill the lack of a three-dimensional protein network, the incorporation of non-gluten protein during batter preparation has been also proposed. Milk proteins and egg albumins from animal source or plant-based proteins from soybean and pea have been used for the nutritional and technological improvement of the gluten-free breads (Ziobro et al., 2016). Water uptake increase, higher dough stability and extended shelf-life have been associated with the addition of milk protein. In fact, after heating, β-lactoglobulin up-folds and starts to aggregate with whey protein via disulphide interactions and hydrophobic bonding, giving rise to a three-dimensional structure (Arendt & Moore, 2006). Marco and Rosell (2008) reported a hardness increase followed by the individual addition of pea protein isolate, while the opposite trend was observed when protein isolate from soybean were used in the rice-based formulation. A valid alternative approach for the processing of gluten-free breads is offered by enzymatic treatments. The main enzymes utilized in gluten-free technology are acting on starch or proteins to improve the batter-handling properties and to enhance the final baking quality. Regarding those that act on the starch substrate, amylase, cyclodextrin glycosyltransferases, and glucose oxidase are the most explored ones. The latter catalyzes the oxidation of β-D-glucose to D-gluconolactone or D-glucon acid and hydrogen peroxide. After that, this strong oxidizing agent reacts with the thiol groups of the proteins by forming disulphide bonds and consequently the rheological properties of the batter change (Rosell, 2009a). Gujral and Rosell (2004) observed that increasing the addition of glucose oxidase levels, the viscous and elastic moduli increased too; moreover, the specific volume of the rice-based breads resulted higher than the control. Regarding protein, transglutaminase enzymes have been successfully applied for the creation of a protein network that allowed a better gas retention and higher specific volumes of the gluten-free breads. This enzyme catalyzes the cross-linking of a wide group of proteins, its association with skim milk and egg powder lead to a protein network formation that could be responsible for the improvement of loaf volume and textural characteristics of gluten-free made by rice flour (Moore et al., 2006). But on the opposite side, even proteases have been proposed for making gluten free breads, owing the protein hydrolysis results in hydrophobicity reduction (Renzetti & Rosell, 2016) (Fig. 2.6).

5 Bread Quality Aspects

Bread quality characteristics can range from crumb and crust properties all the way to bread flavor and aroma. The American Association of Cereal Chemists – International (AACCI) standard methods determine bread quality characteristics such as loaf volume, specific volume, crumb grain and texture, loaf symmetry, crust and crumb color. These characteristics are especially applicable for pan bread-type

Fig. 2.6 Gluten-free bread with seeds

products. These methods are frequently used in industry as well as experimental baking to determine the impact of different ingredients and/or processing methods on the end-product quality. In this case, bread is produced either through the straight-dough baking method or through the sponge-and-dough method, after which the previously stated characteristics are assessed as per the guidelines of the standard methods. Other baking methods can also be used for this purpose and will depend on the focus on the manufacture.

The assessment of flat bread quality varies from that of pan bread, however, to evaluate dough properties of flat bread, methods such as Farinograph, mixograph, extensograph and alveograph can be used. For example, Azizi et al. (2003) assessed the impact of additives on dough quality using Farinograph and extensograph, while flat bread quality was assessed used parameters such as, shape, crust formation, upper and bottom surface properties, firmness and chewiness. Similarly, Panghal et al. (2019) assessed the quality of *chapatti* considering properties such as, height immediately after puffing, softness, brown spots on the surface, mouthfeel, chewiness and texture. For Chinese steamed bread, properties such as spread ratio, specific volume and texture can be tested as outlined by Sim et al. (2015).

To assess the quality of gluten-free bread and sourdough bread and other pan bread-type products, similar characteristics as those described in the AACCI standard methods can be used, as the focus is on characteristics such as loaf volume, appearance and texture.

The quality of bread depends on a multitude of factors ranging from raw ingredient properties to storage. Overall, all ingredients and process steps in the production of bread impacts bread quality; therefore, it is essential to determine optimum ingredient and process characteristics to produce a loaf of bread with favorable end-product quality. It is also important to note that end-product quality characteristics may differ dependent upon the type of product developed, therefore quality needs to be assessed accordingly.

6 Consumer Preferences

Consumers are increasingly seeking food products that are clean-label and nutrient dense. With the growing demand for protein-enriched foods, consumers are attracted to bread with considerable protein content. As described above, one way to achieve enhanced protein nutritional quality is through the addition of pulse flours. Moreover, the production of plant protein concentrates and isolates is skyrocketing, allowing more choice for the bakery industry in terms of the most suitable protein ingredients in bread formulations. In addition, high fiber products are increasingly appealing to consumers, who are aware of the benefits of dietary fiber in terms of maintaining gut health and non-communicable diseases. In this context, consumers are seeking products that are either whole grain and/or fortified with fiber ingredients such as beta-glucan.

Clean-label products have limited number of chemical names that consumers cannot identify and contain more natural or readily found ingredients that consumers are familiar with. With the drive for clean-label products, the bakery industry is increasingly aware of the impact of flour protein and quality on end-product quality as preferably the use of additives needs to be minimized.

Bakery products such as artisan bread and sourdough bread are also increasing in popularity. In fact, the artisan bread market has been growing at a CAGR of 4.1% since 2019 and is expected to reach 4.5 billion USD by 2024 (Market Research Future, 2021). The same is true for sourdough bakery product market, which was valued at 2.4 billion USD in 2018, and is expected to grow at a rate of 6.6% (Grand View Research, 2019). The demand for gluten-free products is also increasing which has led to innovative products in this sector.

Overall, consumer preference regarding bread products is evolving with more consumers seeking nutrient dense, clean-label products. As such the global bakery industry is in need of innovation, especially on the raw ingredient formulation end, to address this demand and produce high quality bread and bakery products which meet consumer needs.

7 Conclusions

As it turns out, bread cannot be only considered as a simple foodstuff, but it is a real sign of identity in many cultures. The easy accessibility and the high affordability of its ingredients make bread a staple food giving it a strong symbolic value in both great megalopolis and developing countries alike. There is not a single breadmaking process, over the years, depending on the raw materials availability and consumers' preferences, bread production has shown all its flexibility, successfully adapting to the changes. Nowadays, the combination of traditional recipes with different unit operations has allowed to diversify the market; typical breads from a particular geographical area can be easily found in other part of the globe, sometimes with different characteristics to fit the local market requirements. The challenge for the future could be based on the nutritional enrichment of this product which is at the base of many people's diets. Lately, a growing trend provides the incorporation of plant-based ingredients to obtain new breads nutritionally enriched. Nevertheless, further research is needed to shed light on the ingredient interactions and their impact on the digestibility.

References

Abdel-Kader, Z. M. (2000). Enrichment of Egyptian 'Balady' bread. Part 1. Baking studies, physical and sensory evaluation of enrichment with decorticated cracked broadbeans flour (Vicia faba L.). *Food/Nahrung, 44*(6), 418–421. https://doi.org/10.1002/1521-3803(20001201)44:6<418::AID-FOOD418>3.0.CO;2-U

Ali, A., Shehzad, A., Khan, M. R., Shabbir, M. A., & Amjid, M. R. (2012). Yeast, its types and role in fermentation during bread making process-A. *Pakistan Journal of Food Sciences, 22*(3), 171–179.

AlliedMarketResearch. (2022). *Flatbread market by product type (Tortilla, Naan, Pita, Focaccia, Fry Bread, and Others) and distribution channel (Business-to-business and business-to-consumers): Global opportunity analysis and industry forecast, 2019–2026*. Retrieved 4 April 2022 from https://www.alliedmarketresearch.com/flatbread-market

Alvarez-Jubete, L., Auty, M., Arendt, E. K., & Gallagher, E. (2009). Baking properties and microstructure of pseudocereal flours in gluten-free bread formulations. *European Food Research and Technology, 230*(3), 437–445. https://doi.org/10.1007/s00217-009-1184-z

Arendt, E. K., & Moore, M. M. (2006). Gluten-free cereal-based products. *Bakery Products: Science and Technology, 2006*, 471–496.

Armero, E., & Collar, C. (1996). Antistaling additives, flour type and sourdough process effects on functionality of wheat doughs. *Journal of Food Science, 61*(2), 299–303. https://doi.org/10.1111/j.1365-2621.1996.tb14180.x

Arora, K., Ameur, H., Polo, A., Di Cagno, R., Rizzello, C. G., & Gobbetti, M. (2021). Thirty years of knowledge on sourdough fermentation: A systematic review. *Trends in Food Science & Technology, 108*, 71–83. https://doi.org/10.1016/j.tifs.2020.12.008

Arranz-Otaegui, A., Gonzalez Carretero, L., Ramsey Monica, N., Fuller Dorian, Q., & Richter, T. (2018). Archaeobotanical evidence reveals the origins of bread 14,400 years ago in northeastern Jordan. *Proceedings of the National Academy of Sciences, 115*(31), 7925–7930. https://doi.org/10.1073/pnas.1801071115

Aussenac, T., Carceller, J.-L., & Kleiber, D. (2001). Changes in SDS solubility of Glutenin polymers during dough mixing and resting. *Cereal Chemistry, 78*(1), 39–45. https://doi.org/10.1094/CCHEM.2001.78.1.39

Azizi, M. H., Rajabzadeh, N., & Riahi, E. (2003). Effect of mono-diglyceride and lecithin on dough rheological characteristics and quality of flat bread. *LWT – Food Science and Technology, 36*(2), 189–193. https://doi.org/10.1016/S0023-6438(02)00201-3

Baker, J. C., Parker, H. K., & Mize, M. (1943). The pentosans of wheat flour. *Cereal Chemistry, 20*, 267.

Barrera, G. N., Bustos, M. C., Iturriaga, L., Flores, S. K., León, A. E., & Ribotta, P. D. (2013). Effect of damaged starch on the rheological properties of wheat starch suspensions. *Journal of Food Engineering, 116*(1), 233–239. https://doi.org/10.1016/j.jfoodeng.2012.11.020

Basman, A., & Köksel, H. (2001). Effects of barley flour and wheat bran supplementation on the properties and composition of Turkish flat bread, yufka. *European Food Research and Technology, 212*(2), 198–202. https://doi.org/10.1007/s002170000241

Belz, M. C., Ryan, L. A., & Arendt, E. K. (2012). The impact of salt reduction in bread: A review. *Critical Reviews in Food Science and Nutrition, 52*(6), 514–524. https://doi.org/10.1080/10408398.2010.502265

Berk, Z. (2018). *Food process engineering and technology*. Academic.

Bevier, I. (1913). *Some points in the making and judging of bread*. University of Illinois.

Borsuk, Y., Arntfield, S., Lukow, O. M., Swallow, K., & Malcolmson, L. (2012). Incorporation of pulse flours of different particle size in relation to pita bread quality. *Journal of the Science of Food and Agriculture, 92*(10), 2055–2061. https://doi.org/10.1002/jsfa.5581

Boukid, F., Zannini, E., Carini, E., & Vittadini, E. (2019). Pulses for bread fortification: A necessity or a choice? *Trends in Food Science & Technology, 88*, 416–428. https://doi.org/10.1016/j.tifs.2019.04.007

Bourekoua, H., Benatallah, L., Zidoune, M. N., & Rosell, C. M. (2016). Developing gluten free bakery improvers by hydrothermal treatment of rice and corn flours. *LWT – Food Science and Technology, 73*, 342–350. https://doi.org/10.1016/j.lwt.2016.06.032

Brooker, B. (1993). The stabilisation of air in foods containing fat-a review. *Food Structure, 12*(1), 12.

Cauvain, S. (2015). Breadmaking processes. In *Technology of breadmaking* (pp. 23–55). Springer. https://doi.org/10.1007/978-3-319-14687-4_2

Cauvain, S. P., & Young, L. S. (2007). Principles of dough formation. In S. P. Cauvain & L. S. Young (Eds.), *Technology of breadmaking* (2nd ed., pp. 299–332). Springer.

Cauvain, S. P., & Young, L. S. (2009). *Bakery food manufacture and quality: Water control and effects*. Wiley.

Cook, W. H., & Rose, R. C. (1934). Solubility of gluten. *Nature, 134*(3384), 380–381. https://doi.org/10.1038/134380c0

De Bellis, P., Rizzello, C. G., Sisto, A., Valerio, F., Lonigro, S. L., Conte, A., Lorusso, V., & Lavermicocca, P. (2019). Use of a selected Leuconostoc Citreum strain as a starter for making a "yeast-free" bread. *Food, 8*(2). https://doi.org/10.3390/foods8020070

De Vuyst, L., & Neysens, P. (2005). The sourdough microflora: Biodiversity and metabolic interactions. *Trends in Food Science & Technology, 16*(1–3), 43–56. https://doi.org/10.1016/j.tifs.2004.02.012

De Vuyst, L., Van Kerrebroeck, S., & Leroy, F. (2017). Microbial ecology and process technology of sourdough fermentation. *Advances in Applied Microbiology, 100*, 49–160. https://doi.org/10.1016/bs.aambs.2017.02.003

Decock, P., & Cappelle, S. (2005). Bread technology and sourdough technology. *Trends in Food Science & Technology, 16*(1), 113–120. https://doi.org/10.1016/j.tifs.2004.04.012

Delcour, J., & Hoseney, R. C. (2010). *Principles of cereal science and technology*. American Association of Cereal Chemists.

Delcour, J. A., Joye, I. J., Pareyt, B., Wilderjans, E., Brijs, K., & Lagrain, B. (2012). Wheat gluten functionality as a quality determinant in cereal-based food products. *Annual Review of Food Science and Technology, 3*(1), 469–492. https://doi.org/10.1146/annurev-food-022811-101303

Della Valle, G., Chiron, H., Jury, V., Raitière, M., & Réguerre, A. L. (2012). Kinetics of crust formation during conventional French bread baking. *Journal of Cereal Science, 56*(2), 440–444. https://doi.org/10.1016/j.jcs.2012.05.004

Demirkesen, I., Mert, B., Sumnu, G., & Sahin, S. (2010). Utilization of chestnut flour in gluten-free bread formulations. *Journal of Food Engineering, 101*(3), 329–336. https://doi.org/10.1016/j.jfoodeng.2010.07.017

Dura, A., & Rosell, C. M. (2016). Enzymes in baking. In R. C. Ray (Ed.), *Microbial enzyme technology and food applications* (pp. 295–314). CRC Press Taylor & Francis Group.

Edwards, W. P. (2007). *The science of bakery products*. Royal Society of chemistry.

Farooq, Z., Rehman, S. U., & Khan, M. U. I. (2014). Effect of association of dough characteristics with storage proteins of wheat and gram flours on texture of the leavened flat bread (naan). *JAPS: Journal of Animal & Plant Sciences, 24*(2).

Fellows, P. J. (2009). 4 – Mixing and forming. In P. J. Fellows (Ed.), *Food processing technology* (3rd ed., pp. 157–187). Woodhead Publishing. https://doi.org/10.1533/9781845696344.2.157

Fenn, D., Lukow, O. M., Humphreys, G., Fields, P. G., & Boye, J. I. (2010). Wheat-legume composite flour quality. *International Journal of Food Properties, 13*(2), 381–393. https://doi.org/10.1080/10942910802571729

Ferrero, C. (2017). Hydrocolloids in wheat breadmaking: A concise review. *Food Hydrocolloids, 68*, 15–22. https://doi.org/10.1016/j.foodhyd.2016.11.044

Finnie, S. M., Jeannotte, R., & Faubion, J. M. (2009). Quantitative characterization of polar lipids from wheat whole meal, flour, and starch. *Cereal Chemistry, 86*(6), 637–645. https://doi.org/10.1094/cchem-86-6-0637

Gan, Z., Ellis, P. R., & Schofield, J. D. (1995). Gas cell stabilisation and gas retention in bread dough. *Journal of Cereal Science, 21*(3), 215–230. https://doi.org/10.1006/jcrs.1995.0025

Garzón, R., Hernando, I., Llorca, E., & Rosell, C. M. (2018). Understanding the effect of emulsifiers on bread aeration during breadmaking. *Journal of the Science of Food and Agriculture, 98*(14), 5494–5502. https://doi.org/10.1002/jsfa.9094

Garzon, R., Skendi, A., Antonio Lazo-Velez, M., Papageorgiou, M., & Rosell, C. M. (2021). Interaction of dough acidity and microalga level on bread quality and antioxidant properties. *Food Chemistry, 344*, 128710. https://doi.org/10.1016/j.foodchem.2020.128710

Gobbetti, M. (1998). The sourdough microflora: Interactions of lactic acid bacteria and yeasts. *Trends in Food Science & Technology, 9*(7), 267–274. https://doi.org/10.1016/S0924-2244(98)00053-3

Gocmen, D., Inkaya, A. N., & Aydin, E. (2009). Flat breads. *Bulgarian Journal of Agricultural Science, 15*(4), 298–306.

Gómez, M., Del Real, S., Rosell, C. M., Ronda, F., & Blanco, C. A. (2004). Functionality of different emulsifiers on the performance of breadmaking and wheat bread quality. *European Food Research and Technology, 219*(2), 145–150. https://doi.org/10.1007/s00217-004-0937-y

GrandViewResearch. (2019). *Sourdough market size, share & trends analysis report by type, by application (breads, cookies, cakes, waffles, pizza), by region, competitive landscape, and segment forecasts, 2019–2025*. Retrieved April 16 from https://www.grandviewresearch.com/industry-analysis/sourdough-market

Greer, E. N., & Stewart, B. A. (1959). The water absorption of wheat flour: Relative effects of protein and starch. *Journal of the Science of Food and Agriculture, 10*(4), 248–252. https://doi.org/10.1002/jsfa.2740100409

Grenier, D., Rondeau-Mouro, C., Dedey, K. B., Morel, M.-H., & Lucas, T. (2021). Gas cell opening in bread dough during baking. *Trends in Food Science & Technology, 109*, 482–498. https://doi.org/10.1016/j.tifs.2021.01.032

Guarda, A., Rosell, C. M., Benedito, C., & Galotto, M. J. (2004). Different hydrocolloids as bread improvers and antistaling agents. *Food Hydrocolloids, 18*(2), 241–247. https://doi.org/10.1016/S0268-005X(03)00080-8

Gujral, H. S., & Rosell, C. M. (2004). Improvement of the breadmaking quality of rice flour by glucose oxidase. *Food Research International, 37*(1), 75–81. https://doi.org/10.1016/j.foodres.2003.08.001

Hall, C. (2021). Pulses: Milling and baking applications. In D. S. Gupta, S. Gupta, & J. Kumar (Eds.), *Breeding for enhanced nutrition and bio-active compounds in food legumes* (pp. 211–228). Springer International Publishing. https://doi.org/10.1007/978-3-030-59215-8_9

Hoseney, R. (1984). Functional properties of pentosans in baked foods. *Food Technology, 106*, 114–117.

Hoseney, R., & Faubion, J. (1981). A mechanism for the oxidative gelation of wheat flour water-soluble pentosans. *Cereal Chemistry, 58*(5), 421–424.

Huang, S., & Miskelly, D. (2016). Chapter 2 – Classification of steamed breads and buns. In S. Huang & D. Miskelly (Eds.), *Steamed breads* (pp. 13–21). Woodhead Publishing. https://doi.org/10.1016/B978-0-08-100715-0.00002-1

Jha, P. K., Chevallier, S., Cheio, J., Rawson, A., & Le-Bail, A. (2017). Impact of resting time between mixing and shaping on the dough porosity and final cell distribution in sandwich bread. *Journal of Food Engineering, 194*, 15–23. https://doi.org/10.1016/j.jfoodeng.2016.07.016
Junge, R., & Hoseney, R. (1981). Effect of surfactants on air incorporation in dough and the crumb grain of bread [Size and distribution of air cells in dough after mixing]. *Cereal Chemistry*.
Kamal-Eldin, A. (2012). 9 fermented cereal and Fermentation: Effects on food properties, 209.
Karizaki, V. M. (2017). Ethnic and traditional Iranian breads: Different types, and historical and cultural aspects. *Journal of Ethnic Foods, 4*(1), 8–14. https://doi.org/10.1016/j.jef.2017.01.002
Kezih, R., Bekhouche, F., & Merazka, A. (2014). Some traditional Algerian products from durum wheat. *African Journal of Food Science, 8*(1), 30–34. https://doi.org/10.5897/AJFS2013.1095
Kiszonas, A. M., Fuerst, E. P., Luthria, D., & Morris, C. F. (2015). Arabinoxylan content and characterisation throughout the bread-baking process. *International Journal of Food Science & Technology, 50*(8), 1911–1921. https://doi.org/10.1111/ijfs.12829
Kornbrust, B. A., Forman, T., & Matveeva, I. (2012). 19 – Applications of enzymes in breadmaking. In S. P. Cauvain (Ed.), *Breadmaking* (2nd ed., pp. 470–498). Woodhead Publishing. https://doi.org/10.1533/9780857095695.2.470
Kuraishi, C., Yamazaki, K., & Susa, Y. (2001). Transglutaminase: Its utilization in the food industry. *Food Reviews International, 17*(2), 221–246. https://doi.org/10.1081/FRI-100001258
Kusunose, C., Fujii, T., & Matsumoto, H. (1999). Role of starch granules in controlling expansion of dough during baking. *Cereal Chemistry, 76*(6), 920–924. https://doi.org/10.1094/cchem.1999.76.6.920
Larré, C., Denery-Papini, S., Popineau, Y., Deshayes, G., Desserme, C., & Lefebvre, J. (2000). Biochemical analysis and rheological properties of gluten modified by transglutaminase. *Cereal Chemistry, 77*(2), 121–127. https://doi.org/10.1094/CCHEM.2000.77.2.121
Larsson, H. (2002). Effect of pH and sodium chloride on wheat flour dough properties: Ultracentrifugation and rheological measurements. *Cereal Chemistry, 79*(4), 544–545.
León, A. E., Ribotta, P. D., Ausar, S. F., Fernández, C., Landa, C. A., & Beltramo, D. M. (2000). Interactions of different carrageenan isoforms and flour components in breadmaking. *Journal of Agricultural and Food Chemistry, 48*(7), 2634–2638. https://doi.org/10.1021/jf991340a
Levent, H., & Bilgiçli, N. (2012). Evaluation of physical, chemical and sensory properties of Turkish flat breads (bazlama and yufka) supplemented with lupin, buckwheat and oat flours. *International Journal of Food Science and Nutrition Engineering, 2*(5), 89–95.
López-Tenorio, J. A., Rodríguez-Sandoval, E., & Sepúlveda-Valencia, J. U. (2015). The influence of different emulsifiers on the physical and textural characteristics of gluten-free cheese bread. *Journal of Texture Studies, 46*(4), 227–239. https://doi.org/10.1111/jtxs.12121
Lyons, D. (2016). Bread in Africa. In *Encyclopaedia of the history of science, technology, and medicine in non-western cultures* (pp. 963–974) Springer.
Macritchie, F. (1979). Flour lipids – Theoretical aspects and functional properties. *Cereal Foods World, 24*(9), 442–442.
MacRitchie, F. (1984). Baking quality of wheat flours. In *Advances in food research* (Vol. 29, pp. 201–277). Elsevier.
Marco, C., & Rosell, C. M. (2008). Functional and rheological properties of protein enriched gluten free composite flours. *Journal of Food Engineering, 88*(1), 94–103. https://doi.org/10.1016/j.jfoodeng.2008.01.018
Market Research Future. (2021). *Artisan bakery market global research report by product type (breads & buns, cakes & pastries, biscuits & cookies and others), by category (conventional and gluten free), by distribution channel (store-based (supermarket & hypermarket, specialty store, and others), and non-store-based) and region—Forecast till* 2027. Retrieved April 10 from https://www.marketresearchfuture.com/reports/artisan-bakery-market-3143
Martin, P. J., Tassell, A., Wiktorowicz, R., Morrant, C. J., & Campbell, G. M. (2008). Chapter 20 – Mixing bread doughs under highly soluble gas atmospheres and the effects on bread crumb texture: Experimental results and theoretical interpretation. In G. M. Campbell, M. G. Scanlon,

& D. L. Pyle (Eds.), *Bubbles in food 2* (pp. 197–206). AACC International Press. https://doi.org/10.1016/B978-1-891127-59-5.50024-9
Matz, S. A. (1992). *Cookie and cracker technology*. Springer.
McCann, T. H., Small, D. M., Batey, I. L., Wrigley, C. W., & Day, L. (2009). Protein–lipid interactions in gluten elucidated using acetic acid fractionation. *Food Chemistry, 115*(1), 105–112. https://doi.org/10.1016/j.foodchem.2008.11.070
Meerts, M., Cardinaels, R., Oosterlinck, F., Courtin, C. M., & Moldenaers, P. (2017). The impact of water content and mixing time on the linear and non-linear rheology of wheat flour dough. *Food Biophysics, 12*(2), 151–163. https://doi.org/10.1007/s11483-017-9472-9
Mehta, K., Scanlon, M., Sapirstein, H., & Page, J. (2009). Ultrasonic investigation of the effect of vegetable shortening and mixing time on the mechanical properties of bread dough. *Journal of Food Science, 74*(9), E455–E461. https://doi.org/10.1111/j.1750-3841.2009.01346.x
Moore, M. M., Heinbockel, M., Dockery, P., Ulmer, H. M., & Arendt, E. K. (2006). Network formation in gluten-free bread with application of transglutaminase. *Cereal Chemistry, 83*(1), 28–36. https://doi.org/10.1094/CC-83-0028
Morreale, F., Garzón, R., & Rosell, C. M. (2018). Understanding the role of hydrocolloids viscosity and hydration in developing gluten-free bread. A study with hydroxypropylmethylcellulose. *Food Hydrocolloids, 77*, 629–635. https://doi.org/10.1016/j.foodhyd.2017.11.004
Nabiev, A. (2019). Physical and chemical processes developing in the mass of components during dough mixing. *Food Science and Technology, 13*(3), 10–17. https://doi.org/10.15673/fst.v13i3.1451
Neela, S., & Fanta, S. W. (2020). Injera (an ethnic, traditional staple food of Ethiopia): A review on traditional practice to scientific developments. *Journal of Ethnic Foods, 7*(1), 1–15. https://doi.org/10.1186/s42779-020-00069-x
Nicolas, V., Vanin, F., Grenier, D., Lucas, T., Doursat, C., & Flick, D. (2016). Modeling bread baking with focus on overall deformation and local porosity evolution. *AICHE Journal, 62*(11), 3847–3863. https://doi.org/10.1002/aic.15301
Nishitsuji, Y., Whitney, K., Nakamura, K., Hayakawa, K., & Simsek, S. (2020). Changes in structure and solubility of wheat arabinoxylan during the breadmaking process. *Food Hydrocolloids, 109*, Article 106129. https://doi.org/10.1016/j.foodhyd.2020.106129
Nunes, M. H. B., Moore, M. M., Ryan, L. A. M., & Arendt, E. K. (2009). Impact of emulsifiers on the quality and rheological properties of gluten-free breads and batters. *European Food Research and Technology, 228*(4), 633–642. https://doi.org/10.1007/s00217-008-0972-1
Ooms, N., & Delcour, J. A. (2019). How to impact gluten protein network formation during wheat flour dough making. *Current Opinion in Food Science, 25*, 88–97. https://doi.org/10.1016/j.cofs.2019.04.001
Pahwa, A., Kaur, A., & Puri, R. (2016). Influence of hydrocolloids on the quality of major flat breads: A review. *Journal of Food Processing, 2016*, 1–9. https://doi.org/10.1155/2016/8750258
Panghal, A., Chhikara, N., & Khatkar, B. S. (2019). Characterisation of Indian wheat varieties for chapatti (flat bread) quality. *Journal of the Saudi Society of Agricultural Sciences, 18*(1), 107–111. https://doi.org/10.1016/j.jssas.2017.02.005
Pareyt, B., Finnie, S. M., Putseys, J. A., & Delcour, J. A. (2011). Lipids in bread making: Sources, interactions, and impact on bread quality. *Journal of Cereal Science, 54*(3), 266–279. https://doi.org/10.1016/j.jcs.2011.08.011
Pasqualone, A. (2017). Origine, diffusione e metodi di cottura dei pani piatti. *Tecnica Molitoria, 68*, 522–540.
Pasqualone, A. (2018). Traditional flat breads spread from the Fertile Crescent: Production process and history of baking systems. *Journal of Ethnic Foods, 5*(1), 10–19. https://doi.org/10.1016/j.jef.2018.02.002
Pasqualone, A., Delcuratolo, D., & Gomes, T. (2011). Focaccia Italian flat fatty bread. In *Flour and breads and their fortification in health and disease prevention* (pp. 47–58). Elsevier. https://doi.org/10.1016/B978-0-12-380886-8.10005-4

Pico, J., Bernal, J., & Gómez, M. (2015). Wheat bread aroma compounds in crumb and crust: A review. *Food Research International, 75*, 200–215. https://doi.org/10.1016/j.foodres.2015.05.051

Poutanen, K. (2014). Chapter 1 – Rye and Rye bread—An important part of the north European bread basket. In K. Poutanen & P. Åman (Eds.), *Rye and health* (pp. 1–6). AACC International Press. https://doi.org/10.1016/B978-1-891127-81-6.50001-8

Randez-Gil, F., Sanz, P., & Prieto, J. A. (1999). Engineering baker's yeast: Room for improvement. *Trends in Biotechnology, 17*(6), 237–244. https://doi.org/10.1016/S0167-7799(99)01318-9

Renzetti, S., & Rosell, C. M. (2016). Role of enzymes in improving the functionality of proteins in non-wheat dough systems. *Journal of Cereal Science, 67*, 35–45. https://doi.org/10.1016/j.jcs.2015.09.008

Ringsted, T., Siesler, H. W., & Engelsen, S. B. (2017). Monitoring the staling of wheat bread using 2D MIR-NIR correlation spectroscopy. *Journal of Cereal Science, 75*, 92–99. https://doi.org/10.1016/j.jcs.2017.03.006

Rosell, C. M. (2009a). Enzymatic manipulation of gluten-free breads. In E. Gallagher (Ed.), *Gluten-free food science and technology* (pp. 83–98). Blackwell.

Rosell, C. M. (2009b). Trends in bread making: Low and subzero temperatures. In M. L. Passos & C. L. Ribeiro (Eds.), *Innovation in food engineering: New techniques and products* (pp. 59–79). Taylor and Francis, CRC Press.

Rosell, C. M. (2011a). Chapter 1 – The science of doughs and bread quality. In V. R. Preedy, R. R. Watson, & V. B. Patel (Eds.), *Flour and breads and their fortification in health and disease prevention* (pp. 3–14). Academic. https://doi.org/10.1016/B978-0-12-380886-8.10001-7

Rosell, C. M. (2011b). The science of doughs and bread quality. In *Flour and breads and their fortification in health and disease prevention* (pp. 3–14). Elsevier.

Rosell, C. M., & Collar, C. (2009). Effect of temperature and consistency on wheat dough performance. *International Journal of Food Science and Technology, 44*(3), 493–502. https://doi.org/10.1111/j.1365-2621.2008.01758.x

Rosell, C. M., Rojas, J. A., & Benedito de Barber, C. (2001). Influence of hydrocolloids on dough rheology and bread quality. *Food Hydrocolloids, 15*(1), 75–81. https://doi.org/10.1016/S0268-005X(00)00054-0

Rosentrater, K. A., & Evers, A. D. (2018). Chapter 8 – Bread-baking technology. In K. A. Rosentrater & A. D. Evers (Eds.), *Kent's Technology of Cereals* (5th ed., pp. 565–622). Woodhead Publishing. https://doi.org/10.1016/B978-0-08-100529-3.00008-6

Sadot, M., Cheio, J., & Le-Bail, A. (2017). Impact on dough aeration of pressure change during mixing. *Journal of Food Engineering, 195*, 150–157. https://doi.org/10.1016/j.jfoodeng.2016.09.008

Sakandar, H. A., Huang, W., Kubow, S., Sadiq, F. A., & Imran, M. (2020). Comparison of bacterial communities in gliadin-degraded sourdough (Khamir) sample and non-degraded sample. *Journal of Food Science and Technology, 57*(1), 375–380. https://doi.org/10.1007/s13197-019-04030-y

Schiraldi, A., & Fessas, D. (2012). The role of water in dough formation and bread quality. In S. P. Cauvain (Ed.), *Breadmaking: Improving quality, 2nd edition* (pp. 352–369).

Serna-Saldivar, S. O., & Chuck-Hernandez, C. (2019). Food uses of lime-cooked corn with emphasis in tortillas and snacks. In *Corn* (pp. 469–500). Elsevier. doi:https://doi.org/10.1016/B978-0-12-811971-6.00017-6.

Shewry, P. (2019). What is gluten—Why is it special? *Frontiers in Nutrition, 6*. https://doi.org/10.3389/fnut.2019.00101

Siepmann, F. B., Ripari, V., Waszczynskyj, N., & Spier, M. (2018). Overview of sourdough technology: From production to marketing. *Food and Bioprocess Technology, 11*(2), 242–270. https://doi.org/10.1007/s11947-017-1968-2

Sim, S. Y., Aziah, A. A. N., & Cheng, L. H. (2015). Quality and functionality of Chinese steamed bread and dough added with selected non-starch polysaccharides. *Journal of Food Science and Technology-Mysore, 52*(1), 303–310. https://doi.org/10.1007/s13197-013-0967-1

Simons, C. W., Hunt-Schmidt, E., Simsek, S., Hall, C., & Biswas, A. (2014). Texturized pinto bean protein fortification in straight dough bread formulation. *Plant Foods for Human Nutrition, 69*(3), 235–240. https://doi.org/10.1007/s11130-014-0421-1

Singh, S., & Singh, N. (2013). Relationship of polymeric proteins and empirical dough rheology with dynamic rheology of dough and gluten from different wheat varieties. *Food Hydrocolloids, 33*(2), 342–348. https://doi.org/10.1016/j.foodhyd.2013.04.007

Sluimer, P., & Sluimer, P. (2005). *Principles of breadmaking: Functionality of raw materials and process steps*. American Association of Cereal Chemists.

Sroan, B. S., & MacRitchie, F. (2009). Mechanism of gas cell stabilization in breadmaking. II. The secondary liquid lamellae. *Journal of Cereal Science, 49*(1), 41–46. https://doi.org/10.1016/j.jcs.2008.07.004

Statista. (2022). *Bread*. Retrieved 14 March 2022 from https://www.statista.com/outlook/cmo/food/bread-cereal-products/bread/worldwide

Struyf, N., Van der Maelen, E., Hemdane, S., Verspreet, J., Verstrepen, K. J., & Courtin, C. M. (2017). Bread dough and baker's yeast: An uplifting synergy. *Comprehensive Reviews in Food Science and Food Safety, 16*(5), 850–867. https://doi.org/10.1111/1541-4337.12282

Tara, K. A., & Bains, G. S. (1972). Determination of damaged starch content of flours of Indian wheats. *Starch – Stärke, 24*(5), 159–161. https://doi.org/10.1002/star.19720240505

Waniska, R. D. (2015). Chapter 5 – Processing of flour tortillas. In L. W. Rooney & S. O. Serna-Saldivar (Eds.), *Tortillas* (pp. 125–145). AACC International Press. https://doi.org/10.1016/B978-1-891127-88-5.50005-0

Wieser, H. (2007). Chemistry of gluten proteins. *Food Microbiology, 24*(2), 115–119. https://doi.org/10.1016/j.fm.2006.07.004

Wu, C., Liu, R., Huang, W., Rayas-Duarte, P., Wang, F., & Yao, Y. (2012). Effect of sourdough fermentation on the quality of Chinese northern-style steamed breads. *Journal of Cereal Science, 56*(2), 127–133. https://doi.org/10.1016/j.jcs.2012.03.007

Wu, T., Wang, L., Li, Y., Qian, H., Liu, L., Tong, L., et al. (2019). Effect of milling methods on the properties of rice flour and gluten-free rice bread. *LWT – Food Science and Technology, 108*, 137–144. https://doi.org/10.1016/j.lwt.2019.03.050

Zaroug, M., Orhan, I. E., Senol, F. S., & Yagi, S. (2014). Comparative antioxidant activity appraisal of traditional Sudanese kisra prepared from two sorghum cultivars. *Food Chemistry, 156*, 110–116. https://doi.org/10.1016/j.foodchem.2014.01.069

Zhang, P., He, Z., Chen, D., Zhang, Y., Larroque, O. R., & Xia, X. (2007). Contribution of common wheat protein fractions to dough properties and quality of northern-style Chinese steamed bread. *Journal of Cereal Science, 46*(1), 1–10. https://doi.org/10.1016/j.jcs.2006.10.007

Zhao, C. J., Kinner, M., Wismer, W., & Gänzle, M. G. (2015). Effect of glutamate accumulation during sourdough fermentation with Lactobacillus reuteri on the taste of bread and sodium-reduced bread. *Cereal Chemistry, 92*(2), 224–230. https://doi.org/10.1094/CCHEM-07-14-0149-R

Zhu, F. (2014). Influence of ingredients and chemical components on the quality of Chinese steamed bread. *Food Chemistry, 163*, 154–162. https://doi.org/10.1016/j.foodchem.2014.04.067

Zhu, F. (2019). Glycemic control in Chinese steamed bread: Strategies and opportunities. *Trends in Food Science & Technology, 86*, 252–259. https://doi.org/10.1016/j.tifs.2019.02.038

Zhu, J., Huang, S., Khan, K., & O'Brien, L. (2001). Relationship of protein quantity, quality and dough properties with Chinese steamed bread quality. *Journal of Cereal Science, 33*(2), 205–212. https://doi.org/10.1006/jcrs.2000.0358

Ziobro, R., Juszczak, L., Witczak, M., & Korus, J. (2016). Non-gluten proteins as structure forming agents in gluten free bread. *Journal of Food Science and Technology, 53*(1), 571–580. https://doi.org/10.1007/s13197-015-2043-5

Zurayk, R., & Rahman, S. A. (2008). *From 'Akkar to Amel'Lebanon's food trail*. Slow Food.

Chapter 3
Biscuits

Antima Gupta and Satinder Kaur

1 Introduction

Owing to accessibility, good sensory appeal and equitable price; bakery products are most acceptable and popular food items among Indian and world population. Among all these bakery products like cake, bread, pastries etc., biscuits are most popular and essential part of society as biscuits are available in varieties. Biscuit is ready to eat, conventional food. The word 'biscuit' means to be 'cooked twice' in Latin; as traditionally, biscuits were first cooked and then allowed to hardening by cooling. Wheat flour is used as major constitute owing to gluten content that provides elasticity and overall acceptability to product. The word biscuit has been originated from modern French whereas in Italian it is known to be *Biscotti*, *Zweiback* and *Beschuit* in German and Dutch respectively. Besides versatility of perishable goods, there is need to easily available, easy to carry food items having good shelf life for various purpose such as military missions and long-distance sea journeys (Anderson, 2010). There are various types of biscuit available in market which generally lack in fiber and protein content but are good source of fat and carbohydrates. Biscuits are accepted as good snack aby all age groups due to its taste and longer shelf-life and also it is considered good because of its easy improvement in nutritional quality and fortification. Major ingredient used in manufacturing of biscuits are refined wheat flour, sugar, milk and other are eggs, margarine, butter etc. Owing to continuous development of new technologies, awareness and knowledge in baking, and various innovative equipment; type and quality of biscuit has been increased drastically. Biscuit manufacturers become aware about consumers expectations and needs thus shifted towards involvement of new and innovative technology. Now a

A. Gupta (✉) · S. Kaur
Department of Food Science and Technology, Punjab Agricultural University, Ludhiana, Punjab, India
e-mail: antimagupta@pau.edu

M. A. Shah et al. (eds.), *Cereal-Based Food Products*,
https://doi.org/10.1007/978-3-031-40308-8_3

days, emergence of new and healthier biscuits by lowering sugar and fat content is in trend and also available in luxurious form in chocolate covering. The study reported that different plant species are used for fortification of wheat-based biscuits (Preedy et al., 2011). Besides fluid, a diet should contain well balanced macro and micronutrients in healthy diet. Globally there are different models used for description of proportions to be involved in healthy diet. Such as United States Department of Agriculture (USDA) use food pyramid, Eatwell/food well plate and the food guide rainbow are used for UK and Canada respectively. Vegetable and fruit rich starchy food stuffs are recommended by health professionals which include potato, pasta, rice etc. These sections of food group should cover around two-third part of daily diet. And one-third remaining part of diet should be composed of non-dairy products which are rich source of protein like fish, meat, egg etc. in addition to reduced fat and sugar content. On the other hand, dairy product like milk, curd etc. should be included in daily diet. Fluid formed nourishment should be around 1.5–2.5 liters daily. As biscuits and other sweet have higher energy value because of presence of higher content are covered under different food group. In the context of energy expenditure, the energy obtained from this food group should be sufficient to meet energy. This balance gives description of energy in = energy out that is see-saw affect; as energy consume by an individual in daily diet needs to expend that in daily exercise and activities. Owing to this, daily requirement of energy is different among individuals and vary according age, gender, level of physical activity and weight. WHO i.e. World Health Organization states that biscuits and cakes included in diet leads to obesity due to high energy as higher fat content is present. Besides of energy source, biscuits are consumed as source of enjoyment on social meet and greet which is also a source of overall well-being. Due to good texture, mouthfeel, easy to carry, lightweight biscuit is considered as good snack. Due to high sugar and fat content recommendation of biscuit is in small quantity in diet. Scientists recently reports that biscuits can be included to diet for making it a proper balanced diet (Astrup et al., 2006; Field et al., 2004). Beneficial output of biscuit taking as snack are listed below;

- Snacking leads to overall well-being by providing mental health as it provides enjoyment and improve concentrations.
- It prevents overeating by fulfil the requirement of frequent cravings.
- It provides nutrients as biscuits are good source of energy.

When individual works hard, brain needs concentration which involves expenditure of 20% of energy level from total body requirement of body. These hard work leads to reduction in total blood sugar level which can be completed by consumption of snacks like biscuits instead of overeating and prevent high glycemic index which cause dizziness. Brain is about only 2% of total body weight but it requires 20% of glucose requirement as it does not store and supply continuous energy for work. So, brain requires about 100 g of glucose to meet the continuous requirement of glucose level. Biscuits as good source of glucose in form of carbohydrates; works easiest and affordable source of energy to brain frequently in working institutions and in our daily routine. In recent years, government public health department turn their focus on reduction of sugar and fat level in biscuits. Such as Food Standard Agency (FSA) of UK, target energy reductions and voluntary saturated fats by publishing in 2010 (FSA, 2008). Beneficial outcomes are emerging in biscuits such as McVitie's lightweight and digestive biscuits is reported as reduced fat content about 80%, McVitie's Hobnobs is reported as 75% reduction of saturated fat. The main impact on food innovation is basically a connection which is between health and food. As Indonesian children prefer various types of biscuits, this ready-to-eat snacks made from composite flour in local market was selected as food carrier for RUSF i.e., ready-to-use supplementary food (Sekiyama et al., 2012; Inayati, 2011). At present, concept of functional food is most popular. To overcome national insecurity and problem of malnutrition, nutritionally rich underdeveloped foods become more popular (Dandin & Kishna Kumar, 2016; Durst & Bayasgalanbat, 2014). There is complete focus on development of health-oriented products; as proper amount of micro and macro nutrients in addition with sufficient amount of energy is required for development of human body (Lazzerini et al., 2013; Guimón & Guimón, 2012). Among the obese people eating of high calorie food is common practice and they are unable to abstain the consumption of high fat products due to their cravings. Therefore, better product which is healthier as compared to that of conventional product need to develop but without major change in taste, texture and other sensory appeal.

2 Position of Biscuit in Nutrition

Traditional rearing of livestock leads to increment in production of animal protein which ultimately leads to unsustainable elevation in greenhouse emissions (Gerber et al., 2013). Decrease in livestock is advantageous for environment on the other hand it leads to increase in malnutrition incidences as 10.7% of total world's population i.e., 815 million is suffering from chronic diseases (FAO et al., 2017). Food insecurity is another cause of malnutrition. A study reported that unavailability of well-balanced food sources leads to deficiency of various important nutrients in population such as folate, iron, protein and vitamin A ((Ayensu et al., 2017; Saaka et al., 2017)). This evokes for the requirement of eco-friendly and well-balanced methods and techniques for production of nutritional food items. As negative impact of malnutrition cannot be underrated; therefore, Various food sources have easy accessibility, cheap, convenient with essential nutrients is requirement of present scenario and The Nutritional requirement of population on global level is varies drastically according to their specific and personal level. Beside all these, some common needs are shared by particular groups of people for solution of common issues. These issues may be religious in which people follow many prohibitions, vegans, vegetarians, undesirable food reactions-gluten and lactose intolerance, infant nutrition, lifestyle. Others include allergy from various allergens present in different food items such as; crustaceans, lupin, molluscs, nuts, peanuts, fish etc. (Table 3.1)

Table 3.1 Global religious practices and food prohibitions. (Waibel, 2010)

Religion	Prohibition or practices
Eastern orthodox Christianity	• Few prohibitions for fish and meat
Buddhism	• Beef is restricted. • Vegetarian diet is preferred.
Islam	• Some birds and pork are restricted.
Protestants Rastafarianism	• Vegetarian diet is preferred. • Salt, condiments and preservative is prohibited.
Hinduism	• Vegetarian diet preferred. • Beef is restricted.
Roman Catholicism Seventh-day Adventist	• Pork is restricted.

It's very common practice to utilize biscuits as a way of exercise for teething in infant and various types of biscuits are good source of nutrition for infants and good in taste. Biscuits are preferred by all age groups and a good source of nutrition specially when fortified with calcium, iron vitamin A, zinc and fluorides. A study reported that for 2 years children, daily intake of fluorides that is contributed by biscuits is 25% of maximum daily intake of fluorides i.e., 0.07 mg/body weight (Buzalaf et al., 2004). Biscuit's consumption provides high molecular weight carbohydrates, some minerals, vitamins (by fortification) and vegetable proteins. Biscuits and rusk for infants are Strictly under control by legislation in order to keep check on quality, both in case of nutrition and ingredients. Both these products provide softness when broken in pieces and dipped into milk (for infants and elderly). Generally, biscuits are known for its contribution of proving high energy owing to presence of high sugar and fat content. In addition to empty calories, snacking has insignificant role in dietary fibers and some other nutrients. One fifth portion of energy intake in young and children and one-eighth portion of energy intake in adults is contributed by various confectionary and bakery products such as pastries, cakes, cookies etc. In various feeding programs, biscuits are used as positive vector for fortification. Therefore, different kinds of fortification like protein and dietary fiber fortification are in trend. Snacking has no direct significant effect on human health but there are indirect factors like frequency and type of food in involved in diet (Table 3.2).

3 Classification of Biscuits

Biscuits have three major classification system- BIS, UK and USA classification (Fig. 3.1). BIS classified biscuits based on the sensory attributes as following:

TYPE I Sweet,
TYPE II Semi-sweet,
TYPE III Crackers,
TYPE IV Cookies,
TYPE V Specialty Biscuits.

Table 3.2 Nutritional composition of standard biscuits available in market

	Standard biscuits (Per 100 g)
Energy	448 kcal
Carbohydrates	68.8, out of which sugars are about 32.5
Protein	4.8
Fat	16.7
Sodium	0.2
Fibre	1.7
Poly-saturates	1.9
Mono-saturates	6.8
Vitamin B6	–
Vitamin 12	–
Vitamin C	–
Folic acid	–

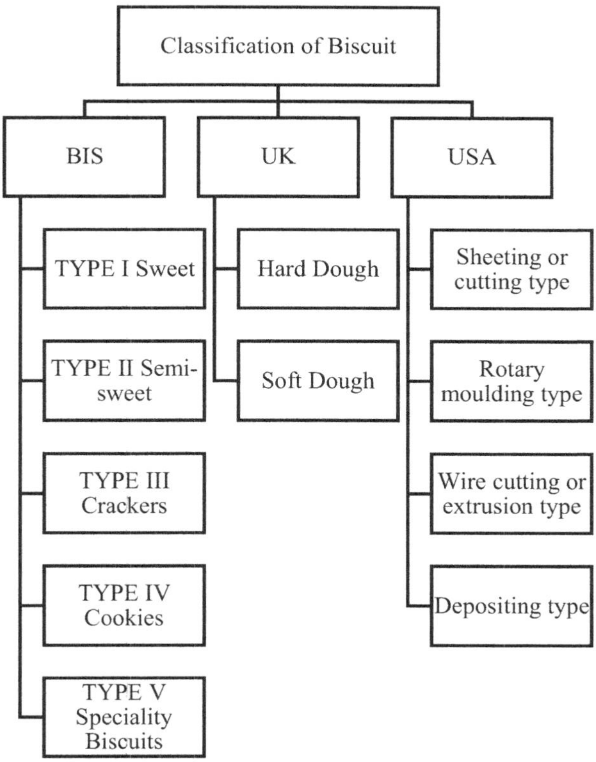

Fig. 3.1 Classification of biscuits

Second classification of biscuit is by UK, based on the formulation as:

- ***Hard dough biscuit:*** They are made from bread wheat having higher protein content, in this we allow the gluten formation during mixing with very low fat and sugar resulting in hard textured biscuit example crackers, sweet and semi-sweet biscuits.
- ***Soft dough biscuit:***They are made soft wheat flour; in this we don't allow gluten formation along with higher fat and sugar resulting in tender biscuits.

Third classification USA based is on the method of dough piece formation:

- ***Sheeting or cutting type:***It involves continuous compression of dough through laminator to form sheet of desired thickness. This sheet is then passed through a short series of gauge rolls to reduce the thickness further followed by passing through the cutter of various size and shape and placed on an oven band for baking. In these majorly hard doughs are used which exhibits some extensibility. Example: Cream cracker, puff cracker, marie gold, krackjack, Monaco etc.
- ***Rotary moulding type:***It is the principal method used to make dough pieces from short dough. Rotary moulding is not suitable for very soft dough or for doughs which contains larger particles like nuts, chocochips etc. In thisdough directly fed into rotator moulder with dies of various size and shape. Example of rotary type biscuit are glucose, good day etc.
- ***Wire cutting or extrusion type:***It is the third type and in this soft short dough is passed through holes in extruder and these doughs are then wire cut in pieces which are directly dropped into oven band. In this co-extrusion can also be done. Example of wire cutting are chocolate cookies, butter cookies etc.
- ***Depositing type:***The dough is very soft, due to high fat and containing egg. The extrusions are intermittent through a row of nozzles mounted in a depositor head which rises and falls and discrete amounts of the dough form as pieces to be baked. The nozzle may rotate or oscillate giving interesting shapes to the deposits formed. Example: cream treat biscuits available in the marker are made by depositor type.

According to food technologists, it is of utmost importance to categorize biscuits on the basis of their internal and external parameters which is beneficial in making and baking procedures of desirable end product such as dough pieces molded, cut or extruded. Formation methods are limited whereas various patterns are formed during baking. Low sugar and fat containing doughs have less impression as compared to doughs which are rich source of sugar and fat. Internal structure in dough have higher amount of sugar and fat is more open, irregular and crumbly whereas gluten provides viscosity in dough of many biscuits. Basically, dough consistency which is directly related to water requirement as it acts as catalyst in biscuits making, depends upon level of various alkaline leavening agent used, flour type a different type of cereals. Water is added to dough and it get absorbed during baking. During this adding and absorbing cycle, it performs various functions as follows;

- It causes hydration of flour starch and protein.

- Hydration cause gluten formation which provides proper structure to dough and thereby end product.
- Water acts as solvent for dissolution of various ingredients such as leavening chemicals, sugar, salt.
- Acts as dispersion medium for fat and others.

If lower quantity of water is used in recipe, fat quantity is increased automatically which leads to reduction in flour hydration and starch gelatinization. This impact marks overall sensory appeal of biscuits. Non-gelatinized starch tends to produce biscuits have soft texture. Therefore, biscuits under groups namely; pizza, bread, and bread crisp, cream crackers, water biscuits and soda crackers, cabin biscuits, semi-sweet/hard-sweet and savory crackers are soft to eat with exceptional case of having high sugar content which leads to hardness of biscuits. In microstructure of hard-sweets types and crackers show microstructure in which starch is embedded in protein matrix and has relatively thin walls as compared to others. Examples like slicing of frozen dough of co-extrusion which engineers emerges in technology of biscuits both in cream and dough. Various cereal and vegetable-based snacks confined to small shapes and flavored with various flavors form crisp type foods. It has been found that flat strips formed by extrusion in medium size cutting can be accepted as biscuits. Wafer biscuits are special category of baked goods owing to its formation which is between pair of hot plates rather than baking band or wires. It is made up from pumpable, fluid able batter has low sugar and fat content.

4 Ingredients and Their Functions

Primarily soft wheat flour, fat, sugar and water are being used in the preparation of biscuit. Moreover, other ingredients like skimmed milk powder, baking powder and emulsifiers are also added for leavening and givingflavour and texture to the biscuit. The quantity and nature of ingredients determines the quality of biscuit (Fig. 3.2).

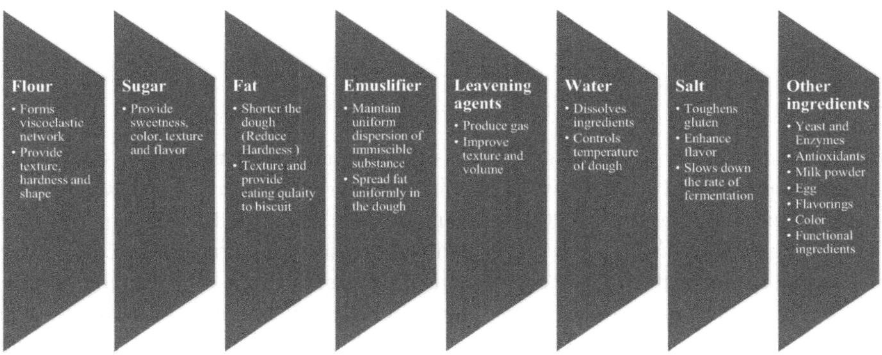

Fig. 3.2 Role of ingredients in biscuits

Wheat Flour

For the preparation of nearly all kinds of biscuit, soft wheat flour is generally being usedowing toits low protein content and soft kernel texture that ultimately results in tender crumb product (Hoseney et al., 1994). Biscuit flour contains around 8–11% protein, 70–75% starch and 14% moisture (Mamat & Hill, 2018). Additionally, it is also composed of fat, some fibres, ash, minerals, vitamins and phytochemicals. Soft wheat flour is obtained from the soft wheat by removing pericarp, seed coat, aleurone layer and germ from the endosperm. Typically, for biscuit flour preparation, the endosperm obtained is milled to the extraction rate of 70–75% with an average particles size of around 50 μm in which less than 10% have a particle size of more than 130 μm (Manley, 2000). The soft wheat ground finer than hard wheat due to soft texture of endosperm which is primarily due to the presence of friabilin protein. Friabilin constitutes puroindolines a and puroindolines b along with grain softness protein (*GSP-1*) (Bettge & Morris, 2000). These finer particles have larger contact area that leads to higher water absorption and produce dough of required consistency. In addition to particle size level of damaged starch and protein decides the amount of water absorption. Kulp et al. (1991) reported that starch in the biscuit system doesn't play any role in structure forming as it doesn't gelatinize during baking due to high levels of sugar and fat rather it acts as a filler in the matrix (Arepally et al., 2020). Low protein content of biscuit flour is considered ideal as minimal gluten network is desirable for sheeting and forming of biscuit dough.

In soft wheat flour gluten is present which are responsible for celiac disease. Thereby, wheat flour is substituted partially or completely for the preparation of nutritionally rich biscuit. Several researches have been conducted to improve the nutritional quality and sometime to make it gluten free biscuit such as buckwheat, sorghum and lentil flour blend (Cairano et al., 2022), rice and carob flour blend (Skaltsi et al., 2021), hybrid corn (Hussein et al., 2020), partially substituted mungbean biscuits (Fathonah et al., 2020), rice and chickpea flour (Benkadri et al., 2018), oat flour (Duta & Culetu, 2015), water chestnut, makhana and potato mixture (Mishra et al., 2015) etc.

Sugar

Sucrose is generally being used in the biscuit preparation that provides sweetness, color, crispiness and crunchiness to the biscuit. However, sucrose is not solely used in the biscuit preparation, 10–20% reducing sugar is also added in the form of syrup containing 70–80% dry solids. The crunchiness majorly governs by the crystal size and speed of dissolution. Quantity of sugar also describes the organoleptic properties of the end product. Higher sugar content reduces the hardness of dough as well as restrict the gluten formation by competing with water resulting in less tough and crumbly biscuits (Gallagher et al., 2003). Sugar also increases the gelatinization temperature thus allowing the dough to make more time to rise during baking. Hoseney et al. (1994) reported that during baking, undissolved crystals melt progressively and results in spreadability of the dough. During cooling of biscuits, the melted sugar converts into non-crystalline glass like structure that give crispiness and crunchiness to the biscuit.

During baking, sugar combines with amino acids and contributes to the color of biscuit by caramelization reaction. Nowadays people are more prone to diabetes thereby bakers replace the sucrose with low calorie sweetener like, saccharine, sucralose, cyclamate, aspartame, acesulfame-K, raftilose, etc. in biscuit preparation (Savitha et al., 2008). Among these, Acesulfame-K is mostly used for biscuit preparation due to its stability during baking (Zoulias et al., 2000). However, these artificial sweeteners are limited to sweetness only, they don't provide any other functionality like texture, color and structure like sucrose. Thereby, bulking agents like polyols are employed along with sweeteners (Zoulias et al., 2000; Olinger & Velasco, 1996).

Fat

Fat is an important ingredient in achieving tenderness, mouthfeel, andbite of the biscuit (O'Brien et al., 2003). Fat acts a stabilizer, lubricating agent and give good organoleptic properties to the biscuit. Mostly, fats with melting point below blood temperature (36.9 °C) is preferred as it avoids waxy mouthfeel of the biscuits. Typically, hydrogenated vegetable oil, high oleic sunflower or palm olein is used. Vegetable oil like cottonseed and soybean are regularly used while corn, groundnut and coconut oil are used to a limited extend. Fat like vegetable fat/oil and butter are generically termed as shortenings. Shortening lubricates during sheeting of dough by hindering the adherence of starch and gluten and results in tender, softer and uniform shape of the biscuit (Mert & Demirkesen, 2016). Moreover, it also provides aeration, spreadability and organoleptic properties. Type of fat, composition of fat and quantity of fat influences the dough and biscuit quality (Mamat & Hill, 2014).

During mixing, the fat envelops the flour particle this restricts the hydration and interrupt the gluten development. Thegluten development is least desirable in the biscuit dough as it results in elastic dough and ultimately results in hard biscuit. It also tends to inhibits the leavening action of the carbon dioxide diffusion in the dough during baking by forming a layer around the flour particles, resulting in softer and finer texture (Davidson, 2016). Moreover, fat also influences the dough machinability, the dough spread after cutting, and textural properties of biscuit after baking. Fat interacts with other ingredients in the dough to give excellent mechanical, rheological and organoleptic properties. Despite the key role ingredient in the biscuit, it is necessary to reduce the fat content due to its associated health risk with it like obesity, arthrosclerosis, and coronary heart disease etc.

In order to combat the health risk, fat can be replaced or substituted with fat replacers. Fat replacer are generally categorized as fat substitutes and mimetics. Fat substitutes are those that have a chemical structure somewhat similar to fats and have similar physiological properties, replace fat on gram-to-gram basis (example- Olestra, capenin, salatrim etc.). Fat mimetics are those that have distinctly different chemical structures from fat and mimic few characteristics of fat (example- carbohydrate based, lipid based, protein based, emulsifier based and synthetic based). Zoulias et al. (2002) studied the effect of fat replacement on the texture of cookies. They recommended maltodextrin and simplesse could be used as a fat replacer to give more tender low-fat cookies in comparison with polydextrose or dairytrim. Sudha et al. (2007) also reported similar result in biscuit, replacement of fat with maltodextrin improves the biscuit texture in comparison to polydextrose. The substitution of fat have greater impact on the textural attributes of biscuit which are still unresolved.

In addition to fat percentage, types of fat especially trans fats are also major point of consideration. Mostly, biscuits available in market contains trans fats ranging from 0.1 to 3.2 g/100 g and in the butter biscuit it accounts for 17.1 g/100 g of fat reported in one survey report by Kala in 2014. They can be replaced totally by using vegetable oils rich in unsaturated fats, but problem of rancidity will be there. Thereby, Tarancón et al. (2015) developed oleogels from olive oil as an alternate to fat replacer. The advantage of oleogels is the health aspect.

Other Ingredients

Emulsifiers are the amphiphilic molecules with hydrophilic head and hydrophobic tail, that reduces the interfacial tension between water and fat and achieves greater stability. Moreover, addition of emulsifier beside providing stability it distributes the fat uniformly, prevent moisture migration and improves texture of finished product (Manohar & Rao, 1999). Glycerol monostearate, sorbitol, propylene glycol, lecithin, sodium stearoyl-2 lactylate, saccinylates monoglycerides and mono and diglycerides are the emulsifier generally used in foods. Among them, lecithin is the most commonly used emulsifiers in the biscuit. It improves the eating quality of biscuits.

Leavening agents are the inorganic salts such as baking powder, sodium pyrophosphate, sodium bicarbonate and ammonium bicarbonate. They added to dough to produce gases that leavens he dough and primarily responsible for the texture development in biscuits. Sodium bicarbonate on reacting with acidulants in the dough releases carbon dioxide while ammonium bicarbonate decomposes on heating and produces carbon dioxide. These salts affect the hydroxymythylfurfural (HMF) and acrylamide formation in the biscuit.

Water also is one of the ingredients in all types of biscuit preparation. The hardness of water may affect the quality of dough, hard water gives tightening effect due to the presence of calcium and magnesium while soft water gives smoothening effect to the dough. Additionally, it has a significant effect on the oval quality and palatability of the finished product. Water leavens the products by converting itself into vapors during baking and assist in controlling the temperature of dough (Patel et al., 2003).

Functional Ingredients

As the paradigm shift towards the healthy and convenience food. Biscuit is one of the convenience product mostly consumed on everyday basis and making them healthier is one of the challenges as one change in ingredient can influence whole structure of biscuit. Thereby standardization is required on every enrichment and fortification. Processing waste like bran, oilcake, pomace, peels etc. can be used as a functional ingredient at optimized level to enrich the biscuits and make them healthier with the nutritional ingredients. Moreover, dried fruits and vegetable powders, gluten free flour, multigrain flour, protein powders, fibers etc. can also be added to make healthier biscuits with required health benefits. Galla et al., 2017 fortified the biscuits with 5, 10, and 15% spinach powder and evaluation the effect on nutritional, textural and organoleptic properties. They concluded 5% spinach power enriched biscuits has a highest acceptability index.

5 Preparation Methods of Biscuits

In this section method, process and equipment involved in the manufacturing of biscuits are discussed. The basic unit operation involve in biscuit preparation includes mixing, shaping/forming and baking (Fig. 3.3).

1. **MIXING:** Mixing is commonly defined as the process of blending different ingredient into a uniform homogenous dough within standardized time period.

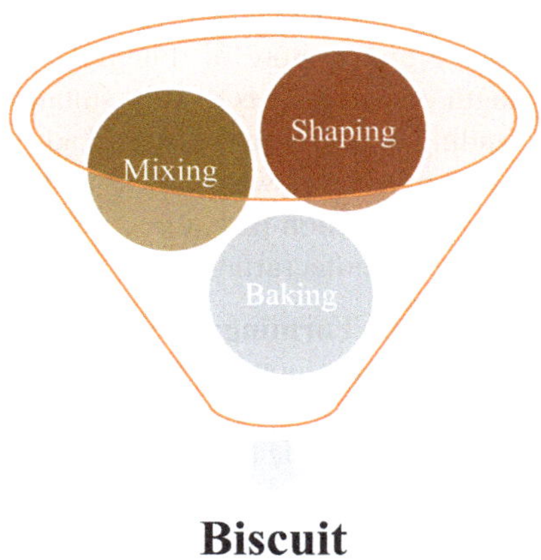

Fig. 3.3 Basic steps in biscuit preparation

Generally, mixing mainly involves blending, dispersing and dissolving solid ingredients into a liquid medium and forming a dough by kneading and pass it to the next step via conveyor belt. Mixing usually takes 15–20 min and depends on the factors such as characteristics of flour, other ingredients and their quantity and temperature of dough during mixing. There are three types of mixers used for mixing ingredients that are *Horizontal mixer, Vertical spindle mixer and continuous mixer.*

Horizontal mixer has bowl mounted on a rigid frame that encloses the drive motor. The dough after mixing can be easily remove just by titling it. Mixer has different blades of variable shape to perform cutting, scraping and kneading action. There are various advantages of horizontal mixer like it works on high speed, continuous dough supply to the processing line is possible and uniform mixing is achievable. Horizontal mixer has some disadvantages also like most of the ingredient are fed manually, chances of unhygienic conditions may prevail due to the spillage of ingredients, can't observe process due to the fully enclosed system, only one process like charging, mixing, or discharging can be performed at a time.

Vertical spindle mixers were extensively used before the horizontal mixers. Vertical mixers are typically used by the small bakeries. They also known as the spiral or detachable bowl mixer. The mixer contains spindle and dough is mixed by upward and downward movement. The blades present do the cutting action instead of kneading. There are various advantages of vertical mixer over horizontal like cleaning operation is easy, suitable of all types of dough, available in different capacities, mixing process can be visually monitored and further mixing of other ingredients can be performed and over and undermixing can be avoided. There are several disadvantages with the vertical mixer primarily its low speed which increases the time of mixing and energy, the tub and dough handling equipment is very heavy and mixing at top occurs which results in uneven dough.

Continuous mixer includes uniform mixing which gives continuous and consistent dough stream for the production line. This mixer works in the continuous line and supply dough for the next unit operation on a continuous basis. All kind of dough can be handled by altering different arms and stators. The orifice at the end of the rotor provides desired shape to the dough. Water jackets are attached to control the temperature of dough and mixing speed can be adjusted by varying the length of barrel. It is mostly suitable for the doughs that requires minimum or no standing time. There are several advantages of continuous mixer, it can operate with minimum man power, uniform and homogeneous dough, minimum wastage, and lower supervision is required. Major limitation with continuous mixer is that it can used for manufacturing single product as changing the machine parts is difficult task.

2. **Shaping/Forming:** In biscuit preparation the mixing and baking facilities can be used in common but for each different type of biscuit forming process is different. After the ingredients are mixed and converted into dough, the dough is rested for some time to allow fermentation or gluten development. At this stage the time for resting and temperature of resting are critical in controlling the final quality of the product. The forming process is classified into three categories:

Sheeting or cutting, rotary moulding and extrusion through the die(Fig. 3.4). For hard dough with good extensibility due to gluten development is generally uses sheeting or cutting forming process while for soft dough with less gluten and extensibility rotary process is acceptable. For very soft dough generally extrusion forming process is used.

*Sheeting or cutting forming process*is usually preferred for the cracker and cookie dough pieces. In this fully developed dough is feeded through the hopper and followed by passing through the three- or four- roller sheeter that will make compression and make sheets of desired thickness (4–5 nm) (Fig. 3.4a). In three rollers, first roller having the grooves that facilitates the drawing of the dough from the hopper into the sheeter and rest successive roller are smooth also known as gauging rollers there to smoothen the dough of desired thickness. The distance between the rollers are set as per the required thickness of the sheet. Once the dough is sheeted it is discharged through conveyor belt either in front or a back discharge. Additionally, for the production of laminated products such as crackers, hard-sweet biscuit and baked snacks laminator is used that will give light and crisp texture product. Laminator repairs the sheet that was formed during simple rollers, uniform stress can be achieved, and repetitive cycles of rolling and folding cause more working of dough results in crispiness in the end product.

*Rotary moulding forming process*is most suitable for short dough type biscuits and sandwich cookies. It consists of a set of three rollers arranged in a triangular fashion and placed below the dough hopper. First roller also known as forcing roller will take the dough to the mountain roller which second have engraved surface, helps to form pieces of dough followed by extraction roller. Excess dough from the forced roller is scrapped by the scrapper which located between forcing and mounting rollers. Extraction roller is pressed against the forcing roller to extract the dough pieces. The dough is then collected in catchment tray and transfers them to baking belt.

*Extrusion forming process*is used for very soft short dough which passes through the orifice by means of pump or rollers. It includes wirecut and rout presses machines. The extruded pieces of desired shape and size are directly falls on the baking band.

3. **Baking:** It is the important step in manufacturing, as the final product quality and shelf life depends on the effectiveness of the baking process. Several changes occur during baking like change in density, structure becomes porous, evaporation of moisture, and change in color due to the browning reaction. Small bakery generally uses simple oven while for industrial baking traveling and band oven is used. Baking profile varies for almost all types of biscuits. Time varies from 3 to 12 min depending upon the product, shortest time is usually for crackers and highest for cookies. Temperature varies from 140 to 240 °C, depending upon product. The freshly baked biscuits need to be cooled before packaging as if we pack warm biscuits, they will get soggy as well as packaging material will also shrink as biscuit leaves oven at 100 °C. The biscuits that are packed in primary packaging may cooled upto 40–45 °C while those are packed in secondary packaging are cooled to 18–26 °C.

Mixing (Horizontal/Vertical/Continuous Mixer)

↓

Dough

↓

Making thin sheet of desired thickness (4-5 nm) ←

↓ Scrap dough

Cutting into desired size and shape →

↓

Baking

↓

Cooling/ Packaging

(A) SHEETING AND CUTTING METHOD

Mixing (Horizontal/Vertical/Continuous Mixer)

↓

Dough ←

↓

Feed roller will pull down the dough Scrap dough

↓

Forcing of dough into cavity of the engraved roller

↓

Scraping of excess dough →

↓

Extraction roller will extract the dough pieces

↓

Fig. 3.4 Methods of biscuit formation (**a**) sheeting and cutting method (**b**) rotary moulding method (**c**) extrusion method

Conveyor belt

↓

Baking

↓

Cooling/ Packaging

(B) ROTARY MOULDING METHOD

Mixing (Horizontal/Vertical/Continuous Mixer)

↓

Dough

↓

Forcing dough into the pressure chamber

↓

Extrusion of very short soft dough through orifice

↓

Cutting of extruded dough by wire cut

↓

Baking

↓

Cooling/ Packaging

(C) EXTRUSION METHOD

Fig. 3.4 (continued)

6 Quality Aspects

The different biscuits prepared are checked for its quality parameters using different test for quality like flour test, physical parameters, rheological parameters, dough test, baking test, texture, sensory and many more as specific to product are conducted. General flour tests are conducted before the biscuit preparation like moisture, protein, fat, ash, carbohydrates, particle size, hydration capacity etc. are studied.

To study the rheological properties alveograph is used that measures the dough extensibility and resistance to expansion which ultimately decides the suitability of flour for different biscuits, crackers and cookies. In this a thin sheet of dough is clamped between two metal plate, upper plate holds the dough and allow it to expand while the lower plate is provided with air valve. The manometer attached records the rapid initial rise in the pressure as the dough sheet resist to expand, then the decline in pressure notices due to the thinning and stretching of sheet ultimately forming bubble which burst on excess pressure. The sudden loss in pressure is recorded and come in graphical form. Extensiograph is another instrument to measure the resistance to stretching of the dough.

Dough test is conducted by farinograph that will estimate the extend of water absorption by flour during mixing, the consistency of the dough and time required for the dough development to the maximum stiffness and the resistance from mixing to breakdown. These parameters will help to know the suitability of flour for specific product. Baking test is another test to check the flour quality and its applicability for the end use. Simulated conditions were provided to check the suitability.

The important physical parameter to be tested for bread is a spread ratio, spread ratio is affected by quality and quantity of sugar and fat. The spread ratio is indirectly affected by the flour type. It is the ratio of diameter is to thickness. Ojha and Thapa (2017) reported decreased spread ratio with increased level of mandarin peel powder from 3 to 9%. Nassar et al. (2008) also reported decreased spread ratio of biscuits supplemented with orange peel. They suggested protein quality difference and disturbance in water absorption capacity influences the spread ratio of biscuits. Ikuomola et al. (2017) reported increase in malted barley bran in cookies @ 0–50% increased the spread ratio from 7.97 to 8.68.

The prepared biscuits were further checked for textural properties by texture analysed by the texture analyser, to know whether the new ingredient or any new treatment applied to the system will influence the textural properties to what extent. They will be check for color using hunter color lab. Sensory characteristics of biscuits like appearance, color, texture, taste, flavor and overall acceptability of biscuit also evaluated for mass consumer acceptability before introducing the product into the market.

7 Sustainability in biscuit Industry

Sustainable agriculture is an integrated system of plant and animal production methods that can meet the long-term food and fibre needs of people while also preserving and improving the environment and natural resources both for now and future population. There is a need to adopt agroecological and other innovative approaches for sustainable agriculture and food systems that enhance food security and nutrition and causes minimum impact on the resources. For the sustainability of the food processing industry, the challenge is to develop strategies that improve social equity, environmental health, and economic profitability. Food demand makes up 26% of the global ecological footprint.

Multiple variety of biscuits are available in the market among them chocolate-coated biscuits are least sustainable where low fat/sugar biscuits are healthier and more sustainable than any other biscuits. The highest (1.81 kg Carbon dioxide eq./kg) global warming potential was found for chocolate-coated biscuit and lowest (1.27 kg Carbon dioxide eq./kg) was for low fat/sugar biscuits. The raw material (41–61%) and manufacturing process (24–38%) especially biscuit baking are the main hotspot for the global warming. In the chocolate-coated biscuits, milk powder is the most significant contributor accounting 30% global warming potential (Konstantas et al., 2019). The environmental problems like ozone depletion was most impacted by raw material used for biscuit making like wheat flour for the cultivation of wheat which requires water and land with full of nutrients. Fresh water and marine eutrophication was mainly due to the wheat cultivation, palm/kernel oil, sugar and cocoa powder.

As the lifestyle changed to sedentary thereby low fat/sugar biscuits can be preferred as they are healthy. Any waste products from industry can be processed and converted into functional food ingredient for the development of healthy biscuits which will increase the sustainability of biscuit industry.

8 Conclusion and Future Trends

Biscuits have appealing qualities like widespread consumption, extended shelf life, and accessibility in a variety of tastes and textures. The nutritional, textural, and physical characteristics of the biscuit are greatly influenced by the ingredients; therefore, they must be used in the right proportion to provide the desired quality. Due to the energy-intensive process of baking, research is being done in finding novel ways, such as microwave and RF drying, to minimise the energy consumption of the industrial continuous baking oven to make the process sustainable. Biscuit industry has evolved from labour intensive to well mechanized industry, minimizing the energy consumption while maintaining the product quality is a big challenge for the industry as well as to the scientists.

In future due to the paradigm shift in the consumer preference towards the healthier foods, new functional ingredients need to be explore to make healthier biscuit while maintaining the organoleptic properties. It is still necessary to understand the different interactions in the biscuit dough system. New fat replacers, sweeteners can be identified and their effect on the structural properties can be studied. Food scientist still need to identify healthier substitute and their inter-relation with the basic ingredients to develop nutritionally healthy biscuits.

References

Anderson, K. (2010). *Hardtack*. Available from http://www.kenanderson.net/hardtack/

Arepally, D., Reddy, R. S., Goswami, T. K., & Datta, A. K. (2020). Biscuit baking: A review. *LWT – Food Science and Technology, 131*, 109726. https://doi.org/10.1016/j.lwt.2020.109726

Astrup, A., Bovy, M. W. L., Nackenhorst, K., & Popova, A. E. (2006). Food for thought for food? – A stakeholder dialogue around the role of snacking industry in addressing the obesity epidemic. *Obesity Reviews, 7*, 303–312.

Ayensu, J., Edusei, A., Oduro, I., & Larbie, C. (2017). Status of some antioxidant micronutrient and pregnancy outcomes in Ghanaian adolescents attending antenatalclinic in urban (Suntreso) and rural (Mampong) hospitals. *European Journal of Nutrition & Food Safety, 7*(2), 120–127.

Benkadri, S., Salvador, A., Zidoune, M. N., & Sanz, T. (2018). Gluten-free biscuits based on composite rice–chickpea flour and xanthan gum. *Food Science and Technology International., 24*, 607–616. https://doi.org/10.1177/1082013218779323

Bettge, A. D., & Morris, C. F. (2000). Relationships among grain hardness, pentosan fractions, and end-use quality of wheat. *Cereal Chemistry, 77*, 241–247.

Buzalaf, M. A., De Almeida, B. S., Cardoso, V. E., Olympio, K. P., De Furlani, T., & A. (2004). Total and acid-soluble fluoride content of infant cereals, beverages and biscuits from Brazil. *Food Additives and Contaminants, 21*(3), 210–215.

Cairano, M. D., Condelli, N., Cela, N., Sportiello, L., Caruso, M. C., & Galgano, F. (2022, January 15). Formulation of gluten-free biscuits with reduced glycaemic index: Focus on *in vitro* glucose release, physical and sensory properties. *LWT, 154*, 112654.

Dandin, S. B., & Kishna Kumar, N. K. (2016). Neglected and underutilized fruit species—An insurance against global mal and under nutrition. In *Regional expert consultation on scoping, prioritizing and mapping of neglected and underutilized crop species in Asia* (pp. 1–40). FAO Regional Office.

Davidson, I. (2016). *Biscuit baking technology: Processing and engineering manual*. Elsevier.

Durst, P., & Bayasgalanbat, N. (2014). *Promotion of underutilized indigenous food resources for food security and nutrition in Asia and the Pacific*; FAO Regional Office. ISBN 9789251082379.

Duta, D. E., & Culetu, A. (2015). Evaluation of rheological, physicochemical, thermal, mechanical and sensory properties of oat-based gluten free cookies. *Journal of Food Engineering, 162*(10), 1–8.

FAO, IFAD, UNICEF, WFP, & WHO. (2017). *The state of food insecurity and nutrition in the world 2017. Building resilience for peace and food security*. FAO.

Fathonah, S., Rosidah, A. B., & Humaizah, S.. (2020). The formulation of alternative gluten-free mung bean biscuits. Journal of Physics: Conference Series. The 8th engineering international conference 2019 1444, 012004.

Field, A., et al. (2004). Snack food intake does not predict weight change among children and adolescents. *International Journal of Obesity, 28*, 1210–1216.

FSA. (2008). *FSA unveils first steps of saturated fat and energy intake programme*. http://www.food.gov.uk/news/pressreleases/2008/feb/satfatprog

Galla, N. R., Pamidighantam, P. R., Karakala, B., Gurusiddaiah, M. R., & Akula, S. (2017). Nutritional, textural and sensory quality of biscuits supplemented with spinach (Spinacia oleracea L.). *International Journal of Gastronomy and Food Science, 7*, 20–26.

Gallagher, E., O'Brien, C. M., Scannell, A. G. M., & Arendt, E. K. (2003). Evaluation of sugar replacers in short dough biscuit production. *Journal of Food Engineering, 56*(2–3), 261–263.

Gerber, P. J., Steinfeld, H., Henderson, B., Mottet, A., Opio, C., Dijkman, J., et al. (2013). *Tackling climate change through livestock: A global assessment of emissions and mitigation opportunities*. Food and Agriculture Organization of the. United Nations (FAO).

Guimón, J., & Guimón, P. (2012). How ready-to-use therapeutic food shapes a new technological regime to treat child malnutrition. *Technological Forecasting and Social Change, 79*, 1319–1327. [CrossRef].

Hoseney, R. C., Rogers, D. E., & Faridi, H. (1994). *Mechanism of sugar functionality in cookies. The science of cookie and cracker production*. AACC.

Hussein, A. M., Yaseen, A. A., Esmail, R. M., et al. (2020). Gluten-free biscuits produced from new drought tolerant corn hybrids: Processing and evaluation. *Bulletin of the National Research Centre, 44*, 25. https://doi.org/10.1186/s42269-020-0279-3

Ikuomola, D. S., Otutu, O. L., & Oluniran, D. D. (2017). Quality assessment of cookies produced from wheat flour and malted barley (Hordeumvulgare) bran blends. *Cogent Food & Agriculture, 3*(1), 1293471.

Inayati, D. A. (2011). *Nutritional interventions among mildly wasted children on Nias Island, Indonesia: Impact on children's nutritional status and knowledge-practice of caregivers*. Ph.D. Thesis,. University of Hohenheim.

Konstantas, A., Stamford, L., & Azapagic, A. (2019). Evaluation of environmental sustainability of biscuits at the product and sectoral levels. *Journal of Cleaner Production, 230*, 1217e1228.

Kulp, K., Olewink, M., & Lorenz, K. (1991). Starch functionality in cookies systems. *Starch/Starke, 43*(2), 53–57.

Lazzerini, M., Rubert, L., & Pani, P. (2013). Specially formulated foods for treating children with moderate acute malnutrition in low and middle-income countries. *Cochrane Database of Systematic Reviews, 6*, 1–126.

Mamat, H., & Hill, S. E. (2014). Effect of fat types on the structural and textural properties of dough and semi-sweet biscuit. *Journal of Food Science and Technology, 51*(9), 1998–2005.

Mamat, H., & Hill, S. E. (2018). Structural and functional properties of major ingredients of biscuit. *International Food Research Journal, 25*(2), 462–471.

Manley, D. (2000). *Technology of biscuits, crackers and cookies*. Woodhead Publishing Limited.

Manohar, R. S., & Rao, P. H. (1999). Effect of emulsifiers, fat level and type on the rheological characteristics of biscuit dough and quality of biscuits. *Journal ofthe Science of Food and Agriculture, 79*(10), 1223–1231.

Mert, B., & Demirkesen, I. (2016). Reducing saturated fat with oleogel/shortening blends in a baked product. *Food Chemistry, 199*(5), 809–816.

Mishra, A., Devi, M., & Jha, P. (2015, July). Development of gluten free biscuits utilizing fruits and starchy vegetable powder. *Journal of Food Science and Technology, 52*(7), 4423–4431.

Nassar, A. G., Abdel-Hamied, A. A., & El-Naggar, E. A. (2008). Effect of citrus by-products flour incorporation on chemical, rheological and organoleptic characteristics of biscuits. *World Journal of Agricultural Sciences, 4*(5), 612–616.

O'Brien, C. M., Chapman, D., Neville, D. P., Keogh, M. K., & Arendt, E. K. (2003). Effect of varying the microencapsulation process on the functionality of hydrogenated vegetable fat in short dough biscuits. *Food Research International, 36*, 215–221.

Ojha, P., & Thapa, S. (2017). Quality evaluation of biscuit incorporated with mandarin peel powder. *Scientific Study & Research Chemistry & Chemical Engineering, Biotechnology, Food Industry, 18*(1), 19–30.

Olinger, P. M., & Velasco, V. S. (1996). Opportunities and advantages of sugar replacement. *Cereal Foods World, 41*(3), 110–117.

Patel, I. N., Rema, S., & Kamaliya, K. B. (2003). Biscuits ingredients. *Procssing Food Indian*, 44–49.

Preedy, V. R., Warson, R. R., & Patel, V. B. (2011). *Flour and breads and their fortification in health and disease prevention* (p. 519). Elsevier.

Saaka, M., Oladele, J., Larbi, A., & Hoeschle-Zeledon, I. (2017). Dietary diversity is not associated with haematological status of pregnant women resident in rural areas of northern Ghana. *Journal of Nutrition and Metabolism, 2017*, 1–10.

Savitha, Y. S., Indrani, D., & Prakash, J. (2008). Effect of replacement of sugar with sucralose and maltodextrin on rheological characteristics of wheat flour dough and quality of soft dough biscuits. *Journal of Texture Studies, 39*(6), 605–616.

Sekiyama, M., Roosita, K., & Ohtsuka, R. (2012). Snack foods consumption contributes to poor nutrition of rural children in West Java, Indonesia. *Asia Pacific Journal of Clinical Nutrition, 21*, 558–567.

Skaltsi, A., Marinopoulou, A., Poriazi, A., Petridis, D., & Papageorgiou, M. (2021). Development and optimization of gluten-free biscuits with carob flour and dry apple pomace. *Journal of Food Processing and Preservation., 46*. https://doi.org/10.1111/jfpp.15938

Sudha, M. L., Srivastava, A. K., Vetrimani, R., & Leelavathi, K. (2007). Fat replacement in soft dough biscuits: Its implications on dough rheology and biscuit quality. *Journal of Food Engineering, 80*(3), 922–930.

Tarancón, P., Salvador, A., Sanz, T., Fiszman, S., & Tárrega, A. (2015). Use of healthier fats in biscuits (olive and sunflower oil): Changing sensory features and their relation with consumers' liking. *Food Research International, 69*(3), 91–96.

Waibel, R. (2010). *Religion and dietary practices*. http://www.faqs.org/nutrition/Pre-Sma/Religion-and-Dietary-Practices

Zoulias, E. I., Piknis, S., & Oreopoulou, V. (2000). Effect of sugar replacement by polyols and acesulfame-K on properties of low-fat cookies. *Journal of the Science of Food and Agriculture, 80*(14), 2049–2056.

Zoulias, E. I., Oreopoulou, V., & Tzia, C. (2002). Textural properties of low-fat cookies containing carbohydrate-or protein-based fat replacers. *Journal of Food Engineering, 55*(4), 337–342.

Chapter 4
Cookies

Manish Tiwari, Mehul Chudasama, Ankita Kataria, and Komal Chauhan

1 Introduction

The Dutch employed a small amount of cake batter for testing oven temperatures before large-scale baking, creating what they called "koekje," meaning "little cake." These were also referred to as "keks" in Germany and "galletas" in Spanish. Italian cookies include varieties such as Amaretti and biscotti (IICA, 2020; Herbest, 2001). The term "cookie" generally denotes a baked product typically composed of three main ingredients: flour, sugar, and fat. Soft wheat flour (Triticum aestivum) is commonly used in cookie preparation, combined with sugar, shortening, and various other ingredients such as raisins, oats, almonds, or chocolate chips to form cookie dough (Aly et al., 2021; Maskey et al., 2021; Mamat et al., 2010). Cookies can be either baked or fried and are often served as snacks, either with or without beverages. Cookies are typically categorized into three types: (a) high sugar cookies, (b) high-fat cookies, and (c) low moisture cookies. Additional ingredients, such as chemical leavening agents, salt, emulsifiers, syrup and others, can be added to the dough at varying levels to improve its textural properties (Pareyt & Delcour, 2008). Cookies are enjoyed worldwide as snacks across all age groups. They can serve as functional foods with added ingredients that impart specific health benefits (Olagunju, 2019; Anitha & Hemavathy, 2015; Arshad et al., 2007). These added ingredients may include alternative flours like rice, millets, legumes, oilseeds, nut seeds, herbal extracts, or bioactive components, aimed at enhancing physicochemical, textural, organoleptic properties, and extending shelf life. Natural antioxidants

M. Tiwari · M. Chudasama · K. Chauhan (✉)
Department of Food Science and Technology, National Institute of Food Technology Entrepreneurship and Management, Sonipat, Haryana, India

A. Kataria
Department of Food Science and Technology, Maharaja Ranjit Singh Punjab Technical University, Bathinda, Punjab, India

M. A. Shah et al. (eds.), *Cereal-Based Food Products*,
https://doi.org/10.1007/978-3-031-40308-8_4

from plant sources (Bhat et al., 2018) can also be incorporated to boost the antioxidative capacity and reduce fat rancidity in cookies.This chapter comprehensively covers all aspects of cookies, including their ingredients, manufacturing methods, and quality characteristics. Special attention is given to unconventional ingredients that can replace wheat flour, thereby creating functional cookies that not only enhance organoleptic qualities but also offer health and nutritional benefits.

2 Ingredients and Their Functions

Baking is a complex process with combination of physical, chemical and biochemical changes in the dough. It involves evaporation of water, volume expansion, porous structure, gelatinization of starch, protein denaturation, browning and formation of crust (Vanin et al., 2009). Cookies are one of majorly consumed baked products prepared fromflour, fat/shortening, and sugar (Chevallier et al., 2000). The other ingredients namely sodium bicarbonate (baking soda), sodium chloride (salt), ammonium bicarbonate and gluten modifying agents are added to improve the physicochemical and functional characteristics of cookies. Several variants of cookies can be prepared by addition of other ingredients at the expense of refined flours (Tacer-Caba et al., 2015).

Major Ingredients

Major ingredients namely flour, sugar and shortening contribute to the base of the cookies. The variations in these ingredients may affect the shape, texture and, flavor to the cookies.

Flour

Flour is obtained from wheat by milling process and is commonly known as *maida* or refined wheat flour (Das et al., 2012). Different fractions of wheat flour can be achieved during milling process and these fractions can affect the final quality of baked products (Shekara et al., 2013). Several factors namely type of flour, water absorption, damaged starch and enzyme catalyzed starch, starch conversions attribute to physicochemical properties and sensory appraisal of cookies. Wheat is classified as hard and soft, the former is more proteinaceous (12–14%) in nature and yields strong flour with higher particle size. The high protein content brings desired leavening owing to entrapment of gases during fermentation. Conversely, soft wheat results in weak flour with poor tolerance to mixing and fermentation (Das et al., 2012; Shekara et al., 2013). The weak flour has low protein content (7–10%) with fine particle size (David et al., 2015). The fine particle size confers to higher rate of water absorption, thereby producing homogeneous dough (Bressiani et al., 2019). Soft wheat flour produces a product with lower height and larger diameter. Spread of cookies can be reduced by using soft wheat flour with higher content of damaged starch (Barrera et al., 2007). The damaged starch controls the water absorption capacity during cookie dough preparation. Furthermore, the broken or damaged starch granules are susceptible to amylolytic enzymatic action resulting into simple sugars. Moreover, the particle size varying between (150–180 μm) along with high content of damaged starch level (5–15%) results in good quality cookies (Belorio et al., 2019; Barak et al., 2014).

Notably flours treated with bleaching agents such as chlorine, chlorine dioxide, benzoyl peroxide, oxides of nitrogen and azodicarbonamide are used in the preparation of bakery products in order to have a soft texture with bright color and voluminous finished product (Saunders et al., 2007). Bleached flours also act as improvers or modifiers in baking industry. Additionally, they control the cookie diameter in order to meet the packaging specifications (Gaines & Tsen, 1980; Slade et al., 2021). Thus, flour is a functional ingredient and responsible for consistency, texture, and overall quality of the cookies.

Sugar

Sugar plays an important role in determining the quality of the dough and cookies besides imparting sweetness, caramelized colour and flavor to cookies. The creaming of sugar with fat affects the dough expansion during baking (Jukic et al., 2022). It also acts as a hardening agent imparting hardness, crispness and crunchiness to cookies. The size of sugar granules affects creaming, spreading and texture of cookies (Boz, 2019). Coarse sugar granules require more mechanical force during creaming which in turn can reduce spreading and surface cracking during baking (Zhou et al., 2014). On the contrary fine sugar granules varying between 150–180 μm require less energy and force during creaming. During dough development of cookie, half of the sugar is dissolved while the remaining is used as syrup for optimal dough expansion during baking (Jukic et al., 2022). On the contrary, fine sugar tends to melt uniformly during baking and give soft textured and higher spread ratio desired cookies (Boz, 2019).

Shortening

Shortening is the key component for structural characterization thereby contributing to the plasticity and spread of the dough. The type and the amount of shortening or fat used affect the viscoelastic properties of cookie dough. During molding, it reduces the elasticity and causes shrinkage of the dough (Maache-Rezzoug et al., 1998). It stabilizes the air cells, generated during creaming and enhances the mouthful, texture, flavor and palatability of baked cookies (Jacob & Leelavathi, 2007;

Jadhav et al., 2022). It must have effective 'plastic' potential with correct solid-to-liquid index during creaming and dough mixing temperatures.

Solid fat index (SFI) is used for determining the solid-to-liquid proportion of fat and product performance at high temperature. High SFI shortening do not have desired oil volume while low SFI shortening do not have ability to hold air until mixing is complete (O'Brien, 2004). Substituting solid fat with liquid oil can affect the stiffness of the dough. Thus, increased content can lead to a softening effect on dough consistency. Oils, owing to their liquid state at room temperature when added to cookies impart hard texture to cookies due to the loss of the incorporated air during baking (Manley, 1998). The liquid oils thus have an adverse effect on handling characteristics of dough and it is essential to use solid fat during preparation of dough (Abboud et al., 1985). The crystalline nature of fat is an important characteristic that determines the quality of the product and is classified as three polymorphs namely α, β and β'. For optimum creaming, fat should be in the form of β' crystal. Size of the β' crystal ranges from 1 to 5 μm (Piska et al., 2006). These β' crystals are smaller and finer which can stabilize more air and liquid components compared to larger α and coarse β crystals (Podmore, 2002).

Minor Ingredients

Salt

In bakery, salt is generally added to modify the intensity and nature of hydrophobic interactions in gluten. Salt promotes aggregation by decreasing the solubility of proteins thereby resulting in increasing the dough strength. Notably salt has little to no effect on hardness of cookies however, can affect the taste and spread of cookies (Panghal et al., 2018).

Sodium Bicarbonate and Ammonium Bicarbonate

Sodium bicarbonate and ammonium bicarbonate are used to leaven the cookies. The two can have synergetic effect on spreading and hardness of cookies. The excess sodium bicarbonate confers alkalinity rendering yellowish crumb colour and unpleasant taste called '*soda bite*' to cookies (Kukurova, 2015). The addition of ammonium bicarbonate generates ammonia which should be driven off before baking otherwise it will impart unpleasant taste to final product (Tiefenbacher, 2017).

Chemical Agents (Additives)

Chemical agents or additives are added to improve physico-functional characteristics and to enhance shelf life of baked products. Several studies demonstrated that the use of surfactants such as sodium stearoyl lactylate (SSL) or sodium stearoyl fumarate (SSF) enhanced cracking at the top surface (called top grain) while the spreadability and quality of the cookies is affected from the type of hard or soft wheat flour used (Majzoobi et al., 2011; Ahmad et al., 2022). SSL is an anionic emulsifier that increases loaf volume besides improving crumb and crust texture. SSL used at 0.25–0.50% (by flour weight) during mixing strengthened the cookie dough and produced fine and uniform surface cracks on cookies. (Sluimer, 2005). The dough containing SSL and SSF increases the spread ratio due to interactions between emulsifier and starch. Lecithin, a natural emulsifier can be added at 0.25–1% of the dough weight, thus reduces the greasiness of shortening. It can produce dough with dry consistency with equivalent moisture and shortening content. It aids in uniform distribution of ingredients in the dough, reduction of mixing time and making the dough more machinable by improving the extensibility of gluten network thereby, producing tender cookies (Perez-Carrillo et al., 2017). Several surfactants namely glycerol monostearate, ethoxylated monoglyceride, sucrose monopalmitate, sucrose mono and di-stearate, sorbitan monostearate, polysorbate 60 and succinylated monoglycerides are also used in baking industry to improve physicochemical and functional properties (Arepally et al., 2020).

Flour modifiers or improvers like sodium metabisulfite, potassium bromates, potassium iodate, L-cysteine, N-ethylmaleimide, glutathione and dithioerythritol effects dough workability and spread factor of the cookies (Pareyt et al., 2010). They also improve the texture, color, and taste of the cookies. Sodium metabisulfite was largely used as a dough plasticizer, however, due to its aftertaste some countries had banned its usage (Ahmad et al., 2022). L-cysteine, an amino acid, naturally present in wheat flour, but can also be synthesized by hydrolysis of cysteine rich proteins is used as a reducing agent in bakery industry. Likewise, N-ethylmaleimide, glutathione and dithioerythritol are used to uncoil the proteins by breaking down disulphide bonds within dough, thereby increasing the extensibility of wheat gluten (Stoica et al., 2010).

Other Ingredients

Water acts as a medium for salt, sugar, and baking powder in cookies. Milk is also used at times as an optional ingredient in cookies to improve color and mouthfeel besides improving the nutritional content of the cookies. Nowadays functional cookies are developed using varied flours, pomace, extracts, peels as ingredients along with major and minor ingredients (Theagarajan et al., 2019; Bhat et al., 2018; Arun et al., 2015). The rice, maize, buckwheat, pseudo cereals, oats, millets, legume flours are widely used to develop gluten free cookies (Rai et al., 2014). In addition, oil seed flours such aschia, flax, pumpkin, melon and alfalfa are also added to enhance therapeutic potential (Duta & Culetu, 2015; Giuberti et al., 2018; Brites et al., 2018; Saric et al., 2019). Nuts flour like chestnut, coconut and konjac flour have also been used to develop gluten free cookies (Paciulli et al., 2018; Akesowan, 2016; Paucean et al., 2016).

3 Preparation of Cookies

The manufacturing of cookies involves sieving of all dry ingredients (wheat flour, skim milk powder, sodium bicarbonate, salt); creaming wherein blending of butter and sugar results in homogenous mixing besides incorporating air for leavening, sheeting, cutting, baking and cooling (Fig. 4.1 and Table 4.1).

Creaming

Creaming involves mixing of fat/shortening with sugar in the ratio of 1:1 to 1:2 as per AACC method (Bettge & Kweon, 2009). The basic principle behind the process is mechanical whipping that softens the fat/shortening and incorporates and traps air

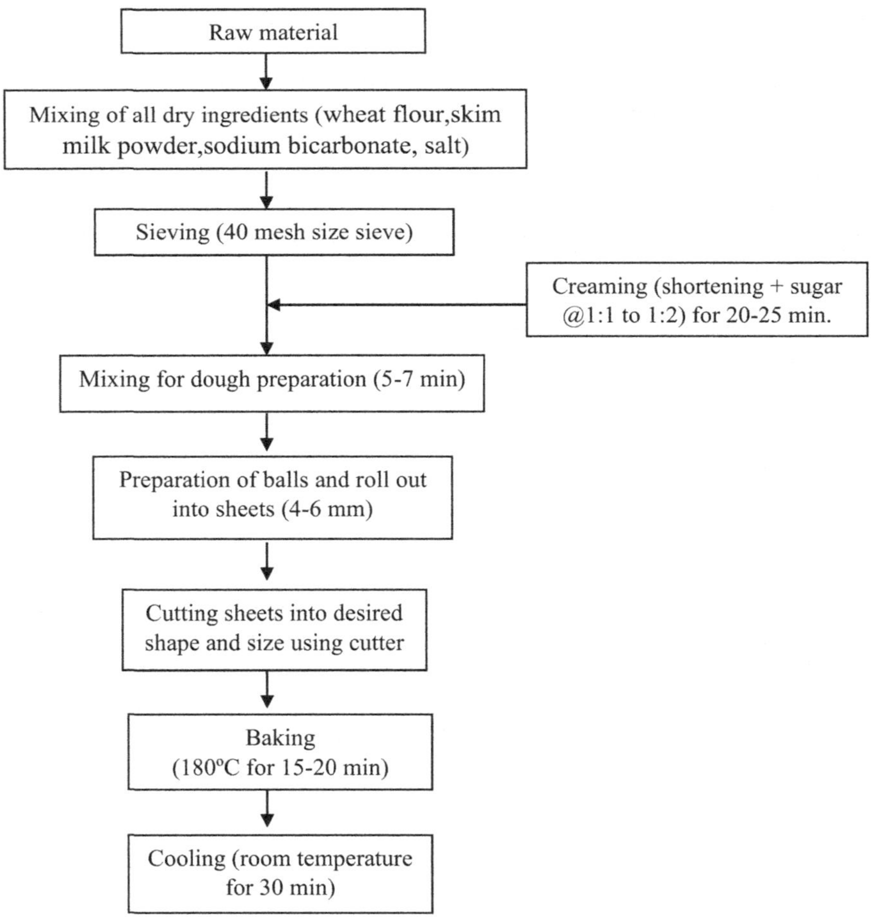

Fig. 4.1 Process flow chart for cookie preparation. Source: (Biljwan et al., 2019)

Table 4.1 Composition of cookies

Ingredients	Amount (g)
Flour	40
Sugar	24
Non-fat dry milk	1.2
Sodium bicarbonate	0.40
Sodium bicarbonate (in sol. A)	0.32 (in 4 ml)
Ammonium chloride (in sol. A)	0.20 (in 2 ml)
Salt (in sol. B)	0.18
Shortening	12.0
Water (based on moisture content of flour)	2.7 ml

Solution A *– 0.95 M sodium bicarbonate (79.8 g dissolved in water to make 1 L)*
Solution B *– 1.9 M ammonium chloride/1.52 M sodium chloride (101.6 g and 88.8 g respectively, dissolved in water to make 1 L)*
Source: (Bettge & Kweon, 2009)

in air pockets in the dough. Initially creaming is carried out at slow speed for 30 sec followed by scrapping of the contents, and then further continued at a medium speed for 20–25 min. Optimum creaming attributes to balanced colour and texture to the dough.

Liquid Incorporation

Addition of the liquid namely water or milk makes mixing easier; improves homogenization and further aerates the dough thereby helps in maintaining the texture of the cookies.

Inclusion of Dry Ingredients

The final stage involves addition or folding in of the dry ingredients/flour to the creamed mixture. These are added in small amounts to prevent air bubble formation and achieve a consistent dough.

Dough Formation

The dough can be placed in covered water-jacketed mixing bowl with water circulating at 26.7 °C for 5–7 min. This improves biting and textural properties of cookies. Dough can also be refrigerated for 10–15 min prior to baking as cold dough slows down the spreading by forming crust around the cookies.

Sheeting and Cutting of Cookies

The dough prepared is subjected to sheeting of 4-6 mm thickness and cut into desired shapes using different moulds. The sheeting can be done manually or through automated sheeter. The former involves keeping the dough balls on a cooking sheet or tray; pressing gently with palm and rolling flat by rolling pin to desired thickness. The latter involves passing the dough through sheeter. The sheeting tray is greased with oil to prevent cookies from sticking. The process is repeated thrice with scraping in between to get uniform layer. Thereafter, the sheeted dough of desired thickness is subjected to cutting to develop cookies of desired size and shape followed by baking.

Baking

Baking is a culinary process that transforms dough into a variety of delicious and enjoyable products, including cookies. The ingredients and techniques used in baking can vary widely, resulting in a diverse array of baked goods with different sensory attributes and enhanced shelf life. The conversion of dough into finished product depends on the baking temperature and time. The cookies are baked in preheated oven at 180 °C for 15–20 min. The said temperature results in desired browning and also aids in reducing the spread of the cookies.

Higher baking temperature involves the complex chemical changes such as melting of fat and sugar, gelatinization of starch, denaturation of proteins, dough expansion and maillard reaction (Chevallier et al., 2002). The spread of the cookie dough is caused by the melting of the sugar and fat in the oven. Higher bakery temperature may decrease the spread factor of the cookies due to the layering of sugar and produce the under baked product at the center. Early protein denaturation and starch gelatinization may increase the dough viscosity and prevent the spread of cookie dough at higher temperature. Due to stress hardening of the dough caused by a higher oven temperature, instant release of gas and vapor from the dough makes the cookies denser and harder. The, Maillard reaction imparts brown colour but reduces the lysine content owing to the reacion with the reducing sugars present within the dough. Baking time influences the open porous structure of cookies. During lower baking time (below 14 min), cookies produced hard texture with reduced spread factor (Panghal et al., 2018).

Cooling

The baked cookies need to be cooled to room temperature. They are removed from oven and allowed to cool at room temperature for 30 min. The cookies are thereafter packed and sealed to prevent any contamination or spoilage.

Table 4.2 Comparitive properties of different types of cookies

Property	Cookies			
	Semisweet	High fat	High sugar	Soft
Dough moisture content (%)	22	9	15	11
Cookie moisture content (%)	1–2	2–3	2–3	3
Dough temperature (°C)	40–42	20	21	21
Critical ingredients	Flour	Fat	Fat and sugar	Fat and sugar
Baking time (min)	5.5	12–25	7	12
Oven type	Wire	Steel	Steel	Steel

Types of Cookies

Cookies generally contain high fat/shortening, sugar and low water content. They can be classified as semi-sweet, high fat, high sugar and soft cookies on the basis of varied ingredients (Table 4.2). High sugar cookies have crisp and firm texture as sugar act as a hardening agent (Zhou et al., 2014). The semi-sweet cookies have highest dough moisture content (22%) while other cookies show a variation between 9–15%. Conversely the moisture content of cookies is low varying between 1–3%. Thebaking temperature of semisweet cookies varies between 40–42 °C while for the other types it ranges between 20–21 °C.

Classification of Cookies

Cookies are also classified based on baking band i.e., rotary molded, wire-cut and cutting machines.

Rotary Molded Cookies

They are produced by forcing dough into mold on rotating rolls. After half a turn of the roller, the dough is removed and placed in baking oven band. These cookies are developed from the dough that contains high sugar and fat but has low moisture content (up to 15% on flour basis) (Fig. 4.2).

Hence it gives a lumpy, stiff and crumbly texture that lacks elasticity. During baking, dough rise or spread is very little and to maintain the size and shape of molded cookies, molds are greased with a little shortening and sugar (Zhou et al., 2014).

Wire-Cut Cookies

The wire cut cookies are produced by extruding very soft dough through a die and cutting into pieces using a reciprocating wire. The cookies are developed from dough containing high sugar, shortening and water (up to 40% on flour basis) that

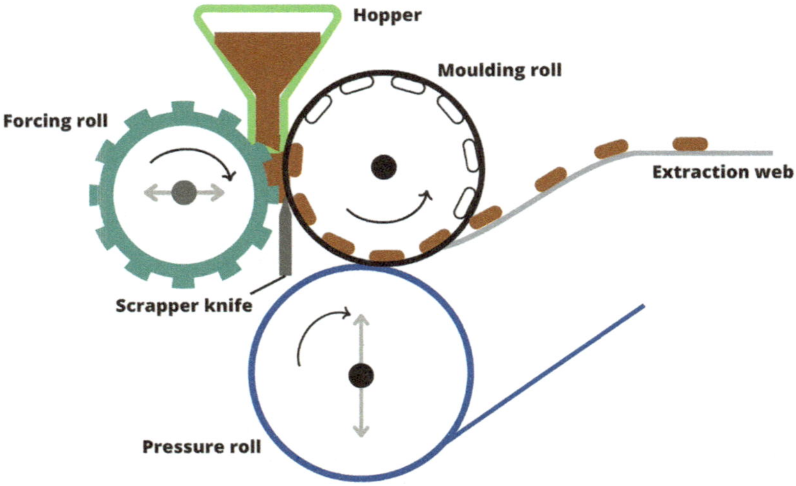

Fig. 4.2 Schematic representation of Rotary mold machine. Source: (Manley, Manley, 2011)

Fig. 4.3 Wire cut cookies

increases spreading of the dough during baking. High moisture content and excessive mixing can lead to gluten development and decrease the tenderness of dough. To retain the tenderness, shortening is incorporated with the flour, prior to addition of water. Chocolate chip cookies are one of the examples of wire-cut cookies (Zhou et al., 2014) (Fig. 4.3).

Cutting Machine

These cookies are produced from dough that contains less sugar and shortening while have high water content (up to 25% on flour basis) compared to rotary molded. Continuous sheet is made by passing the dough through multiple sheeting rolls (Zhou et al., 2014).

Functional Cookies

Rice Flour Cookies

Rice flour is the major substitute for the production of gluten-free cookies. Schober et al. (2003) developed rice flour-based cookies with 70% rice flour, 10% soy flour, 10% potato starch and 10% corn starch. Likewise, wheat and rice flour blends can be used for developing cookies. Substituting 20% wheat flour with rice flour can produce more adhesive and less cohesive cookie dough (Gujral et al., 2003). Furthermore, the type of rice flour affects the quality of cookies. Rice flour-based cookies having high protein content with low cohesiveness and adhesiveness (Ha et al., 2007).

Rice bran can also be used in the formula to increase protein, and fiber content of wheat cookies (Sharma & Chauhan, 2002). It also affects the taste and texture of cookies. The overall acceptability of rice bran cookies has been reported to be high at 10% level of incorporation (Saeed et al., 2009). Protein concentrate extracted from rice bran can also be added to increase the protein content of the cookies (Yadav & Yadav, 2011). Rice bran extracts (Bhanger et al., 2008) or defatted rice bran (Watanabe et al., 2010) have been reported to reduce oxidation of fats in cookies.

Hence rice cookies add on to list ofavailable gluten free alternates for celiac patients. Nonetheless, blend of rice and wheat flour cookies have higher dietary fiber and proteins that can be a nutritious and healthy supplement for different age groups (Sharif et al., 2009).

Millet Based Cookies

Millets are known as the *"nutri-cereals"*owing to their high nutritional content and therapeutic potential. They are gluten free and can be used to prepare products ifor patients with gluten intolerance or celiac disease. Major (sorghum, pearl millet, finger millet, foxtail millet) and minor millet (kodo, kutki, barnyard millet, proso millets) based products are used to prepare gluten free nutrient dense products (Dewan et al., 2022). Millets are rich in bioactive compounds rendering antioxidative potential. Millets have been reported to control the blood pressure, form new blood cells, improves growth by initiating the several chemical reactions and maintain osmotic balance within the body. Dietary fiber of millets restricts the rate of glucose level in and thus possess antidiabetic properties (Dewan et al., 2022). Hence millet based cookies will not only improve the nutritional status but alsoprovide therapeutic benefits to consumers.

Cookies can be made from sorghum flour (Chiremba et al., 2009; Taylor et al., 2006; Morad et al., 1984) as unlike bread and cakes, they are not leavened hence, no leavening agents are required for its manufacturing. However, sorghum cookies tend to be denser and have a harder texture than their wheat counterparts and can have a somewhat gritty mouthfeel (Chiremba et al., 2009). The value addition of cookies with pearl millet, barnyard millet, kodo millet, little millet, finger millet not only increases the nutritional profile but also satisfy the sensorial properties in terms of color, flavor, texture and overall acceptability (Padma & Rajendran, 2016; Wandhekar et al., 2021; Salunke et al., 2019; Mitkal et al., 2021).

Pseudo Cereals Based Cookies

Pseudocereals such as quinoa (*Chenopodium quinoa*), buckwheat (*Fagopyrum esculentum*and *Fagopyrum tartaricum*) and amaranth (*Amaranthus*spp.)contain high-quality proteins, rich in essential micronutrients and bioactive compounds. Like millets, pseudo cereals can be the substitute for preparing gluten free diet food products for people sensitive to gluten allergen. Amaranth grain has high-quality protein, and flour is used in non-gluten formulations to obtain decent quality cookies (de la Barca et al., 2010). Defatted amaranth cookiescontainhigher amount of protein and ash as compared to maize cookies. Due to lower viscosity of amaranth flour than wheat flour, the spread ratio and thickness of cookies increases by 20%. Also, cookies breakage prepared from amaranth flour is comparatively less due to higher protein content (Chavez-Jauregui et al., 2010).

Quinoa cookies are rich in protein, fat and minerals such as potassium, magnesium, calcium, iron and zinc. The incorporation of quinoa flour slightly increases the cookies thickness with decrease in diameter along with hardness up to 30% (Brito et al., 2015; Demir & Kilinç, 2017). Other types of functional cookies also can be made by addition of gluten-free rice or buckwheat or oat flour (Xu et al., 2020).

Legume Based Cookies.

Cookies prepared from legumes are rich in protein, essential amino acids and fiber. The cookies supplemented with jering seed flour (upto 10%) into wheat flour decreases the spread ratio with good rising ability due to higher protein content. Higher protein is proportional to the water absorption capacity of the composite flour (Olapade & Adeyemo, 2014). Composite flour cookie includes chickpea flour, pigeon pea flour, moong bean flour, cowpea flour, and wheat flour at the ratio of 1:1:1:1:6. The composite flour contains good amount of proteins, fats and energy along with appreciable amount of antioxidant activity, and acceptable sensory properties (Thongram et al., 2016).

Multigrain Cookies

Multigrain cookies confer several health benefits owing to presence of different grains. Wheats flour, chickpea flour, barley flour, maize, flour, pearl millet flour, finger millet flour oats, peas, fenugreek seeds and other oil and nut seed flours are widely used as ingredients in multigrain flours. They are good source of dietary fiber and helps in prevention of various chronic diseases such as diabetes, cardiovascular disease, and obesity (Arshad et al., 2014). The cookies prepared with different grain flours not only have high nutritional index but also therapeutic index (Kishorgoliya et al., 2018).

Germinated Flour Cookies

Germinated grains confers several nutritional and health advantages. Germination not only improves nutritional content but also the digestibility as well as bioavailability of nutrients. The cookies prepared from sprouted grains have higher quantities of of micronutrients in bioavailable form than cookies prepared from raw or non-germinated grain flours (Chauhan et al., 2015).

4 Quality Aspects

Quality is defined as the degree of acceptability of product by the consumers for various nutritional, physicochemical parameters and sensory attributes. The quality of cookies can be determined by physical parameters like surface cracking, spread ratio, diameter and sensory evaluation for overall acceptability. The quality also depends on the type of packaging material used which acts as a barrier and protects the product from external factors, contaminants and microbial action.

Physical Aspects

Physical evaluation involves visual inspection of cookies for surface cracking and colorthrough colorimeter. The spread ratio is also an important parameterfor assessing the quality of cookies. Apart from this, hardness and chewiness of cookies can be analyzed using texture analyzer.

Cracking

Flour moisture content and sugar content of cookies affects the cracking of the cookies. Cookies containing varied flours with adjusted moisture content produce different cracking patterns. Lower moisture content of the flour results in cracking while high moisture content prevents cracking irrespective of total water used in the formula. Furthermore, the unique ability of sucrose to recrystallize at the surface during baking is responsible for the cracking pattern. Moreover, the coarse or granular sugar results in rapid drying of the outer surface leading to cracks on spreading (Curley & Hoseney, 1984).

Spread Ratio

Cookies' quality can be predicted by its size i.e., ratio between the height and width as well as diameter. The quality cookies have lower height and larger diameter (Barak et al., 2014). The spread ratio of the cookies prepared from different wheat cultivars ranges from 5.55 to 10.31 (Barak et al., 2013). The other important factors affecting its quality are the type of wheat flour and level of damaged starch. Hard wheat contains 10 to 14% protein content and higher amount of damaged starch which increase its water absorption capacity. Soft wheat flour has protein content of 8–11% and produces fluffy flour with less amount of damaged starch and therefore, is most suitable for the manufacturing of cookies(Manley, 2011). This implies that with an increase in protein content in wheat flour, the spread ratio decreases (Mudgil et al., 2017).

Barak et al. (2014) reported that damaged starch i.e., 12–15%reduces the size of cookie; however, high content of damaged starch reducesthe baking performance (Dexter et al., 1985) owing toincreased water absorption by damaged starch that hinders gluten formation and thus reduces the spread ratio. Furthermore, during fermentation phase, starch degrades causing decrease in dough consistency and loss ofair retention capacity (Barrera et al., 2007).

Damaged starch level in reconstituted flour can affect the diameter of the cookie. Flour and sugar are the main hydrophilic components in cookie formula. When flour absorbs low water content, it provokes sugar to absorb more water which increases the syrup and decrease the dough viscosity during baking. This spread the dough more and produces larger diameter (Slade & Levine, 1994). Water has little to no effect on final cookie diameter and spread ration. Good cookie poorly holds the water (Faridi et al., 1994). Processing conditions and steps also affect the cookie quality i.e., mixing time, creaming time, baking time and temperature and thickness of cookies(Panghal et al., 2018)

Sensory Aspects

Sensory evaluation is widely used to assess the overall acceptability of the cookies using 9-point hedonic scale. Overall acceptability involves the various sensory properties, like appearance, flavor, taste, and texture for cookies. Hedonic scale value of overall acceptability ≥7 represents that the cookies are highly acceptable and achieve all sensory attributes. The cookies made from the various varieties of wheat flours generally possess overall acceptability in the range between 8.0 to 8.7 (Mudgil et al., 2017). The overall acceptability of cookies mainly depends on several factors such as ingredients used, formulations, processing type, baking time, temperature and storage conditions and thus it is required to optimize the process.

Role of Packaging Material on Quality of the Cookies

Packaging material plays an important role in preserving the quality and the shelf life of the product (Galic et al., 2009). It acts as a barrier for light, moisture and oxygen, microbes and chemical contaminants thereby protecting the product from microbial and chemical spoilage. Further, it protects the product from any physical damage during handling and transportation. Besides enhancing the shelf life, the packaging material should be attractive to consumers with all the details mentioned on the label as per labeling norms and regulations. Commonly used packaging material for cookies are polypropylene, metallized orientated polypropylene (OPP) and biaxially oriented polypropylene films (BOPP), laminates, plastic-coated papers, OPP/paper and metallized polylactic acid/paper, biaxially oriented polyethylene terephthalate (BOPET), paperboard, coated paperboard and paperboard/polyethylene/aluminium foil.

5 Conclusion

Cookies are widely consumed throughout the world. Traditionally wheat flour is the dominant ingredient in cookies but due to an increase in demand of functional cookies, the manufacturers are looking towards healthier alternatives. The manufacturing steps for the different types of cookies include uniform mixing, dough preparation, baking, cooling and packaging. The quality of cookies is directly influenced by type of ingredients, their formulations and processing steps. Damaged starch in the flour can affect the cookie quality like spreading and size. Although it helps in size reduction of the cookie, high levels can also reduce the baking quality. Flours other than wheat can also be used individually or as replacement, for preparation of functional cookies. Exhaustive research and planned approach are required to understand the implementation of functional cookies into the market. Knowledge regarding the different ingredients and the variations caused by these during the processing and quality is highly necessary so that functional cookies especially for celiac patients can be provided for regular consumption for the health and nutritional benefit of the consumers.

References

Abboud, A. M., Rubenthaler, G. L., & Hoseney, R. C. (1985). Effect of fat and sugar in sugar snap cookies and evaluation of tests measure cookie flour quality. *Cereal Chemistry, 62*, 124–129.

Ahmad, S., Naz, A., Usman, M., Amjad, A., Pasha, I., & Farooq, U. (2022). Impediment effect of chemical agents (additives) on gluten development in cookie dough. *Journal of Food Science and Technology, 59*(4), 1396–1406.

Akesowan, A. (2016). Influence of konjac flour on foaming properties of milk protein concentrate and quality characteristics of gluten-free cookie. *International Journal of Food Science and Technology, 51*(7), 1560–1569.

Aly, A. A., Zaky, E. A., Mahmoud, H. A., Alrefaei, A. F., Hameed, A. M., Alessa, H., & Kadasah, S. (2021). The impact of addition oats (*Avena sativa*) and cinnamon on cookies and their biological effects on rats treated with cirrhosis by CCL4. *Saudi Journal of Biological Sciences, 28*(12), 7142–7151.

Anitha, D., & Hemavathy, V. (2015). Organoleptic evaluation of five cereal based cookies and commercial cookies. *International Journal of Innovative Research in Science, Engineering and Technology, 4*, 6966–6970.

Arepally, D., Reddy, R. S., Goswami, T. K., & Datta, A. K. (2020). Biscuit baking: A review. *Lwt, 131*, 109726.

Arshad, M. U., Anjum, F. M., & Zahoor, T. (2007). Nutritional assessment of cookies supplemented with defatted wheat germ. *Food Chemistry, 102*, 123–128.

Arshad, U., Anjum, F. M., Rehman, U. S., & Sohaib, M. (2014). Development and characterization of multigrain cookies. *Pakistan Journal of Food Sciences, 24*(1), 1–5.

Arun, K. B., Persia, F., Aswathy, P. S., Chandran, J., Sajeev, M. S., Jayamurthy, P., & Nisha, P. (2015). Plantain peel-a potential source of antioxidant dietary fibre for developing functional cookies. *Journal of Food Science and Technology, 52*(10), 6355–6364.

Barak, S., Mudgil, D., & Khatkar, B. S. (2013). Effect of composition of gluten proteins and dough rheological properties on the cookie-making quality. *British Food Journal, 115*(4), 564–574.

Barak, S., Mudgil, D., & Khatkar, B. S. (2014). Effect of flour particle size and damaged starch on the quality of cookies. *Journal of Food Science and Technology, 51*(7), 1342–1348. https://doi.org/10.1007/s13197-012-0627-x

Barrera, G. N., Pérez, G. T., Ribotta, P. D., & León, A. E. (2007). Influence of damaged starch on cookie and bread-making quality. *European Food Research and Technology, 225*(1), 1–7.

Belorio, M., Sahagún, M., & Gómez, M. (2019). Influence of flour particle size distribution on the quality of maize gluten-free cookies. *Food, 8*(2), 1–9.

Bettge, A., & Kweon, M. (2009). Collaborative study on updated method 10-52: Baking quality of cookie flour-micro method (sugar-snap cookie). *Cereal Foods World, 54*(2), 70–73.

Bhanger, M. I., Iqbal, S., Anwar, F., Imran, M., Akhtar, M., & Zia-Ul-Haq, M. (2008). Antioxidant potential of rice bran extracts and its effects on stabilisation of cookies under ambient storage. *International Journal of Food Science and Technology, 43*, 779–786.

Bhat, N. A., Hamdani, A. M., & Masoodi, F. A. (2018). Development of functional cookies using saffron extract. *Journal of Food Science and Technology, 55*(12), 4918–4927.

Biljwan, M., Naik, B., Sharma, D., Singh, A., & Kumar, V. (2019). Recent development in dough based bakery products: A mini review. *The Pharma Innovation Journal, 8*(5), 654–658.

Boz, H. (2019). Effect of flour and sugar particle size on the properties of cookie dough and cookie. *Czech Journal of Food Sciences, 37*(2), 120–127.

Bressiani, J., Oro, T., Da Silva, P., Montenegro, F., Bertolin, T., Gutkoski, L., & Gularte, M. (2019). Influence of milling whole wheat grains and particle size on thermo-mechanical properties of flour using Mixolab. *Czech Journal of Food Sciences, 37*(4), 276–284.

Brites, L. T. G. F., Ortolan, F., Silva, D. W. D., Bueno, F. R., Rocha, T. D. S., Chang, Y. K., & Steel, C. J. (2018). Gluten-free cookies elaborated with buckwheat flour, millet flour and chia seeds. *Food Science and Technology, 39*, 458–466.

Brito, I. L., de Souza, E. L., Felex, S. S. S., Madruga, M. S., Yamashita, F., & Magnani, M. (2015). Nutritional and sensory characteristics of gluten-free quinoa (*Chenopodium quinoa* Willd)-based cookies development using an experimental mixture design. *Journal of Food Science and Technology, 52*(9), 5866–5873.

Chauhan, A., Saxena, D. C., & Singh, S. (2015). Total dietary fibre and antioxidant activity of gluten free cookies made from raw and germinated amaranth (Amaranthus spp.) flour. *LWT-Food Science and Technology, 63*(2), 939–945.

Chavez-Jauregui, R. N., Santos, R. D., Macedo, A., Chacra, A. P. M., Martinez, T. L., & Arêas, J. A. G. (2010). Effects of defatted amaranth (*Amaranthus caudatus* L.) snacks on lipid metabolism of patients with moderate hypercholesterolemia. *Food Science and Technology, 30*, 1007–1010.

Chevallier, S., Colonna, P., Buléon, A., & Della Valle, G. (2000). Physicochemical behaviors of sugars, lipids, and gluten in short dough and biscuit. *Journal of Agricultural and Food Chemistry, 48*, 1322–1326.

Chevallier, S., Della Valle, G., Colonna, P., Broyart, B., & Trystram, G. (2002). Structural and chemical modifications of short dough during baking. *Journal of Cereal Science, 35*(1), 1–10.

Chiremba, C., Taylor, J. R. N., & Duodu, K. G. (2009). Phenolic content, antioxidant activity and consumer acceptability of sorghum cookies. *Cereal Chemistry, 86*, 590–594.

Curley, L. P., & Hoseney, R. C. (1984). Effect of sugar type and flour moisture on surface cracking of sugar-snap cookies. *Cereal Chemistry, 62*(3), 274–279.

Das, A., Raychaudhuri, U., & Chakraborty, R. (2012). Cereal based functional food of Indian subcontinent: A review. *Journal of Food Science and Technology, 49*(6), 665–672.

David, O., Arthur, E., Kwadwo, S. O., Badu, E., & Sakyi, P. (2015). Proximate composition and some functional properties of soft wheat flour. *International Journal of Innovative Research in Science, Engineering and Technology, 4*(2), 753–758.

de la Barca, A. M. C., Rojas-Martínez, M. E., Islas-Rubio, A. R., & Cabrera-Chávez, F. (2010). Gluten-free breads and cookies of raw and popped amaranth flours with attractive technological and nutritional qualities. *Plant Foods for Human Nutrition, 65*(3), 241–246.

Demir, M. K., & Kilinç, M. (2017). Utilization of quinoa flour in cookie production. *International Food Research Journal, 24*(6), 2394–2401.

Dewan, A., Tiwari, M., Chhikara, N., & Khatkar, B. S. (2022). Millet based functional food. *Functional Foods*, 91–160.

Dexter, J. E., Preston, K. R., Tweed, A. R., Kilborn, R. H., & Tipples, K. H. (1985). Relationship of flour starch damage and flour protein to the quality of Brazilian-style hearth bread and remix pan bread produced from hard red spring wheat. *Cereal Foods World, 30*, 511–514.

Duta, D. E., & Culetu, A. (2015). Evaluation of rheological, physicochemical, thermal, mechanical and sensory properties of oat-based gluten free cookies. *Journal of Food Engineering, 162*, 1–8.

Faridi, H., Gaines, C., & Finney, P. (1994). Soft wheat quality in production of cookies and crackers. *Wheat*, 154–168. https://doi.org/10.1007/978-1-4615-2672-8_11

Gaines, C. S., & Tsen, C. C. (1980). A baking method to evaluate flour quality for rotary-molded cookies. *Cereal Chemistry, 57*, 429–433.

Galic, K., Ćurić, D., & Gabrić, D. (2009). Shelf life of packaged bakery goods—A review. *Critical Reviews in Food Science and Nutrition, 49*(5), 405–426.

Giuberti, G., Rocchetti, G., Sigolo, S., Fortunati, P., Lucini, L., & Gallo, A. (2018). Exploitation of alfalfa seed (*Medicago sativa* L.) flour into gluten-free rice cookies: Nutritional, antioxidant and quality characteristics. *Food Chemistry, 239*, 679–687.

Gujral, H. S., Mehta, S., Singh, S. I., & Goyal, P. (2003). Effect of wheat bran, coarse wheat flour, and rice flour on the instrumental texture of cookies. *International Journal of Food Properties, 6*(2), 329–340.

Ha, M., Roh, Y., Hong, K., Kang, Y., Jung, D., Kim, K., Park, S., Ha, S., & Bae, D. (2007). Textural properties of processed foods produced from newly developed non-glutinous rice cultivars. *Food Science and Biotechnology, 16*, 789–795.

Herbst, S. T. (2001). *The new food lover's companion: Comprehensive definitions of nearly 6,000 food, drink, and culinary terms* (3rd ed., p. 385). Barron.

IICA. (2020). Cookies and Biscuits, Lesson-2, *The Coconut Pyramids*. (Accessed at 11 May 2022)

Jacob, J., & Leelavathi, K. (2007). Effect of fat-type on cookie dough and cookie quality. *Journal of Food Engineering, 79*(1), 299–305.

Jadhav, H. B., Gogate, P. R., & Annapure, U. S. (2022). Understanding the beneficial effects of using designer lipids in the formulation of cookies. *Journal of Food Processing and Preservation, 46*(5).

Jukic, M., Nakov, G., Komlenić, D. K., Vasileva, N., Šumanovac, F., & Lukinac, J. (2022). Quality assessment of cookies made from composite flours containing malted barley flour and wheat flour. *Plants, 11*(6), 761.

Kamel, B. S. (1994). Creaming, emulsions, and emulsifiers. In H. Faridi (Ed.), *The science of cookie and cracker production*. Chaman & Hall.

Kasapis, S. (2012). Relation between the structure of matrices and their mechanical relaxation mechanisms during the glass transition of biomaterials: A review. *Food Hydrocolloids, 26*(2), 464–472.

Kawai, K., Toh, M., & Hagura, Y. (2014). Effect of sugar composition on the water sorption and softening properties of cookie. *Food Chemistry, 145*, 772–776.

Kishorgoliya, N., Mehra, M., & Goswami, P. (2018). Nutritional quality of the developed multigrain flour and cookies. *Journal of Pharmacognosy and Phytochemistry, 7*, 2886–2888.

Kozlowska, M., Zbikowska, A., Marciniak-Lukasiak, K., & Kowalska, M. (2019). Herbal extracts incorporated into shortbread cookies: Impact on color and fat quality of the cookies. *Biomolecules, 9*(12), 858.

Kukurova, K. (2015). Effect of inorganic salts on acrylamide formation in cereal matrices. *Acrylamide in food: Analysis, content and potential health effects, 27*, 377–392.

Maache-Rezzoug, Z., Bouvier, J. M., Allaf, K., & Patras, C. (1998). Effect of principal ingredients on rheological behavior of biscuit dough and on quality of biscuits. *Journal of Food Engineering, 35*, 23–42.

Majzoobi, M., Farahnaky, A., & Amiri, S. (2011). Physicochemical characteristics of starch component of wheat flours obtained from fourteen Iranian wheat cultivars. *International Journal of Food Properties, 14*, 685–696.

Mamat, H., Abu-Hardan, M. O., & Hill, S. E. (2010). Physicochemical properties of commercial semi-sweet biscuit. *Food Chemistry, 121*(4), 1029–1038.

Manley, D. (1998). *Biscuit dough piece forming, manual 3*. Woodhead Publishing Ltd..

Manley, D., & Clark, H. (2011). Rotary moulding in biscuit manufacture. In *Manley's Technology of Biscuits, Crackers and Cookies,* 453–466.

Manley, D. (Ed.). (2011). *Manley's technology of biscuits, crackers and cookies*. Elsevier.

Maskey, B., Paudel, A., & Shrestha, N. K. (2021). Effect of incorporation of oats and malted barley flours on the quality of cookies. *Himalayan Journal of Science and Technology, 5*(01), 67–74.

Mitkal, K. T., Kotecha, P. M., Godase, S. N., & Chavan, U. D. (2021). Studies on nutritional quality of Kodo millet cookies. *International journal of Chemical Studies, 9*(1), 1669–1674.

Morad, M. M., Doherty, C. A., & Rooney, L. W. (1984). Effect of sorghum variety on baking properties of US conventional bread, Egyptian Pita "Balady" bread and cookies. *Journal of Food Science, 49*(4), 1070–1074.

Mudgil, D., Barak, S., & Khatkar, B. S. (2017). Cookie texture, spread ratio and sensory acceptability of cookies as a function of soluble dietary fiber, baking time and different water levels. *LWT, 80*, 537–542.

O'Brien, R. D. (2004). *Fats and oils. Formulating and processing for applications*. CRC Press.

Olagunju, A. (2019). Influence of whole wheat flour substitution and sugar replacement with natural sweetener on nutritional composition and glycemic properties of multigrain bread. *Preventive Nutrition and Food Science, 24*(4), 456.

Olapade, A. A., & Adeyemo, M. A. (2014). Evaluation of cookies produced from blends of wheat, cassava and cowpea flours. *International Journal of Food Studies, 3*(2), 175–185.

Paciulli, M., Rinaldi, M., Cavazza, A., Ganino, T., Rodolfi, M., Chiancone, B., & Chiavaro, E. (2018). Effect of chestnut flour supplementation on physico-chemical properties and oxidative stability of gluten-free biscuits during storage. *LWT, 98*, 451–457.

Padma, A., & Rajendran, R. (2016). Development and evaluation of spirulina incorporated little millet cookies. *Madras Agricultural Journal, 103*(6), 199–201.

Panghal, A., Chhikara, N., & Khatkar, B. S. (2018). Effect of processing parameters and principal ingredients on quality of sugar snap cookies: A response surface approach. *Journal of Food Science and Technology, 55*(8), 3127–3134.

Pareyt, B., & Delcour, J. A. (2008). The role of wheat flour constituents, sugar and fat in low moisture cereal-based products: A review on sugar-snap cookies. *Critical Reviews in Food Science and Nutrition, 48*, 1–16.

Pareyt, B., Brijs, K., & Delcour, J. A. (2010). Impact of fat on dough and cookie properties of sugar snap cookies. *Cereal Chemistry, 87*(3), 226–230.

Paucean, A., Man, S., Muste, S., & Pop, A. (2016). Development of gluten free cookies from rice and coconut flour blends. *Bulletin of University of Agricultural Sciences and Veterinary Medicine Cluj-Napoca Food Science and Technology, 73*(2), 163–164.

Perez-Carrillo, E., Frías-Escobar, A., Gutiérrez-Mendívil, K., Guajardo-Flores, S., & Serna-Saldívar, S. O. (2017). Effect of maize starch substitution on physicochemical and sensory attributes of gluten-free cookies produced from nixtamalized flour. *Journal of Food Processing*, 1–6.

Piska, I., Zarubova, M., Louzecky, T., Karami, H., & Filip, V. (2006). Properties and crystallization of fat blends. *Journal of Food Engineering, 77*(3), 433–438. https://doi.org/10.1016/j.jfoodeng.2005.07.010.OUR

Podmore, J. (2002). Bakery fats. In K. K. Rajah (Ed.), *Fats in food technology*. Sheffield Academic Press.

Rai, S., Kaur, A., & Singh, B. (2014). Quality characteristics of gluten free cookies prepared from different flour combinations. *Journal of Food Science and Technology, 51*(4), 785–789.

Saeed, G. S. M., Arif, S., Ahmed, M., Ali, R., & Shih, F. (2009). Influence of rice bran on rheological properties of dough and in the new product development. *Journal of Food Science and Technology, 46*, 62–65.

Salunke, P., Chavan, U., Kotecha, P., & Lande, S. (2019). Studies on nutritional quality of barnyard millet cookies. *International Journal of Chemical Sciences, 7*(4), 651–657.

Saric, B., Dapcevic-Hadnadev, T., Hadnadev, M., Sakac, M., Mandic, A., Misan, A., & Skrobot, D. (2019). Fiber concentrates from raspberry and blueberry pomace in gluten-free cookie formulation: Effect on dough rheology and cookie baking properties. *Journal of Texture Studies, 50*(2), 124–130.

Saunders, J. A., Rosentrater, K. A., & Krishnan, P. G. (2007). Potential bleaching techniques for use in distillers grains. In *2007 ASAE annual meeting* (p. 1). *American Society of Agricultural and Biological Engineers.*

Schober, T., O'Brien, C., McCarthy, D., Darnedde, A., & Arendt, E. (2003). Influence of gluten-free flour mixes and fat powders on the quality of gluten-free biscuits. *European Food Research and Technology, 216*, 369–376.

Sharif, M. K., Butt, M. S., Anjum, F. M., & Nawaz, H. (2009). Preparation of fiber and mineral enriched defatted rice bran supplemented cookies. *Pakistan Journal of Nutrition, 8*(5), 571–577.

Sharma, H. R., & Chauhan, G. S. (2002). Effects of stabilized rice bran: Fenugreek blends on the quality of breads and cookies. *Journal of Food Science and Technology, 39*(3), 225–233.

Sharma, S., & Riar, C. S. (2020). Effect of storage period and packaging materials on textural, phenolic, antioxidant properties of cookies made from raw and germinated minor millet blends flour. *Annals Food Science and Technology, 21*, 74–85.

Shekara, P., Kumar, V. P., & Hosamane, G. G. (2013). Gravity flow operated small electricity generator retrofit kit to flour mill industry. *Journal of Food Science and Technology, 50*(5), 1006–1011.

Slade, L., & Levine, H. (1994). In H. Faridi (Ed.), *The science of cookie and cracker production* (pp. 23–141). Chapman & Hall/AVI.

Slade, L., Kweon, M., & Levine, H. (2021). Exploration of the functionality of sugars in cake-baking, and effects on cake quality. *Critical Reviews in Food Science and Nutrition, 61*(2), 283–311.

Sluimer, P. (2005). *Principles of breadmaking* (Vol. 212). American Association of cereal Chemists, Inc..

Stoica, A., Popescu, C., Bărăscu, E., & Iordan, M. (2010). L-cysteine influence on the physical properties of bread from high extraction flours with normal gluten. *Annual Review of Food Science and Technology, 11*, 6–10.

Tacer-Caba, Z., Nilufer-Erdil, D., & Ai, Y. (2015). Chemical composition of cereals and their products. *Handbook of Food Chemistry*, 301–329.

Taylor, J. R. N., Schober, T. J., & Bean, S. R. (2006). Novel food and non-food use for sorghum and millets. *Journal Cereal Science, 44*, 252–271.

Theagarajan, R., Malur Narayanaswamy, L., Dutta, S., Moses, J. A., & Chinnaswamy, A. (2019). Valorisation of grape pomace (cv. Muscat) for development of functional cookies. *International Journal of Food Science and Technology, 54*(4), 1299–1305.

Thomson, T. C., Franko, D., & Alberton, D. (2010). *Cereal intake is associated with an improved nutrient intake profile among ood insecure children in the state*. NHANES.

Thongram, S., Tanwar, B., Chauhan, A., & Kumar, V. (2016). Physicochemical and organoleptic properties of cookies incorporated with legume flours. *Cogent Food & Agriculture, 2*(1), 1172389.

Tiefenbacher, K. F. (2017). Technology of minor ingredients for wafers and waffles. In *The Technology of Wafers and Waffles I. Operational Aspects* (pp. 227–311). Academic.

Vanin, F. M., Lucas, T., & Trystram, G. (2009). Crust formation and its role during bread baking. *Trends in Food Science and Technology, 20*(8), 333–343.

Wandhekar, S. S., Bondre, N., & Swami, A. (2021). Formulation and quality evaluation of finger millet (Ragi) fortified cookies. *International Journal of Innovation, Leadership, Society and Sustainability, 1*(1), 1–8.

Watanabe, Y., Hashimoto, K., Omori, A., Uda, Y., & Nomura, M. (2010). Suppressive ability of defatted rice bran against lipid oxidation in cookies containing iron. *Bioscience, Biotechnology, and Biochemistry, 74*, 262–265.

Xu, J., Zhang, Y., Wang, W., & Li, Y. (2020). Advanced properties of gluten-free cookies, cakes, and crackers: A review. *Trends in Food Science and Technology, 103*, 200–213.

Yadav, R. B., Yadav, B. S., & Chaudhary, D. (2011). Extraction, characterization and utilization of rice bran protein concentrate for biscuit making. *British Food Journal, 113*, 1173–1182.

Zhou, W., Hui, Y. H., De Leyn, I., Pagani, M. A., Rosell, C. M., Selman, J. D., & Therdthai, N. (Eds.). (2014). *Bakery products science and technology* (pp. 391–407). Wiley.

Chapter 5
Cakes and Muffins

Tahira Mohsin Ali, Marium Shaikh, Tooba Mehfooz, and Abid Hasnain

1 Introduction

Bakery products are popular worldwide due to easy availability, nutritional quality and improved palatability. Consequently, a wide range of bakery products are nowadays available on super market shelves including breads, biscuits, doughnuts, cakes, rusks, buns, muffins and many others. Among these, cakes represent the most diverse category when compared with other baked products. By definition, cake is a semi-dry foamed bakery product in which air bubbles are entrapped into the starch and protein matrix that is held in aqueous phase of batter (Khanam et al., 2019). Besides this simple explanation, a great deal of variations arises when ingredients are blended in different mixing orders and proportions forming different assorted products. There are certain features which distinguish cakes from other bakery products. Cakes belong to intermediate moisture content class of baked products with average moisture content in the range of 22–28% (Cauvain & Young, 2008) which is lower than bread (35–40%) and higher than biscuits (1–5%) (Sani et al., 2014). Carrot cake and cheese cake, however, belong to high moisture content food groups. These are characterized by a thin yellow-brown layer on surface known as crust (thinner than bread crust) that surrounds the internal sponge like part known as crumb. Cakes are generally formulated from batter which is in contrast from bread and biscuits prepared generally from dough. Additionally, cakes are unique in the sense that the crust can be decorated with colorful frostings, nuts, chocolate chips, edible sprinkles etc. giving attractive aesthetics.

T. M. Ali (✉) · M. Shaikh · A. Hasnain
Department of Food Science and Technology, University of Karachi, Karachi, Pakistan
e-mail: tahira.mohsin@uok.edu.pk

T. Mehfooz
Department of Human Nutrition and Dietetics, Faculty of Health Sciences, Iqra University, Karachi, Pakistan

M. A. Shah et al. (eds.), *Cereal-Based Food Products*,
https://doi.org/10.1007/978-3-031-40308-8_5

This chapter deals with the major types of cakes, role of ingredients, preparation methods of cakes and muffins, development and the challenges associated with gluten free cakes and muffins and quality aspects of cakes and muffins.

2 Types of Cakes

Cake is known to be a batter-based bakery product. Batter is principally a water based mixture of flour with either sugar for making sweet products like cake, cupcake, pancake, etc. or with seasoning to make coated deep-fried products. Cake batter is a sweet viscoelastic mixture in which air bubbles are initially trapped in the fat phase, and remaining dry ingredients are dissolved and/or dispersed in the continuous water phase. This type of arrangement in batter gives rise to an oil-in-water emulsion. Cake batter can be further categorized into emulsion-based and foam based batters. The emulsion type batter is generally regarded as a high fat containing batter whereas, the foam type batter is a 'no' or 'low-fat batter'. In emulsion type batter, air bubbles are first incorporated into the fat phase which generates a non-homogenous system in which air cells are trapped inside emulsified fat phase. This mixture is eventually added into the aqueous phase forming oil in water emulsion. The foam type batter is formulated by incorporating air bubbles directly into the aqueous medium thus forming liquid foam.

Cakes are believed to be the most versatile baked goods in terms of ingredient utilization (Fig. 5.1). However, the basic recipe contains wheat flour, sugar, shortening, eggs, and leavening agent (Sumnu & Sahin, 2008). Some cake recipes also include certain additives like emulsifiers, flavoring agents and enzymes (amylase) in order to enhance stability of batter as well as flavor, volume and shelf life of final product (Pycarelle & Delcour, 2021). Cakes are often classified on the basis of two aspects: (1) type and level of ingredients used in batter and (2) method of batter

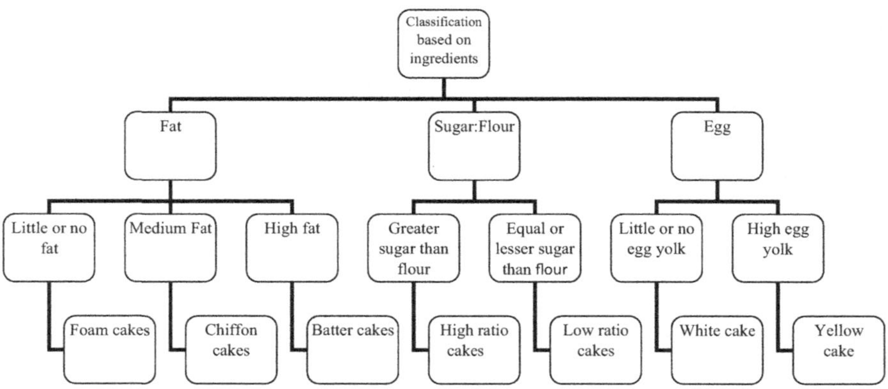

Fig. 5.1 Classification of cakes according to different ingredients

formulation. Ingredient based classification involves utilization of high and low levels of two major components of recipe i.e. fat and sugar. Some recipes mention the use of egg white alone while some cakes are prepared by using whole egg. According to the lipid content and preparation method of batter, cakes are distinguished as foam style, batter style and chiffon style. Foam type cake contains very little or sometimes no fat content (0.1–10) %, based on batter weight) in the recipe like angel food cake and sponge cake. Batter type cake is a high fat cake (10–30%, based on batter weight) and includes cream cake and layer cake (Pycarelle & Delcour, 2021). A third classification is relatively combination of batter type as well as foam type cakes and is known as chiffon cake. A chiffon cake batter is composed of both an emulsion and foam and constitutes the lightest type of cake. The lipid content of chiffon cake is an intermediate between foam and batter style cakes.

Cakes are also classified on the basis of sugar level in recipe as high ratio and low ratio cakes. A recipe which has higher sugar to flour ratio belongs to the former type while lesser or equal sugar to flour ratio in recipe represents the latter type. Yellow and white cakes have a major differentiation in usage of egg contents. Yellow cake is prepared by using whole egg in recipe which gives yellowish shade to cake due to presence of egg yolk. On the other hand, white cake is formulated with egg white only and forms lighter colored crumb.

Type of Cakes Based on Fat Content

Foam Type Cakes

The principal ingredients of foam style cakes are wheat flour, sugar, and eggs (either whole egg or egg white only) usually in the proportion of approximately 1:1:1 (Der Sman & Renzetti, 2021). These are known as unshortened cakes as the formula does not include any added shortening. For this reason, the major leavening action in this category of cakes is derived by maximum incorporation of air cells into the beaten egg white phase which give rise to foam like appearance. Whole egg is made up of two parts: egg yolk and egg white. The proteins in egg white are crucial for the development of a foam which contains dispersed air bubbles (Çelik et al., 2007). Some batter formulations indicate use of acidic ingredient which lowers the pH and further expedite the whipping ability of egg white. The ultimate volume and texture of these cakes depend on beating time as well as amount and size distribution of added air bubbles (Godefroidt et al., 2019). These are relatively softer and lighter (less dense) than batter style cakes (Suas, 2012). Their recipe usually does not mention any chemical leavening agent and are prepared either by single step mixing method or multi step mixing method. Common examples of foam style cakes are angel food, sponge, meringue, dacquoise, genoise, and roulade cakes. Sponge cake is the simplest type of foam cake and can be made from only three ingredients: wheat flour, sugar, and whipped egg white. Foam based cakes are lighter, soft, and airy due to sufficient whipping of egg whites.

Batter Type Cakes

These are also known as shortened cakes as high amount of added fat/shortening is the integral recipe component which is crucial for incorporating air into the batter. Other major ingredients consist of wheat flour, sugar, eggs and leavening agents. The leavening in batter style cakes is achieved by addition of air bubbles either into the fat phase of cream (mixture of sugar and fat) or directly into the aqueous phase (Delcour & Hoseney, 2010; Godefroidt et al., 2019). The former method refers to multistage batter mixing while the latter method corresponds to one stage batter mixing. Their recipe usually includes chemical leavening agent like baking powder or baking soda for the final development of volume and structure (Labensky et al., 2009). Batter style cakes are richer and denser than foam-based cakes. Pound cake, cream cake, yellow and white layer cake are some of the examples of shortening based cakes.

Chiffon Cakes

Chiffon cake is basically a hybrid between foam style and batter style cakes. The key ingredients of chiffon cake are wheat flour, sucrose, eggs, oil, leavening agent and water (Sirisoontaralak et al., 2017). These are termed as unshortened or low fat containing cakes with use of liquid oil in place of solid shortening (Kim et al., 2014). Their recipe involves high amount of eggs which primarily mimics the lightness feature of foam style cakes (Sinha, 2007). The uniqueness of chiffon style cake batter is that, it is formed of vegetable oil-egg yolk emulsion containing air bubbles that is carefully folded into the beaten foam phase. The resultant product forms a super soft and airy textured crumb which appears like chiffon.

Types of Cakes Based on Sugar/Flour Ratio (High-Ratio and Low-Ratio Cakes)

Based on sugar to flour ratio, cakes are categorized as "high-ratio" and "low- ratio" cakes (Table 5.1). The basic difference between a high-ratio and a low-ratio cake is the quantity of sugar used in batter recipe. The high ratio product contains greater amount of sugar than flour content (sugar: flour ratio > 1), while the low ratio product is formulated with equal or less sugar proportion as compared to flour weight (sugar: flour ratio ≤ 1). The external appearance like volume, crust color and character, symmetry of formation and bloom of both types of cakes is almost similar. However, differences exist in the internal crumb structure. The high-ratio cakes possess uniform and fine crumb grain. These also offer tender eating quality partially due to higher moisture content in the final baked product. High ratio cakes usually contain added shortening in the formula that is responsible for entrapping maximum air bubbles in cream, surrounding starch and protein molecules to impart further crumb tenderization, soft

Table 5.1 Difference between high-ratio and low-ratio cake characteristics

Recipe component	Low-ratio	High-ratio	References
Sugar	Lesser or equal quantity than the weight of flour	Higher quantity than the weight of flour	Kim (1994)
Water (added + liquid containing ingredients)	Lesser or equal quantity than the weight of flour	Higher quantity than the weight of flour	Cauvain and Young (2008)
Flour treatment	Plain	Dry heated or chlorine treated	Gough et al. (1978)
Fat	Without added emulsifier	With added emulsifier	Cauvain and Young (2008)

mouth feel, greater emulsification of recipe liquids to enhance crumb moisture content and ultimately forms softer cakes (Zhou et al., 2011).

Types of Cakes Based on Egg/Egg Components (Yellow & White Cakes)

The use of whole egg and/or egg contents separately in cake formulation give rise to another classification based on color as yellow and white cakes. Yellow cakes are prepared by adding whole eggs in recipe in which the yellow color of yolk (due to xanthophylls) contributes to the golden-yellow appearance of cake crumb. Alternatively, white cakes are formed with egg white only and possess lighter/white colored crumb due to absence of egg yolk. Angel food cake is a low fat and high ratio white cake while sponge cake is a low fat and low-ratio yellow cake (Brown, 2018; Godefroidt et al., 2019; Pycarelle et al., 2019). In absence of leavening agent, angel food cake relies completely on the foaming capacity of egg white proteins for the final volume and crumb structure.

3 Cake Ingredients and Their Functions

Flour, fat and sugar are some common ingredients in cakes. While, eggs, surface active and leavening agents are also used in the preparation of cakes. These ingredients are associated with the development of specific texture, flavor and mouth feel of cake.

Wheat Flour

Wheat flour is the foremost important ingredient which makes (30–40)% dry basis of the cake and ultimately defines the texture, volume, color and flavor of the cake (Dewaest et al., 2017).. Ideally, low protein and ash content refined/milled flour derived from soft wheat variety is used for the production of cakes (Edmund & Perry, 2008). This flour is also known as cake flour and contains protein content in the range of (7–9)% and is majorly extracted from the endosperm of wheat flour (Al-Dmoor & El-Qudah, 2016; Al-Dmoor, 2013). The general composition for a cake flour is moisture (13.5–14.5%), protein (7–8)%, starch (72–74)%, sugars (1–2)%, lipids (0.4–0.6)%, cellulose 0.1% and ash (0.2–0.5)% (Al-Dmoor, 2013; Al-Dmoor & El-Qudah, 2016). The particle size of flour is also very important for the development of appropriate cake volumes and crumb textures by influencing batter viscosity and density (Moiraghi et al., 2013). An inverse relation was observed by Choi and Baik (2013) and Nakamura et al. (2010) between cake flour particle size and viscosity of flour-batter system. According to Choi and Baik (2013), the finer particle size of flour subsides the negative impact of higher protein and damaged starch content. The cake flours with particle size less than 55 micrometers produces cakes of higher volumes primarily due to reduction in batter viscosity and increase in surface area of flour which allows entrapment of more air bubbles during mixing of batter (Choi & Baik, 2013). Cake flours are also characterized with a lower water retention capacity (Moiraghi et al., 2013).

The major functional ingredients of cake flour which dictates processing and physical traits of finished product are "gluten" and "starch" (Wilderjans et al., 2013). However, unlike hard wheat products, the high protein content and gluten development is undesirable in cake mixes as it will show high degree of elasticity which prevents expansion and demotes spread-ability of cake during leavening and will result in cakes with lower volumes along with coarser and harder textures (Al-Dmoor, 2013). In contrast to bread doughs, the gluten development in cake batters is very weak due to the presence of high amount of sugar and fat which prevents hydration of wheat proteins. Also, dilution of wheat proteins due to higher amount of water and shorter mixing times lead to weak gluten development (Al-Dmoor, 2013; Wilderjans et al., 2008). But, still appreciable amount of protein content (7–9)% is nevertheless important to bind water and confer elasticity along with egg proteins to develop the spongy cellular structure of the cake as baking crosslinks/aggregates wheat and egg proteins (Wilderjans et al., 2008).

The starch as constituent of wheat flour has a more dominant role in cake making compared to protein. In cake batters, starch exists in ungelatinized form. However, during baking it absorbs water from the aqueous batter and swells leading to gelatinization of starch in the temperature range of (94–97)°C (Marcotte et al., 2004). The elevated gelatinization temperature of starch is due to presence of sugars and other interfering agents like fat which delay/retard loss of crystallinity. The absorption of water by starch converts the liquid batter into a solid foam as evident by increase in storage moduli and viscosity (Wilderjans, Luyts, et al., 2010b) of batter which helps in retention of air bubbles resulting in higher cake volume. This thermal transition (starch gelatinization) is followed by denaturation of egg white proteins (Deleu et al., 2016) which altogether prevents the collapse of cake structure due to expanding air-bubbles and thus thermosets the internal porous crumb texture of cake. The starch is embedded in the denatured protein network and reinforces the cellular structure (van der Sman, 2018a, b) and this was confirmed by Rodríguez-García et al. (2014) using Confocal Laser Scanning Microscopy. The presence of high amount of sugars and fats are responsible for the delay in gelatinization of starch (Van Der Sman, 2018a, b). Sucrose by competing for water with starch elevates gelatinization temperature and this phenomenon is more pronounced in high ratio cakes. Higher starch swelling allows the bridging of starch granules and thereby gives a stronger crumb cellular texture which prevents collapse both during baking and cooling when gas contracts (Wilderjans et al., 2008). Damaged starch is also detrimental for the finished quality of cake. The gelatinization is a crumb related phenomenon whereas due to lower availability of water in crust, the starch here remains fully or partially ungelatinized (Hesso et al., 2015). However, since crust has higher temperature (>180 °C) than crumb due to direct heat exposure therefore, dextrinization or caramelization of starch takes place giving brownish crust color (Cauvain & Young, 2008; Wilderjans, Kerckhofs, et al., 2010a). Milling of soft wheat varieties result in lower amount of damaged starch compared to hard-wheat varieties owing to lower amount of energy required to mill the former. Damaged starch has very high degree of water absorption (Saari et al., 2016). Cake volume is negatively correlated with damaged starch (Kahraman et al., 2008) as it affects batter rheology by competing for water with undamaged starch and other hydrophilic ingredients (Saari et al., 2016). Besides stabilizing air-water interface during baking by increasing the storage modulus of batter, the leached out amylopectin chains retrogrades or recrystallizes resulting in hardening of crumb and thus contributes to staling of cakes (Luyts et al., 2013).

Cake flours are also bleached with chlorine as it results in better and lighter crumb color and also decreases the gelatinization temperature of starch resulting in earlier thermosetting which subsequently prevents the escape of gas during early baking stages. Bleached flour is popularly used for high ratio cakes where sucrose drastically elevates the gelatinization temperature of starch (Finnie et al., 2006). Bleaching of flour is allowed as per 21CFR137.105 (CFR, 2022). However, alternative treatments with xanthan gum and L-cysteine also resulted in similar cake volumes and crumb scores as observed for chlorinated or bleached flours (Finnie et al., 2006).

Eggs

Egg gives tenderness, yellow color and adds protein to the cake (Yazici & Ozer, 2021). The yellow color of cakes is due to carotenes and xanthophylls present in egg yolk (Wu, 2014). However, the functional role of eggs in cake depends on sponge or batter type of cake. But, majorly it may act as foaming/whipping, surface active and thermosetting agent.

Role of Egg in Sponge Type Cakes

Role of egg as whipping agent is mainly for sponge type cakes. Maximum air is incorporated in aqueous phase of sponge cake batters by beating egg white as no further air bubbles will be formed later down the process (Conforti, 2006). The extra-ordinary foaming capacity of egg white is due its albumins and globulins and particularly the interaction of the former with lysozyme protein present in egg white. The ovalbumin orients itself at the air-water interface such that the (-SH) groups are exposed which then later form disulphide (-SS) bridges which confers stability to foam (Godefroidt et al., 2019). However, addition of cream of tartar further reduces the pH of egg-white bringing it close to its isoelectric point where electrostatic repulsive forces are minimized due to equal positive and negative charge on proteins. The reduction in repulsive forces promotes protein aggregation which further reinforces foam stability (Jarpa-Parra et al., 2017; Oldham et al., 2000). For sponge cakes made via multi-stage mixing process, first egg-sugar mixture is whipped to incorporate air followed by addition of egg yolk, flour and fat. Leavening agents must be added in case if sponge cakes are high in fat or contains egg yolk which restricts air incorporation during beating phase (Pycarelle et al., 2019). For sponge type cakes made via single-stage mixing procedure, development of stable foam is manifested via addition of surfactants/emulsifiers which gives stability at the air-water interface. Whereas, in case of multi-stage mixing, overall stabilization of foam/air (prevention of bubble coalescence) is due to egg white proteins and surface active low density lipoproteins present in egg yolk (Godefroidt et al., 2019).

When the cake batter is placed in oven, the egg proteins denature simultaneously with starch gelatinization. The structure of cake relies on gelatinized starch granules which increase storage modulus of batter and forms brick like structure in cakes. However, these bricks are adhered through denatured gluten proteins (acting as mortar between starch bricks) (Pyler & Gorton, 2008). The unfolding of egg protein (both ovalbumin and yolk proteins) at high temperature is due to formation of covalent disulfide bridges. The denatured egg proteins aggregate with wheat proteins (glutenins). The formation of this protein network enveloping grainy starch granules make stronger cell walls which sustain expanding gas pressure during baking and thus prevents the collapse of cake volume due to its weaker structure (Wilderjans et al., 2008, 2013).

Role of Egg in Batter Type Cakes

During mixing stage, the role of eggs in batter type cakes is emulsion stabilization in contrast to aeration for sponge type cakes. For batter type cakes made via multi-stage mixing procedure which involves creaming (beating of sugar and fat together) step followed by addition of other ingredients, the yolk proteins of the eggs convert water-in-oil emulsion into oil-in-water emulsion where sugar is present in the aqueous phase (Shepherd & Yoell, 1976). The yolk proteins actually orient itself at the oil-water interface and form a thin film around the fat droplets which subsequently prevents their coalescence or aggregation. Since, air is also present in the fat phase of batter as immobilized gas cells therefore, egg proteins acts as both foam and emulsion stabilizing agent (Mine, 2002).

During baking stage, both egg white and yolk proteins coagulate at 65 °C and 85 °C, respectively and thermosets to give a cellular gel like structure. This denatured protein network along with starch contributes to foam like structure of cake (Kiosseoglou, 2003, 2004). High temperature of baking melts fat and air cells which then migrate into aqueous phase. Interaction between different egg white proteins develops elastic cellular structure around the air cells which sustain the expanding gas pressure in the aqueous phase during baking. The loss of cellular elasticity can cause collapse of cake structure (Weijers et al., 2006).

Sucrose

Sucrose is a non-reducing sugar responsible for giving sweet flavor to the cake. Also, caramelization of sucrose in the crust part during last stages of baking may also contribute to the brown crust color. Besides caramelization at high temperatures, sucrose may also be thermally decomposed into reducing sugars which then also participate in maillard reaction and contributes to both color and flavor particularly of the cake crust (Der Sman & Renzetti, 2021; Zhang et al., 2012). Sucrose being hydrophilic reduces water activity and thereby also exerts a preservative effect in cake.

Besides preservative effect and flavor development in cake, sucrose also has many other roles during mixing and baking. In batter type cakes, where creaming step is conducted by beating sucrose and fat, the former has an abrasive effect on fat crystals and chops down its size to (1–2) μm. This small size is considered effective for increasing aeration (Shepherd & Yoell, 1976). However, unlike batter type cakes, where sucrose increases aeration and foaming capability by reducing fat crystal size; in foam type cakes, sucrose only stabilizes the already developed foam and it does so by increasing the viscosity of the batter (Zhu & Damodaran, 1994). Effective gluten development is not desirable in cake making, as it will prevent the rise in cake volume and will make cakes harder. However, sucrose being extremely hydrophilic competes for water with wheat proteins and keeps it dehydrated and thereby prevents optimum gluten development (Wilderjans et al., 2013).

Sucrose is also known to increase the gelatinization temperature of starch in both batter and foam type cakes. As earlier discussed, gelatinization of starch is of utmost importance as it is considered a building block in the development of cake's structure along with denatured egg proteins. Different mechanisms have been proposed regarding why sucrose elevates starch gelatinization temperature; the most popular being a) sucrose-water interactions and b) sucrose-starch interactions. Sucrose-water interactions lead to decline in plasticizing effect of pure water and thus higher amount of heat energy is required to cause loss of crystallinity or birefringence in starch leading to elevation in gelatinization temperature of starch (Slade et al.,

2021). Also, sucrose-water interactions limits the availability of water and volumetric density of hydrogen bonds in polymer solution surrounding starch leading to increased melting temperature of gelatinization (Der Sman & Mauer, 2019; Perry & Donald, 2002). According to Chiotelli et al. (2000) and Gonera and Cornillon (2002), sucrose penetrates the amorphous region of starch where it interacts with amylose chains and thereby restricts its mobility leading to increment in starch melting temperature. Too much elevation in starch gelatinization temperature causes delay in thermosetting and may lead to collapse of cake volume. Sucrose though delays gelatinization but increases rate and extend of starch swelling (Bean & Yamazaki, 1978).

Another important thermal transition affected by sucrose is the delay in protein denaturation temperature. It does so my altering the protein hydration leading to stabilization of protein conformation which eventually increases thermal stability and raises the coagulation temperature. It is hypothesized that protein hydrophobic groups are less exposed in sugar-water medium compared to pure water (Kaushik & Bhat, 1998). Another role of sucrose in cake is as anti-staling agent as it retards starch retrogradation (Der Sman & Renzetti, 2021).

Fats

Fats give tenderness, moistness, mouthfeel and flavor to the cake. The commonly used fats are margarine, plastic fats (shortenings), while the premium quality cakes are made with butter. Plastic fats are basically made via hydrogenation of vegetable oils and therefore contain trans fatty acids which increase the risk of cardiovascular diseases (Chopra et al., 2021). Liquid shortening containing emulsifiers and seeds of β-crystals are also developed which have obvious advantages of being trans-free and are easily pump-able at room temperature allowing ease of metering in cake formulations (Zhou et al., 2011). In contrast to shortenings, margarine is an emulsion containing 16% water phase (Andersen & Williams, 2016). The role of fat is very important in batter type cakes, where fat allows incorporation of air cells during creaming (Brooker, 1993, 1996). These air cells then reside in fat phase of the batter and provide the necessary structure/framework for leavening gases and vapors produced during baking. The β′ (Beta prime) fat crystals with smaller size and larger surface area to volume ratio compared to per gram of beta crystals are more preferable as they provide higher interfacial area for stabilization of air cells (Brooker, 1993; Ghotra et al., 2002). Usually, fat crystals having needle shape and of (1–2) μm size allows better aeration. The liquid portion of shortening allows aeration while solid portion stabilizes or immobilizes the incorporated air cells. Oil based cakes are therefore very tender due to air-cell drainage and are unable to withstand processing conditions (Conforti, 2006).

During baking, the fat crystals melt and as a result, the air cells move from fat towards the aqueous phase. These air cells are however, still enveloped with a thin layer of oil which form interface between air and aqueous phase. This interfacial fat layer is further stabilized with egg yolk proteins (Shepherd & Yoell, 1976; Wilderjans et al., 2013). However, later on this melted fat again crystallizes and coats the surface of cake and prevents moisture loss. Fat is also known to coat starch and gluten and thus prevents their close association which can otherwise make cakes harder (Pyler & Gorton, 2008).

Leavening Agents

The beating of air in cream/egg white and addition of baking soda, baking powder, self-raising flour and water vapors produced during baking cause leavening in cakes. Chemical leavening agents are commonly used in batter type cakes. The major chemical leavening agent used is sodium bicarbonate along with an acid source which reacts with each other to produce carbon dioxide gas. However, at higher temperatures, the sodium bicarbonate may also thermally decompose to produce CO_2. This carbon dioxide produced expands already existing air-cells which act as nuclei and thus results in foam like structure of cake after baking. The rate at which carbon dioxide is released is a decisive factor for final cake characteristics. Too early leavening before thermosetting of cake structure will result in coalescence of bubbles which then move on the top and are finally lost from the surface. This type of early leavening will result in lower cake volumes. Similarly, leavening during later stages of baking is also undesirable as starch gelatinization and protein denaturation would have already taken place which sets the structure of cake and will thus resist the expansion of air-bubbles due to release of carbon dioxide gas leading to cracks on the surface of cake (Godefroidt et al., 2019; Heidolph, 1996). Therefore, controlled release of CO_2 is necessary to give proper volume and symmetry to cakes. This is usually done by incorporating fast and slow acting acids. The commonly used acids in baking powder are SAPP (Sodium acid pyrophosphate), CAPP (Calcium acid pyrophosphate), fumaric acid, monocalcium phosphate, glucono-delta-lactone etc. The use of either slow acting or fast acting acid in combination with bicarbonate source leads to production of single acting baking powder whereas using a combination of slow and fast acting acids together may produce a double-acting baking powder (Heidolph, 1996). Another strategy used to control quick release of CO_2 is by encapsulation of sodium bicarbonate. The encapsulation wall melts at higher temperature and thus prevents premature release of CO_2. Aluminum containing leavening agents has been banned by many countries globally. Cream of tartar (CT) is also added in angel cakes where it improves the foaming capacity of egg white by reducing pH which alters protein conformation. The addition of CT also results in white color and uniformly distributed fine crumb grain. The most commonly used acid in baking powder is SAPP which is available in different grades depending on its capability or timing of carbon dioxide released and neutralization value. These different SAPPs have tailored solubility and thus release carbon dioxide at different time and temperature combinations (Heidolph, 1996). In a detailed study conducted by Book et al. (2015) on impact of different leavening acids on cakes, they found that CAPP-S (Calcium acid pyrophosphate, slow acting) produced cakes with the highest volumes and more symmetrical structure in comparison to CAPP-D (double acting), SAPP-28, SAPP-40 and SALP-M (double acting baking powder made with fast acting monocalcium phosphate and heat activated Sodium aluminum phosphate).

Surface Active Agents

Surface active agents added to cakes orient themselves at the air-water interface and thereby stabilizes the foam. Different surface active agents like SSL, polysorbates, lecithin, glycerol monostearate are used in cake batters (Wilde, 2000). The role of surface active agents is more pronounced in foam type cakes compared to batter cakes. For multi-stage mixing in foam type cakes, surface-active agents are not needed. However, foam type cakes made with single-mixing step, it is necessary to add emulsifiers or surfactants to stabilize air-water interface as presence of both surface active lipids and proteins together during single stage mixing competes for air-water interface resulting in less stable foam (Wilde, 2000). The surfactants adsorb at air-water interface and reduce surface tension leading to stabilization of air cells during single stage mixing. Though oils are not used for creaming in cakes because these do not allow similar incorporation of air as observed for shortenings. However, liquids oils along with use of appropriate emulsifiers and hydrocolloids allow air incorporation, stabilization and distribution during cake mixing (Kumari et al., 2011). Usually, hydrophilic emulsifiers are more useful for cakes as these bind water and increase viscosity of the batter and at the same time disperse oil droplets to form oil in water emulsion. Moreover, it is also known to decrease the solubility of amylose chains after starch gelatinization via formation of amylose-lipid complex and thus it decreases strong adherence between starch and protein chains during baking resulting in development of a softer crumb (Kim, 1994).

4 Preparation Methods of Cakes

A cake is prepared by following the fundamental steps shown in Fig. 5.2. First of all, dry ingredients are weighed accurately and mixed with water to form a fluid batter. The batter is then filled into a cake mould/pan and kept in baking oven for predetermined time for conversion of batter into a cake. Finally, the cake is cooled and packed in suitable packaging.

Cakes are categorized according to type and/or level of certain ingredients. The most accepted category is based on the batter lipid content and differentiates cakes broadly into: (a) batter style/shortened cake and (b) foam style/unshortened cake. Both types of cakes are prepared by following the general procedure outlined in Fig. 5.2 i.e., ingredient weighing, mixing, depositing, baking, cooling and packaging. However, their batter mixing methods differ greatly in terms of order of adding

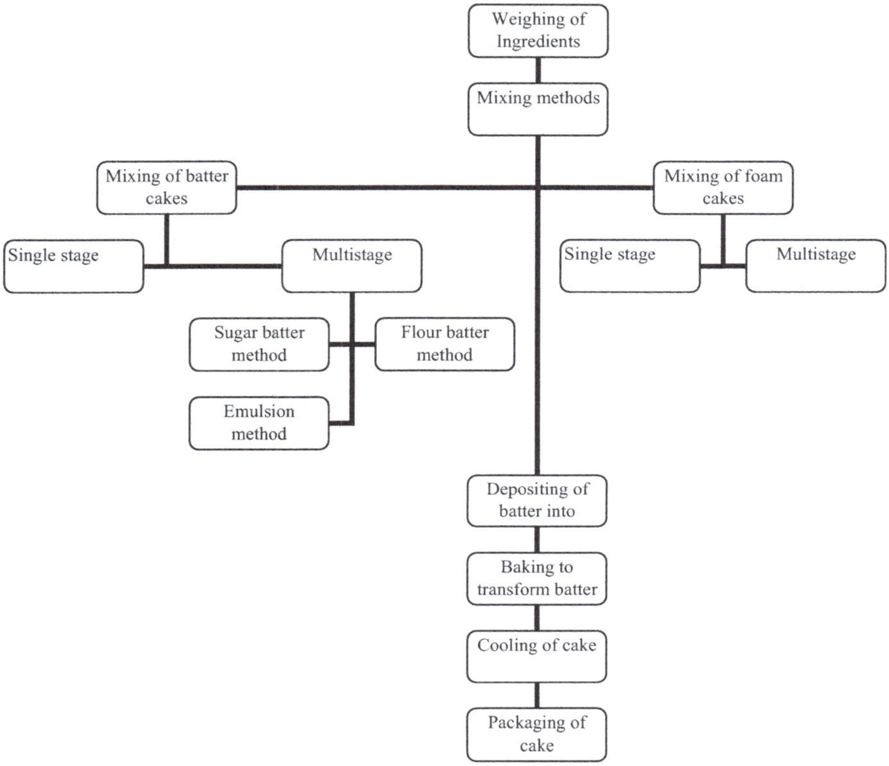

Fig. 5.2 Flow diagram of cake manufacturing

Fig. 5.3 Crust of cake

ingredients, mixing speed and mixing time. Figures 5.3 and 5.4 shows the crust and crumb of the cake, respectively.

Fig. 5.4 Crumb of cake

Mixing

Mixing Methods for Batter Type Cakes

The primary objective of mixing in case of shortened cakes is twofold a) the combination of all ingredients together to form a homogenous fluid batter with stabilized emulsion as this type of batter contains high levels of added fat; b) maximum incorporation of small air cells in the emulsion type batter in order to achieve desirable volume and structure of final baked product. The batter of shortened cakes is prepared by one of the following methods:

(a) Sugar batter or creaming method
(b) Flour batter method
(c) Emulsion method
(d) Single stage or all in one method

5 Sugar Batter or Creaming Method

It is a multistage mixing procedure in which all of the ingredients are not added together but added at different steps in particular order. Firstly, castor/granulated sugar is combined with fat at low to moderate mixing speed to form a uniform aerated mixture known as cream. This method is also known as creaming method due to formation of a homogenous light mixture of sugar and fat that resembles off white colored dairy cream. In the second stage, eggs are added to the cream and mixing action is continued. Finally, other liquid components of recipe (water, milk, essence and color) along with flour are added alternatively in small parts to form the resultant batter in which over development of gluten is prevented. The final batter is a multiphasic system in which air bubbles are trapped in the fat phase while other ingredients are dispersed in the aqueous phase (Wilderjans et al., 2013).

The purpose of cream formation in sugar batter method is to mechanically leaven the system by incorporating as many as possible small air cells into the fat phase of cream via whisking. Consequently, air cells are trapped in the fat making the cream lighter. These air cells are further stabilized by sugar and fat crystals (Der Sman & Renzetti, 2021). The number of air bubbles trapped in the cream and their stabilization is crucial for the final product's volume and structure as no new cells will formed in subsequent mixing steps. The addition of liquid eggs serves to solubilize the sugar crystals as well as the systems gets emulsified holding the fat and liquid fraction uniformly which otherwise would not be possible. The flour is added at the last step so as to minimize gluten formation which is the required feature of cake making. The overall mixing time for this method is approximately 15–20 min in which initial creaming is done for 8–10 min; creaming after egg addition continues for 5–6 min; and final mixing during which flour is combined lasts for 5–6 min. Prolonged mixing accounts for loss in aeration and makes the cake denser.

6 Flour Batter Method

The flour batter method is also an example of a multistage process. In this method, two mixing bowls work simultaneously and deal with different ingredients. In first mixing bowl, equal or slightly less amount of flour and fat are creamed together until it becomes light and fluffy. Sometimes, baking powder and salt are also sifted into flour portion (Sudha et al., 2007). The second bowl carefully whips eggs and same quantity of sugar together at medium speed for about 6 min to form a foamy structure (Bennion & Bamford, 1997). Next, the resultant mixtures of both the bowls are carefully combined together into which milk is gradually added to achieve final batter. The purpose of such dual mixing of ingredients separately is to promote thorough dispersion of fat with minimum chances of batter curdling. Also, the cake prepared by this method possesses highly fine texture. This method also allows the use of higher sugar and liquid content than in the creaming method. The increased running cost of two separate mixing bowls, reduced amount of air incorporation leading to a product with low volume, and greater gluten development which forms a tough cake in comparison to creaming method are the major shortcomings of this method.

7 Emulsion Method

The emulsion method also belongs to multi-stage batter mixing and is suitable for large scale production of cake in high volume mixers. This method is quite similar to creaming method but the order of adding ingredients is different from creaming method (Sumnu & Sahin, 2008). Firstly, sugar and fat are creamed together for about 2–3 min at medium speed to form a uniform mixture. Milk is then added in small parts with continuous beating at low to medium speed for approximately 5 min until butter cream like mixture is formed. The flour is added and mixing is continued for additional 2 min. Finally, eggs are added and mixing is done for 4–5 min to achieve final batter.

8 Single Stage or All in One Method

This cake mixing method combines all ingredients of the recipe at once in a single mixing bowl equipped with a wired-whisk to prepare homogenous batter. The total mixing time is about 10–12 min. The speed of mixing is set as follows: low speed for 1–3 min; medium speed for 3–5 min and finally low speed for 2 min. Contrary to creaming process, in which air cells are added into the fat phase, single step mixing method promotes aqueous phase aeration (Brooker, 1993).

Mixing Methods for Foam Type Cakes

This category of cakes is short of added fat and depends on beaten egg's ability to incorporate and sustain air cells. Therefore, the goal of mixing in unshortened cakes is to include as many as possible air bubbles in whipped eggs by appropriate mixing action which ultimately forms a stable light foam. This foam is responsible for volume and texture of final baked product. The mixing methods for foam style cakes are broadly divided into:

(a) Single step mixing action
(b) Multi step mixing action

9 Single Step Mixing Action

This method is similar to all in one mixing of high fat cake as all ingredients are added together at same time in mixing bowl. However, care must be taken in case of formulating unshortened cake batter in terms of using interface stabilizing agents/emulsifiers. As foam style cakes contain air cells that are dispersed in the aqueous phase and therefore, the water and air interface should be stabilized to confer foam stability. The presence of surface active agent in recipe facilitates incorporation and stabilization of air cells making product more uniform (Richardson et al., 2002). Mixing time is critical in all in one method as prolonged mixing may lead to over development of gluten, forming tough cakes. Rodríguez-García et al. (2014) studied the effect of mixing time (3, 5 and 15 min) on foam style cake quality and suggested that high speed mixing for 3 min resulted in cake with higher volume and tenderness. This kind of method is suitable for large scale manufacturing of cake where ingredient addition and mixing method should be optimized for cost effectiveness and higher production rates.

10 Multi Step Mixing Action

In this method, egg white and half amount of sugar are whipped together to make egg white foam. In another mixing bowl, egg yolk and remaining amount of sugar are homogenously mixed together to form second foam (Godefroidt et al., 2019). These two foams are then combined together and finally folded carefully with remaining ingredients like flour and leavening agents. Care must be taken during folding of foam into dry ingredients so as to assure that no air cells are disrupted.

Depositing

Depositing, in case of cake manufacturing is the process of transferring finished batter from mixing bowl into the container known as a cake pan. A cake pan is made up of any material that is safe to be used in baking oven like plain or enameled metal, silicon or ceramic. However, metal moulds are used preferably because of their higher conductivity. Various styles of cake pans are available ranging from circular, square or rectangular shaped. Some metal pans are also available with nonstick surface to facilitate easy release of cake after baking (Sinha, 2007).

Baking

Baking is responsible for converting wet foam (batter) into dry foam (cake) as a result of heat application. The entire event of cake baking can be typically divided into three distinct segments: early, middle and late baking phase. Early baking phase: occurs up to 50 °C (majorly responsible for melting of shortening), middle baking phase: occurs approximately between 50–65 °C (majorly responsible for expansion of air cells), and late baking phase: occurs above 65 °C (majorly responsible for setting of cake structure due to starch gelatinization and protein coagulation).

Early Baking Phase

As far as the early baking phase is concerned, there are certain variations in baking mechanisms of multi stage and single stage mixed batters. The major transformations occurring in the early baking phase of multi stage creamed batter are related to the initial temperature increment that causes melting of fat molecules (creamed batter contains air bubbles in fat phase) followed by release of the air bubbles from fat part of the batter into the aqueous part. The air cells are then partially covered from inner side by melted fat layer which in turn is surrounded by egg protein layer creating stabilized foam (Brooker, 1993; der Sman & Renzetti, 2021). The flour particles remain dispersed in the water phase that also contains melted fat droplets. The single stage mixed batter contains air cells incorporated into the water phase and shortening is present in the form of fine protein stabilized emulsion (Brooker, 1993). Mixing promotes adsorption of fat crystals onto the surface of air cells that finally creates a protein layer between air-water interface and adsorbed fat molecules. Upon heating, the adsorbed fat starts melting and penetrates inside the air bubbles forming a continuous fat coated layer allowing the smooth air bubbles expansion without rupturing (Wilderjans et al., 2013).

Middle Baking Phase

As the temperature exceeds 50 °C, the size of the air cells start increasing (causing expansion in cell walls) due to accumulation of CO_2 gas generated by leavening agents, water evaporation from aqueous phase, and thermal expansion of the air cells.

Late Baking Phase

This last phase of baking is largely related to the so-called setting of structure which is the final conversion of fluid batter into solid cake like appearance (Pycarelle & Delcour, 2021; Shepherd & Yoell, 1976). The two major events that are greatly responsible for structure setting are gelatinization of starch granules and coagulation of proteins (Mizukoshi, 1979). As the temperature reaches around 65 °C, starch granules start swelling with concomitant coagulation of egg proteins contributing to significant increase in batter viscosity imparting solid like feature to the cake (Wilderjans, Kerckhofs, et al., 2010a). Due to temperature gradient, the base of the cake attains higher temperature earlier than the center and top of the cake and as a result, the cake bottom set first followed by middle and top parts (Wilderjans et al., 2008). Finally, the closed air bubbles containing CO_2 gas and water vapors rupture due to increased pressure of expanding gases and consequently the pressure of ruptured and opened gas cells drops. Usually this step happens after aforementioned starch gelatinization and protein coagulation but if the pressure drops prior to structure setting then it collapses the cake structure (Gough et al., 1978). The last few minutes in oven are also accountable for dark colored crust formation. During the later baking stage, the moisture content of crust reduces significantly leading to glass formation in crust that is crispy and also favors color development via maillard reaction (Ureta et al., 2017).

Cooling, Packaging and Storage

Post baking, the cake pan is gently removed from the oven and cooled down on a shelf for around 5–10 min (Brown, 2018). During cooling, the gases contract which may result in cake collapse (sinking of the cake in the middle). The degree of collapse depends upon the structural strength of crumb which in turn is affected by extent of starch gelatinization and protein coagulation. Usually high ratio cakes are more prone to cake collapse because higher sugar content prevents proper starch gelatinization forming weak structured crumb which upon cooling collapses. Cake collapse during cooling phase is also dependent on the degree of protein coagulation during baking. The greater the protein coagulation, the stronger will be the cell walls of crumb and lower will be the chances of cake collapse (Wilderjans et al., 2008).

After cooling, cake is packaged into appropriate packaging material and stored prior to final consumption. The packaging materials utilized for cake packaging are glassine, grease proof paper, cellophane, polystyrene and polypropylene.

11 Preparation of Muffins

Muffins can be classified as cake-like muffins and bread-like muffins (Cross, 2007). Soft wheat flour with high amount of fat (18–40%) and sugar (50–70%) are used in cake-like muffins. For bread-like muffins, the amount of fat and sugar is 12%. Ingredients found in muffins are common as in cakes. However, some other ingredients like bran, fruit and vegetable chunks, chocolate chips may also be added to the recipe (Ortiz, 2016). The typical picture of plain muffin is shown in Fig. 5.5. While, Fig. 5.6, shows the top and cross-sectional view of centre filled muffin.

Mixing Methods

Muffins can be prepared using two mixing methods i.e. cake method and muffin method. Cake method implicates creaming of sugar and shortening, followed by the incorporation of wet and dry ingredients, respectively. Muffin method is a three step process in which all the dry ingredients are first mixed together. In the second stage,

Fig. 5.5 Muffin

Fig. 5.6 (**a**) Top view of centre-filled Muffin (**b**) Cross-sectional view of centre-filled muffin

shortening/oil and all other liquid materials are combined with each other. Lastly, liquid ingredients are folded into the dry mix (Ortiz, 2016).

Moulding

Muffin size may vary from 1/2 ounce to 5 ounce for mini and large sized muffins, respectively. Two ounce muffin size is considered as standard muffin size (Willyard, 2000). With the aid of dosing machine, 30–45 g of batter is transferred to a mould of size ranging from 55 mm diameter × 30 mm height to 60 mm diameter x 36 mm height (Bender et al., 2017; Herranz et al., 2016).

Baking

According to Khouryieh et al. (2005), Lamdande et al. (2018) and Singh et al. (2020), muffins are usually baked in a pre-heated oven in a temperature range of (110–204) °C from (21–40) min depending upon the type of oven and the size of muffin. The ideal baking temperature for 2 ounce muffin is around 204 °C (Ortiz, 2016).

Cooling

The muffins are allowed to cool at room temperature for at least an hour before further processing. This step ensures reduction in moisture, prevention of condensation after packaging and limits microbial spoilage.

Packaging

Muffins may be packaged in foil/plastic tray or wrapped individually. Plastic or foil wrapped muffins remain shelf stable for 4–7 days. Cake-like muffins are more shelf stable due to the presence of high amount of sugars. Use of preservatives can increase shelf life up to several months.

12 Gluten Free Cakes and Muffins, and Challenges

Over the past few years, people are more interested in gluten free products. Gluten is responsible for unique stretching properties of dough forming a three dimensional structure and also plays a principal role in cake development by giving cohesiveness and promoting the retention of the CO_2 produced during leavening (Salehi, 2019). Hence, it is a great challenge to develop a gluten devoid product without compromising quality. According to the Food and Drug Administration "gluten free" label shall only be used when a product contains less than 20 ppm gluten. The European Union also suggested proper labeling of gluten free products with complete awareness regarding naturally/processed gluten free products (El Khoury et al., 2018). Gluten free products are developed for special class of people called coeliac who are allergic to gluten (Gallagher et al., 2004).

Rice, oats, pseudocereals (Amaranth, Quinoa, Buckwheat), maize, millet, sorghum, chestnut flour, chia flour and legumes are naturally gluten free sources and are possible alternative to wheat flour (Hosseini et al., 2018). Some minor ingredients are also combined with alternative flours to help build and strengthen gluten-free products (e.g. starches, proteins, gums, hydrocolloids, emulsifiers, dietary fibers) (Gao et al., 2018). Texture of cakes is highly dependent on the aeration and stability of gas bubbles produced during baking (Sahi & Alava, 2003). The gas bubble incorporation during mixing of batter also affects volume and texture of crumb. A good quality of gluten free cake is highly dependent on several factors like grain milling, flour type and particle size and additives in the batter. Gluten free flours mostly results in low volume and poor texture for cakes. Rice flour is commonly used for the production of gluten-free foods, as it has hypoallergenic properties due to the absence of gliadin (Gujral & Rosell, 2004). Rice flour also contributes to unique characteristics, such as bland taste, white color, and ease of digestion. Rice flour (particle size $^{<}$ 100 μm) has been widely used in the preparation of gluten free cakes and muffins as it develops small and even bubbles in cakes (De la Hera et al., 2013). Composite flours like (rice flour and quinoa), (rice and soya bean flour), (maize starch, rice flour and buckwheat), (rice flour and green banana flour), (buckwheat flour and maize starch) have been incorporated to develop gluten free muffins (Iovan et al., 2015; Kaur et al., 2017; Man et al., 2014; Miranda-Villa et al., 2019; Zieliński et al., 2017). Bhaduri (2013) replaced rice flour with 75% quinoa flour to produce acceptable gluten free muffins.

Starches are used as gelling agent, thickener, adhesive, moisture-retention, stabilizer, film forming agent and texturizer. In gluten-free products, starch is incorporated into the food formulation to improve one or more of these properties. Starch gelatinization plays an important role in the expansion volume of cake/muffin. Removal of gluten increases the role of starch in providing structure and texture to gluten free products. Starches of alternative (gluten-free) sources have been added to the gluten-free recipes. In order to make further improvements in textural and structural attributes physically and chemically modified starches are also widely used in gluten free products (Korus et al., 2015; Witczak et al., 2016). Ronda et al. (2011) observed that starches like rice, potato and corn produced satisfactory gluten free products.

Apart from non-gluten cereals and starches, gums (hydrocolloid) contribute effectively in developing gluten free structure by further increasing batter viscosity. According to Gómez et al. (2007), increase in batter viscosity slows down the rate of gas diffusion and allows its retention during the early stages of baking. Salehi (2019) reported several gluten-free formulations that have utilized gums to mimic the viscoelastic properties of gluten. Gums comprise of several water-soluble polysaccharides with diverse chemical structures providing a range of functional properties making these suitable for this application. Gums also alter the pasting properties of starch. Itthivadhanapong et al. (2016) reported that the addition of 1% xanthan, guar, hydroxymethylcellulose and carrageenan improved batter quality and the batter demonstrated gel like behavior. According to Majzoobi et al. (2017), use of pectin and xanthan gum along with carrot pomace powder resulted in softer, springy and chewable cake. Xanthan gum is a more popular choice as hydrocolloids as it improves texture, volume, moisture and gas retention by reducing flour sedimentation and also improves the rheological properties (Preichardt et al., 2011).

Emulsifier can provide essential aeration and gas bubble stability during development of gluten free cakes. Turabi et al. (2008) incorporated blend of emulsifiers that resulted in softer and porous rice cakes with higher volume. Emulsifiers are also reported to improve batter viscosity and volume of millet based muffins (Jyotsna et al., 2016). Emulsifier interacts with starch molecules forming starch-lipid complex. In case of steamed rice cakes, addition of emulsifier increased sponginess and decreases hardness (Feng et al., 2021). Gluten-free recipes are usually rich in carbohydrates. Many studies focused on the enrichment of these recipes with protein to fulfill dietary requirements. It has been reported that protein, besides enhancing nutritional density of gluten-free baked product, also played key role in improving the final quality (Geera et al., 2011; Ronda et al., 2011). Proteins are considered as main factors responsible for higher muffin volume and height, as these could swell and denature at high temperature, providing structural support for better gas retention (Ziobro et al., 2013). Shevkani and Singh (2014) introduced kidney bean, field pea and amaranth protein isolates in corn starch-based muffin batters. These protein isolates produced viscoelastic batter, springier muffins which retained more air bubbles. Protein isolates from white cowpea are also reported to enhance product characteristics of rice-based muffins (Shevkani et al., 2015).

Several challenges are associated with gluten free products. The challenge is to prepare a gluten free product that not only satisfies consumers' need but is also pocket friendly for developers as well as processing friendly. Absence of gluten flour makes products inferior in protein content and lacks desirable mouthfeel and texture. Gluten free flours show different behaviour as compared to wheat flour. These differences not only affects the physical properties of the products but also negatively influences sensory attributes.

13 Quality Aspects of Cakes

Cakes may be judged in terms of different quality aspects like flavor, cake volume, symmetry, moistness, tenderness and crumb structure which must be of uniform size with finely distributed grain structure and resilience besides milder brown crust color. The ingredients and processing factors majorly dictate the physical characteristics of the finished product. These factors are evaluated both subjectively using trained/untrained sensory panel and also by using some objective methods described below:

Batter Properties

The cake manufacturers in order to have better control over the process now frequently measure batter properties like density/specific gravity and viscosity as quality control tests before baking. Both simple methods like Bostwick consistometer or viscometers to more sophisticated technologies like rheometer has been reported for the measurement of batter viscosity (Choi & Baik, 2013; Hesso et al., 2015; Rodriguez-Garcia et al., 2013). It is important to understand that cake batters are non-Newtonian and thus their viscosities must be measured at a fixed shear rates for control purpose. Batter viscosity dictates whether the gas will remain entrapped during baking or not. An overly reduced viscous batter may not be able to hold gas pressure during heating (Wilderjans et al., 2008). Determination of batter viscosity can directly impact texture, specific volume, cake height, expansion and crumb cell distribution (Shao et al., 2015) as it affects thermal transitions during baking when liquid batter sets into a porous structure (Hesso et al., 2015). Specific gravity or density is also measured for batters. Usually, egg containing cakes are lighter, fluffy and have lower densities due to higher amount of air incorporation compared to egg-replaced batters which are heavier as hydrocolloids or/and proteins are added as egg substitutes (Agrahar-Murugkar et al., 2018).

Texture

A hard or even semi-hard cake is sign of staling (Kahraman et al., 2008) or imbalance of ingredients during processing and is highly undesirable. The tenderness of the cake has been frequently measured using texture analyzers whereby different parameters like hardness, adhesiveness, cohesiveness and chewiness are measured directly. It is based on a two cyclic compression test whereby the cubical cake is compressed twice up to (25–50)% of its original height. The maximum force observed during first compression cycle highly correlates with the hardness of cake (Abhay Kumar et al., 2017; Díaz-Ramírez et al., 2016; Kahraman et al., 2008; Rodriguez-Garcia et al., 2013). If only hardness of the cake is to be measured then one-cyclic compression test is performed instead of two-cyclic. High volume cakes are said to be negatively correlated with cake hardness while addition of emulsifiers is positively correlated with crumb softness (Lin et al., 2017).

Weight Loss

The difference in weight of cake batter before and soon after baking when it sets into a cake structure divided by original weight of the batter is defined as the percent weight loss or baking loss. A high baking loss will result in lower cake weight

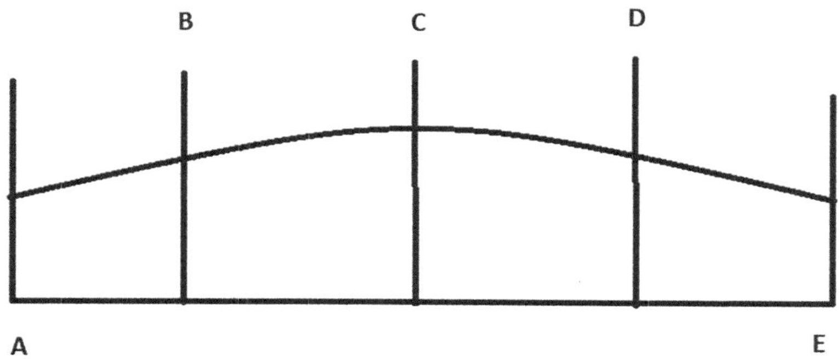

Fig. 5.7 AACC method for measurement of physical characteristics of cakes. (AACC, 2000)

which is an economic loss for the cake manufacturers as net weight is declared on the cake package and thus may fail to comply. Furthermore, it may also make cake hard as baking loss means loss of moisture and volatiles. In contrast, lower baking loss is also undesirable as it may result in extremely moist cake (Geera et al., 2011). Thus, control of net weight of cake after baking is a very important quality control parameter for the processors. The baking loss can be controlled by optimizing fat content and through use of hydrophilic gums and emulsifiers (Seyhun et al., 2003; Shao et al., 2015).

Specific Volume, Volume Index, Symmetry Index and Uniformity Index

The physical characteristics of finished cake is determined in terms of volume index, symmetry index and uniformity Index using method 10–91 (of Cereal Chemists, 2000). This method is used for an 8″ (20.3) cm diameter layer cake for which the cake must neither have peak nor dip at the center. By using the Fig. 5.7, different parameters related to physical characteristics are calculated by measuring diameter from A to E and through measurement of vertical heights B, C and D. Shrinkage value can be calculated using Eq (5.1) and is defined as reduction in diameter of cake batter after baking (of Cereal Chemists, 2000). However, another way of calculating shrinkage volume is as ratio of central height to peripheral height. If the ratio is equal to or greater than one it means no structural collapse while values <1 indicate collapse of cake after baking and cooling (Slade et al., 2021).

$$\text{Shrinkage Volume} = 20.3 - (\text{Diameter A to E}) \tag{5.1}$$

Volume index is calculated by cumulative addition of heights B, C, D to nearest 0.1 cm (Eq. 5.2), (of Cereal Chemists, 2000) whereas specific volume is calculated as ratio of weight to volume of finished cake. The volume of cake must be calculated using seed displacement method (of Cereal Chemists, 2000).

$$\text{Volume Index} = B + C + D \tag{5.2}$$

Volume index (VI) and specific volume (SV) of the cake depends on the amount of entrapped air. Lower VI and SV indicates heavier and denser cakes. Appropriate foaming during preparation of cake batters by using egg proteins or via creaming procedure and its subsequent retention after baking and cooling through starch gelatinization and coagulation/gelation of proteins ultimately decides volume indices of cakes. That is why egg replaced cakes are heavier with low VI and SV values as air-entrapment during mixing and protein coagulation functionalities are both affected in the absence of egg proteins which stabilizes air-cells (Agrahar-Murugkar et al., 2018; Lin et al., 2017). Besides composition, process parameters during mixing and baking may also impact volume of cake (Conforti, 2006).

Similarly, symmetry and uniformity indices are calculated by Eqs. (5.3) and (5.4) (of Cereal Chemists, 2000). Symmetry index is the measure of surface contours. The negative value indicates collapse of centre of cake/muffin or dip at the centre which is a negative trait for cake quality. On the other hand uniformity index (UI) is directly correlated with surface unevenness and higher UI values are therefore undesirable (Gupta et al., 2009; Oliete et al., 2010).

$$\text{Symmetry index} = 2C - B - D \tag{5.3}$$

$$\text{Uniformity Index} = B - D \tag{5.4}$$

Color

Both crust and crumb colors are evaluated for cakes using CIE (L*, a*, b*) values employing colorimeters. 'L' denotes lightness and darkness, 'a*' denotes redness (+) and greenness (−) and 'b*' denotes blueness (−) and yellowness (+). Barium sulphate is taken as the standard for color comparison (Aydogdu et al., 2018).

$$\Delta E^* = \sqrt{\left[\left(L_0 - L^*\right)^2 + \left(a_0 - a^*\right)^2 + \left(b_0 - b^*\right)^2\right]}$$

Where, L_0, a_0 and b_0 are values of barium sulphate taken as standard and L*, a*, b* represent values of cake samples.

Caramelization of sugars and maillard reactions derive the brown color of the crust which is darker than the crumb due to high temperature and low moisture conditions in crust during baking. These two non-enzymatic reactions are also responsible for the taste, flavor and aroma of the cake and are dependent on ingredients like reducing sugars, proteins, baking temperature, baking time, water activity and moisture (Purlis, 2010; Slade et al., 2021). Addition of cream of tartar was found to reduce the pH of batter and resulted in white colored crumb with fine grain structure (Conforti, 2006). The yellow gold and or light brown color is considered desirable for cake crust while the crumb depending upon the type may vary from white to yellowish. Nevertheless, the crust always has lower L* values compared to crumb.

Sensory Analysis

Sensory analysis forms the corner stone for any food product development. Appropriate analysis, free of biasedness can set correct standards for ingredients/process and will result in more likely success rather than failure of the product in market. The importance of sensory analysis has increased manifold in cake industry where new cakes with different health claims designed for specific population continue to strike market like gluten-free cakes, high fiber cakes and eggless cakes. Despite different satisfactory instrumental analysis, a product cannot end-up in any market before successful sensory evaluation. For most of the studies based on major replacement/substitution of any ingredient like development of composite flour cakes or gluten free cakes, usually the panelists are given cake samples monadic sequence so that they could analyze the attributes in isolation (Casas Moreno et al., 2015; Monnet et al., 2022). Both 9-point and 7-point hedonic scales are used in descriptive sensory analysis where different attributes like color, texture, mouthfeel, taste, odor and overall acceptability of cakes are judged (Díaz-Ramírez et al., 2016; Volpini-Rapina et al., 2012). According to AACC (2000), the crumb internal cells must be of uniform size with even distribution and thin walls. Also, the silky textured cells with moist and tender texture are considered the best for cakes.

14 Conclusion and Future Trends

The functionality of different ingredients during mixing and baking has always been a topic of research and diversified studies have been conducted to understand the role of different ingredients in various types of cakes particularly foam and batter type cakes. Also, from the production perspectives, different shortcut methods have been designed which could eliminate multi-stage mixing procedures so as to reduce production time which subsequently results in increased production volumes. Globally, the need for the development of specialty cakes for coeliac and people allergenic to eggs is also needed and how to optimize the process factors and ingredients to develop such cake products commercially.

References

AACC (2000). AACC International approved methods. In *American Association of Cereal Chemists (AACC) international,* St. Paul. The Association.

Abhay Kumar, N., Prasada Rao, U. J. S., Jeyarani, T., & Indrani, D. (2017). Effect of ingredients on rheological, physico-sensory, and nutritional characteristics of omega-3-fatty acid enriched eggless cake. *Journal of Texture Studies, 48*(5), 439–449.

Agrahar-Murugkar, D., Zaidi, A., & Dwivedi, S. (2018). Development of gluten free eggless cake using gluten free composite flours made from sprouted and malted ingredients and its physical, nutritional, textural, rheological and sensory properties evaluation. *Journal of Food Science and Technology, 55*(7), 2621–2630.

Al-Dmoor, H. M. (2013). Cake flour: Functionality and quality. *European Scientific Journal, 9*(3).

Al-Dmoor, H., & El-Qudah, J. M. (2016). Cake flour chlorination and alternative treatments. *Current Research in Nutrition and Food Science, 4*, 127–134.

Andersen, A. J. C., & Williams, P. N. (2016). *Margarine*. Elsevier.

Aydogdu, A., Sumnu, G., & Sahin, S. (2018). Effects of addition of different fibers on rheological characteristics of cake batter and quality of cakes. *Journal of Food Science and Technology, 55*(2), 667–677.

Bean, M. M., & Yamazaki, W. T. (1978). Wheat starch gelatinization in sugar solutions. I. Sucrose: Microscopy and viscosity effects. *Cereal Chemistry, 55*(6), 936–944.

Bender, A. B. B., Speroni, C. S., Salvador, P. R., Loureiro, B. B., Lovatto, N. M., Goulart, F. R., Lovatto, M. T., Miranda, M. Z., Silva, L. P., & Penna, N. G. (2017). Grape pomace skins and the effects of its inclusion in the technological properties of muffins. *Journal of Culinary Science & Technology, 15*(2), 143–157.

Bennion, E. B., & Bamford, G. S. T. (1997). *The technology of cake making*. Springer Science & Business Media.

Bhaduri, S. (2013). A comprehensive study on physical properties of two gluten-free flour fortified muffins. *Journal of Food Processing and Technology, 4*(8).

Book, S., Brill, R., & others. (2015). Effects of chemical leavening on yellow cake properties. *Cereal Foods World, 60*(2), 71–75.

Brooker, B. E. (1993). The stabilisation of air in foods containing fat-a review. *Food Structure, 12*(1), 12.

Brooker, B. E. (1996). The role of fat in the stabilisation of gas cells in bread dough. *Journal of Cereal Science, 24*(3), 187–198.

Brown, A. C. (2018). *Understanding food: Principles and preparation*. Cengage learning.

Cauvain, S. P., & Young, L. S. (2008). *Baked products: Science, technology and practice*. Wiley.

Çelik, İ., Yılmaz, Y., Işık, F., & Üstün, Ö. (2007). Effect of soapwort extract on physical and sensory properties of sponge cakes and rheological properties of sponge cake batters. *Food Chemistry, 101*(3), 907–911.

CFR. (2022). *CFR – Code of federal regulations title 21*. CFR – Code of federal regulations title 21, Volume (2). https://www.accessdata.fda.gov/scripts/cdrh/cfdocs/cfcfr/CFRSearch.cfm?fr=137.105

Chiotelli, E., Rolée, A., & Le Meste, M. (2000). Effect of sucrose on the thermomechanical behavior of concentrated wheat and waxy corn starch- water preparations. *Journal of Agricultural and Food Chemistry, 48*(4), 1327–1339.

Choi, H.-W., & Baik, B.-K. (2013). Significance of wheat flour particle size on sponge cake baking quality. *Cereal Chemistry, 90*(2), 150–156.

Chopra, S., Arora, C., Malhotra, A., Khurana, S. C., & others. (2021). Industrially produced trans fat: Usage, health implications, global and indian regulations. *Indian Journal of Public Health, 65*(1), 71.

Conforti, F. D. (2006). Bakery products: Science and technology. In Y. H. Hui (Ed.), *Cake Manufacture*. (Chap. 22 (pp. 393–410). Blackwell.

Cross, N. (2007). Muffins and bagels. *Handbook of food products manufacturing, volume 1: Principles, bakery, beverages, cereals, cheese, confectionary, fats, fruits, and functional foods, 1*, 279.
de la Hera, E., Martinez, M., & Gómez, M. (2013). Influence of flour particle size on quality of gluten-free rice bread. *LWT-Food Science and Technology, 54*(1), 199–206.
Delcour, J. A., & Hoseney, R. C. (2010). *Principles of cereal science and technology* (pp. 229–235). AACC International..
Deleu, L. J., Wilderjans, E., Van Haesendonck, I., Brijs, K., & Delcour, J. A. (2016). Protein network formation during pound cake making: The role of egg white proteins and wheat flour gliadins. *Food Hydrocolloids, 61*, 409–414.
der Sman, R. G. M., & Mauer, L. J. (2019). Starch gelatinization temperature in sugar and polyol solutions explained by hydrogen bond density. *Food Hydrocolloids, 94*, 371–380.
der Sman, R. G. M., & Renzetti, S. (2021). Understanding functionality of sucrose in cake for reformulation purposes. *Critical Reviews in Food Science and Nutrition, 61*(16), 2756–2772.
Dewaest, M., Villemejane, C., Berland, S., Michon, C., Verel, A., & Morel, M.-H. (2017). Changes in protein size distribution during wheat flour cake processing. *LWT-Food Science and Technology, 79*, 333–341.
Díaz-Ramírez, M., Calderón-Domínguez, G., García-Garibay, M., Jiménez-Guzmán, J., Villanueva-Carvajal, A., de la Paz Salgado-Cruz, M., Arizmendi-Cotero, D., & Del Moral-Ramírez, E. (2016). Effect of whey protein isolate addition on physical, structural and sensory properties of sponge cake. *Food Hydrocolloids, 61*, 633–639.
Edmund, J. T., & Perry, K. W. (2008). Soft wheat quality. In G. S. Servet & S. Serpil (Eds.), *Food engineering aspects of baking sweet goods* (pp. 1–30). CRC Press.
El Khoury, D., Balfour-Ducharme, S., & Joye, I. J. (2018). A review on the gluten-free diet: Technological and nutritional challenges. *Nutrients, 10*(10), 1410.
Feng, W., Zhang, H., Wang, R., Zhou, X., & Wang, T. (2021). Modifying the internal structures of steamed rice cakes by emulsifiers for promoted textural and sensory properties. *Food Chemistry, 354*, 129469.
Finnie, S. M., Bettge, A. D., & Morris, C. F. (2006). Influence of flour chlorination and ingredient formulation on the quality attributes of pancakes. *Cereal Chemistry, 83*(6), 684–691.
Gallagher, E., Gormley, T. R., & Arendt, E. K. (2004). Recent advances in the formulation of gluten-free cereal-based products. *Trends in Food Science & Technology, 15*(3–4), 143–152.
Gao, Y., Janes, M. E., Chaiya, B., Brennan, M. A., Brennan, C. S., & Prinyawiwatkul, W. (2018). Gluten-free bakery and pasta products: Prevalence and quality improvement. *International Journal of Food Science & Technology, 53*(1), 19–32.
Geera, B., Reiling, J. A., Hutchison, M. A., Rybak, D., Santha, B., & Ratnayake, W. S. (2011). A comprehensive evaluation of egg and egg replacers on the product quality of muffins. *Journal of Food Quality, 34*(5), 333–342.
Ghotra, B. S., Dyal, S. D., & Narine, S. S. (2002). Lipid shortenings: A review. *Food Research International, 35*(10), 1015–1048.
Godefroidt, T., Ooms, N., Pareyt, B., Brijs, K., & Delcour, J. A. (2019). Ingredient functionality during foam-type cake making: A review. *Comprehensive Reviews in Food Science and Food Safety, 18*(5), 1550–1562.
Gómez, M., Ronda, F., Caballero, P. A., Blanco, C. A., & Rosell, C. M. (2007). Functionality of different hydrocolloids on the quality and shelf-life of yellow layer cakes. *Food Hydrocolloids, 21*(2), 167–173.
Gonera, A., & Cornillon, P. (2002). Gelatinization of starch/gum/sugar systems studied by using DSC, NMR, and CSLM. *Starch-Stärke, 54*(11), 508–516.
Gough, B. M., Whitehouse, M. E., Greenwood, C. T., & Miller, B. S. (1978). The role and function of chlorine in the preparation of high-ratio cake flour. *Critical Reviews in Food Science & Nutrition, 10*(1), 91–113.
Gujral, H. S., & Rosell, C. M. (2004). Improvement of the breadmaking quality of rice flour by glucose oxidase. *Food Research International, 37*(1), 75–81.

Gupta, M., Bawa, A. S., & Semwal, A. D. (2009). Effect of barley flour incorporation on the instrumental texture of sponge cake. *International Journal of Food Properties, 12*(1), 243–251.
Heidolph, B. B. (1996). Designing chemical leavening systems. *Cereal Foods World (USA), 41*, 118–126.
Herranz, B., Canet, W., Jiménez, M. J., Fuentes, R., & Alvarez, M. D. (2016). Characterisation of chickpea flour-based gluten-free batters and muffins with added biopolymers: Rheological, physical and sensory properties. *International Journal of Food Science & Technology, 51*(5), 1087–1098.
Hesso, N., Loisel, C., Chevallier, S., Marti, A., Le-Bail, P., Le-Bail, A., & Seetharaman, K. (2015). The role of ingredients on thermal and rheological properties of cake batters and the impact on microcake texture. *LWT-Food Science and Technology, 63*(2), 1171–1178.
Hosseini, S. M., Soltanizadeh, N., Mirmoghtadaee, P., Banavand, P., Mirmoghtadaie, L., & Shojaee-Aliabadi, S. (2018). Gluten-free products in celiac disease: Nutritional and technological challenges and solutions. *Journal of Research in Medical Sciences: The Official Journal of Isfahan University of Medical Sciences*, 23.
Iovan, M., Stoin, D., Trașcă, T., & Rădoi, B. (2015). Studies regarding the obtaining and sensory analysis of gluten-free muffins with buckwheat flour addition. *Journal of Agroalimentary Processes and Technologies, 21*(3), 222–226.
Itthivadhanapong, P., Jantathai, S., & Schleining, G. (2016). Improvement of physical properties of gluten-free steamed cake based on black waxy rice flour using different hydrocolloids. *Journal of Food Science and Technology, 53*(6), 2733–2741.
Jarpa-Parra, M., Wong, L., Wismer, W., Temelli, F., Han, J., Huang, W., Eckhart, E., Tian, Z., Shi, K., Sun, T., & others. (2017). Quality characteristics of angel food cake and muffin using lentil protein as egg/milk replacer. *International Journal of Food Science & Technology, 52*(7), 1604–1613.
Jyotsna, R., Soumya, C., Swati, S., & Prabhasankar, P. (2016). Rheology, texture, quality characteristics and immunochemical validation of millet based gluten free muffins. *Journal of Food Measurement and Characterization, 10*(4), 762–772.
Kahraman, K., Sakiyan, O., Ozturk, S., Koksel, H., Sumnu, G., & Dubat, A. (2008). Utilization of Mixolab®to predict the suitability of flours in terms of cake quality. *European Food Research and Technology, 227*(2), 565–570.
Kaur, K., Singh, G., & Singh, N. (2017). Development and evaluation of gluten free muffins utilizing green banana flour. *Bioved, 28*(2), 359–365.
Kaushik, J. K., & Bhat, R. (1998). Thermal stability of proteins in aqueous polyol solutions: Role of the surface tension of water in the stabilizing effect of polyols. *The Journal of Physical Chemistry B, 102*(36), 7058–7066.
Khanam, T., Malik, S., Gururani, P., & Chaudhary, V. (2019). Nutritionally enriched cake using vegetable and fruit waste: A review. *Current Topics in Phytochemistry, 8*(3), 722–725.
Khouryieh, H. A., Aramouni, F. M., & Herald, T. J. (2005). Physical and sensory characteristics of no-sugar-added/low-fat muffin. *Journal of Food Quality, 28*(5–6), 439–451.
Kim, C.-S. (1994). The role of ingredients and thermal setting in high-ratio layer cake sytems. *Journal of the Korean Society of Food Science and Nutrition, 23*(3), 520–529.
Kim, J., Park, S., & Shin, W. (2014). Textural and sensory characteristics of rice chiffon cake formulated with sugar alcohols instead of sucrose. *Journal of Food Quality, 37*(4), 281–290.
Kiosseoglou, V. (2003). Egg yolk protein gels and emulsions. *Current Opinion in Colloid \& Interface Science, 8*(4–5), 365–370.
Kiosseoglou, V. (2004). Interactions and competitive adsorption effects in egg-based products. *World's Poultry Science Journal, 60*(3), 311–320.
Korus, J., Witczak, M., Ziobro, R., & Juszczak, L. (2015). The influence of acorn flour on rheological properties of gluten-free dough and physical characteristics of the bread. *European Food Research and Technology, 240*(6), 1135–1143.
Kumari, R., Jeyarani, T., Soumya, C., & Indrani, D. (2011). Use of vegetable oils, emulsifiers and hydrocolloids on rheological, fatty acid profile and quality characteristics of pound cake. *Journal of Texture Studies, 42*(5), 377–386.

Labensky, S. R., Martel, P., & Van Damme, E. (2009). *On baking: A textbook of baking and pastry fundamentals*. Pearson/Prentice Hall.
Lamdande, A. G., Khabeer, S. T., Kulathooran, R., & Dasappa, I. (2018). Effect of replacement of sugar with jaggery on pasting properties of wheat flour, physico-sensory and storage characteristics of muffins. *Journal of Food Science and Technology, 55*(8), 3144–3153.
Lin, M., Tay, S. H., Yang, H., Yang, B., & Li, H. (2017). Development of eggless cakes suitable for lacto-vegetarians using isolated pea proteins. *Food Hydrocolloids, 69*, 440–449.
Luyts, A., Wilderjans, E., Van Haesendonck, I., Brijs, K., Courtin, C. M., & Delcour, J. A. (2013). Relative importance of moisture migration and amylopectin retrogradation for pound cake crumb firming. *Food Chemistry, 141*(4), 3960–3966.
Majzoobi, M., Vosooghi Poor, Z., Mesbahi, G., Jamalian, J., & Farahnaky, A. (2017). Effects of carrot pomace powder and a mixture of pectin and xanthan on the quality of gluten-free batter and cakes. *Journal of Texture Studies, 48*(6), 616–623.
Man, S., Păucean, A., Muste, S., & Pop, A. (2014). Studies on the formulation and quality characteristics of gluten free muffins. *Journal of Agroalimentary Processes and Technologies, 20*(2), 122–127.
Marcotte, M., Sablani, S., Kasapis, S., Baik, O., & Fustier, P. (2004). The thermal kinetics of starch gelatinization in the presence of other cake ingredients. *International Journal of Food Science and Technology, 39*(7), 807–810.
Mine, Y. (2002). Recent advances in egg protein functionality in the food system. *World's Poultry Science Journal, 58*(1), 31–39.
Miranda-Villa, P. P., Mufari, J. R., Bergesse, A. E., & Calandri, E. L. (2019). Effects of whole and malted quinoa flour addition on gluten-free muffins quality. *Journal of Food Science, 84*(1), 147–153.
Mizukoshi, M. (1979). Model studies of cake baking. I: Continuous observations of starch gelatinization and protein coagulation during baking. *Cereal Chemistry, 56*, 305–309.
Moiraghi, M., de la Hera, E., Pérez, G. T., & Gómez, M. (2013). Effect of wheat flour characteristics on sponge cake quality. *Journal of the Science of Food and Agriculture, 93*(3), 542–549.
Monnet, A. F., Saint-Eve, A., Michon, C., Jeuffroy, M. H., Delarue, J., & Blumenthal, D. (2022). Tailoring the properties of pea-enriched soft cakes using a multiobjective model based on sensory-relevant instrumental characterization. *Food and Bioprocess Technology, 15*(2), 459–473.
Moreno, C., del Mar, M., Barreto-Palacios, V., Gonzalez-Carrascosa, R., Iborra-Bernad, C., Andres-Bello, A., Martínez-Monzó, J., & García-Segovia, P. (2015). Evaluation of textural and sensory properties on typical spanish small cakes designed using alternative flours. *Journal of Culinary Science \& Technology, 13*(1), 19–28.
Nakamura, K., Taniguchi, Y., Taira, M., & Ito, H. (2010). Prediction of specific Japanese sponge cake volume using pasting properties of flour. *Cereal Chemistry, 87*(6), 505–510.
Oldham, A. M., Mccomber, D. R., & Cox, D. F. (2000). Effect of cream of tartar level and egg white temperature on angel food cake quality. *Family and Consumer Sciences Research Journal, 29*(2), 111–124.
Oliete, B., Pérez, G. T., Gómez, M., Ribotta, P. D., Moiraghi, M., & León, A. E. (2010). Use of wheat, triticale and rye flours in layer cake production. *International Journal of Food Science & Technology, 45*(4), 697–706.
Ortiz, D. E. (2016). *Cakes, muffins and bagels*. W. K. Kellogg Institute.
Perry, P. A., & Donald, A. M. (2002). The effect of sugars on the gelatinisation of starch. *Carbohydrate Polymers, 49*(2), 155–165.
Preichardt, L. D., Vendruscolo, C. T., Gularte, M. A., & Moreira, A. S. (2011). The role of xanthan gum in the quality of gluten free cakes: Improved bakery products for coeliac patients. *International Journal of Food Science & Technology, 46*(12), 2591–2597.
Purlis, E. (2010). Browning development in bakery products--A review. *Journal of Food Engineering, 99*(3), 239–249.
Pycarelle, S. C., & Delcour, J. A. (2021). The role and impact on quality of exogenous and endogenous lipids during sponge cake making. *Trends in Food Science & Technology, 114*, 158–166.

Pycarelle, S. C., Winnen, K. L. J., Bosmans, G. M., Van Haesendonck, I., Pareyt, B., Brijs, K., & Delcour, J. A. (2019). Wheat (Triticum aestivum L.) flour free lipid fractions negatively impact the quality of sponge cake. *Food Chemistry, 271*, 401–409.
Pyler, E. J., & Gorton, L. A. (2008). *Baking science & technology: Volume I: Fundamentals & ingredients*. Sosland Publishing.
Richardson, G., Langton, M., Fäldt, P., & Hermansson, A. (2002). Microstructure of α-crystalline emulsifiers and their influence on air incorporation in cake batter. *Cereal Chemistry, 79*(4), 546–552.
Rodriguez-Garcia, J., Puig Gómez, C. A., Salvador, A., & Hernando Hernando, M. (2013). Functionality of several cake ingredients: A comprehensive approach. *Czech Journal of Food Sciences, 31*(4), 355–360.
Rodríguez-García, J., Sahi, S. S., & Hernando, I. (2014). Optimizing mixing during the sponge cake manufacturing process. *Cereal Foods World, 59*(6), 287–292.
Ronda, F., Oliete, B., Gómez, M., Caballero, P. A., & Pando, V. (2011). Rheological study of layer cake batters made with soybean protein isolate and different starch sources. *Journal of Food Engineering, 102*(3), 272–277.
Saari, H., Heravifar, K., Rayner, M., Wahlgren, M., & Sjöö, M. (2016). Preparation and characterization of starch particles for use in Pickering emulsions. *Cereal Chemistry, 93*(2), 116–124.
Sahi, S. S., & Alava, J. M. (2003). Functionality of emulsifiers in sponge cake production. *Journal of the Science of Food and Agriculture, 83*(14), 1419–1429.
Salehi, F. (2019). Improvement of gluten-free bread and cake properties using natural hydrocolloids: A review. *Food Science & Nutrition, 7*(11), 3391–3402.
Sani, N. A., Taip, F. S., Kamal, S. M. M., & Aziz, N. (2014). Effects of temperature and airflow on volume development during baking and its influence on quality of cake. *Journal of Engineering Science and Technology, 9*(3), 303–313.
Seyhun, N., Sumnu, G., & Sahin, S. (2003). Effects of different emulsifier types, fat contents, and gum types on retardation of staling of microwave-baked cakes. *Food/Nahrung, 47*(4), 248–251.
Shao, Y.-Y., Lin, K.-H., & Chen, Y.-H. (2015). Batter and product quality of eggless cakes made of different types of flours and gums. *Journal of Food Processing and Preservation, 39*(6), 2959–2968.
Shepherd, I. S., & Yoell, R. W. (1976). Cake emulsions. In S. Friberg (Ed.), *Food emulsions* (pp. 215–276). Marcel Dekker.
Shevkani, K., & Singh, N. (2014). Influence of kidney bean, field pea and amaranth protein isolates on the characteristics of starch-based gluten-free muffins. *International Journal of Food Science & Technology, 49*(10), 2237–2244.
Shevkani, K., Kaur, A., Kumar, S., & Singh, N. (2015). Cowpea protein isolates: Functional properties and application in gluten-free rice muffins. *LWT-Food Science and Technology, 63*(2), 927–933.
Singh, S., Chauhan, E. S., Agarwal, S., & Rajora, T. (2020). Development of iron rich muffin and its acceptability evaluation. *Journal of Pharmacognosy and Phytochemistry, 9*(1), 104–107.
Sinha, N. K. (2007). *Handbook of food products manufacturing*. Wiley.
Sirisoontaralak, P., Suthirak, P., Papaka, K., & Vongsawasdi, P. (2017). Development of shelf stable chiffon cake using gamma irradiation. *LWT, 75*, 78–84.
Slade, L., Kweon, M., & Levine, H. (2021). Exploration of the functionality of sugars in cake-baking, and effects on cake quality. *Critical Reviews in Food Science and Nutrition, 61*(2), 283–311.
Suas, M. (2012). Advanced bread and pastry. Cengage Learning.
Sudha, M. L., Baskaran, V., & Leelavathi, K. (2007). Apple pomace as a source of dietary fiber and polyphenols and its effect on the rheological characteristics and cake making. *Food Chemistry, 104*(2), 686–692.
Sumnu, S. G., & Sahin, S. (2008). *Food engineering aspects of baking sweet goods*. CRC Press.
Turabi, E., Sumnu, G., & Sahin, S. (2008). Rheological properties and quality of rice cakes formulated with different gums and an emulsifier blend. *Food Hydrocolloids, 22*(2), 305–312.

Ureta, M. M., Olivera, D. F., & Salvadori, V. O. (2017). Influence of baking conditions on the quality attributes of sponge cake. *Food Science and Technology International, 23*(2), 156–165.
van der Sman, R. G. M. (2018a). Progress in understanding of supplemented state diagrams of hydrophilic food materials. *Current Opinion in Food Science, 21*, 32–38.
Van Der Sman, R. G. M. (2018b). Theoretical investigation of the swelling of polysaccharide microgels in sugar solutions. *Food & Function, 9*(5), 2716–2724.
Volpini-Rapina, L. F., Sokei, F. R., & Conti-Silva, A. C. (2012). Sensory profile and preference mapping of orange cakes with addition of prebiotics inulin and oligofructose. *LWT-Food Science and Technology, 48*(1), 37–42.
Weijers, M., van de Velde, F., Stijnman, A., van de Pijpekamp, A., & Visschers, R. W. (2006). Structure and rheological properties of acid-induced egg white protein gels. *Food Hydrocolloids, 20*(2–3), 146–159.
Wilde, P. J. (2000). Interfaces: Their role in foam and emulsion behaviour. *Current Opinion in Colloid & Interface Science, 5*(3–4), 176–181.
Wilderjans, E., Pareyt, B., Goesaert, H., Brijs, K., & Delcour, J. A. (2008). The role of gluten in a pound cake system: A model approach based on gluten--starch blends. *Food Chemistry, 110*(4), 909–915.
Wilderjans, E., Kerckhofs, G., Lagrain, B., Brijs, K., Wevers, M., & Delcour, J. A. (2010a). Baking gradients cause heterogeneity in starch and proteins in pound cake. *Cereal Chemistry, 87*(5), 475–480.
Wilderjans, E., Luyts, A., Goesaert, H., Brijs, K., & Delcour, J. A. (2010b). A model approach to starch and protein functionality in a pound cake system. *Food Chemistry, 120*(1), 44–51.
Wilderjans, E., Luyts, A., Brijs, K., & Delcour, J. A. (2013). Ingredient functionality in batter type cake making. *Trends in Food Science & Technology, 30*(1), 6–15.
Willyard, M. (2000). Muffin technology (update). *Technical bulletin. American Institute of Baking, 22*(10), 1–6.
Witczak, M., Ziobro, R., Juszczak, L., & Korus, J. (2016). Starch and starch derivatives in gluten-free systems–A review. *Journal of Cereal Science, 67*, 46–57.
Wu, J. (2014). Eggs and egg products processing. Food Processing: Principles and Applications, 437–455.
Yazici, G. N., & Ozer, M. S. (2021). A review of egg replacement in cake production: Effects on batter and cake properties. *Trends in Food Science & Technology, 111*, 346–359.
Zhang, Y.-Y., Song, Y., Hu, X.-S., Liao, X.-J., Ni, Y.-Y., & Li, Q.-H. (2012). Effects of sugars in batter formula and baking conditions on 5-hydroxymethylfurfural and furfural formation in sponge cake models. *Food Research International, 49*(1), 439–445.
Zhou, J., Faubion, J. M., & Walker, C. E. (2011). Evaluation of different types of fats for use in high-ratio layer cakes. *LWT-Food Science and Technology, 44*(8), 1802–1808.
Zhu, H., & Damodaran, S. (1994). Proteose peptones and physical factors affect foaming properties of whey protein isolate. *Journal of Food Science, 59*(3), 554–560.
Zieliński, H., Ciesarová, Z., Kukurová, K., Zielinska, D., Szawara-Nowak, D., Starowicz, M., & Wronkowska, M. (2017). Effect of fermented and unfermented buckwheat flour on functional properties of gluten-free muffins. *Journal of Food Science and Technology, 54*(6), 1425–1432.
Ziobro, R., Witczak, T., Juszczak, L., & Korus, J. (2013). Supplementation of gluten-free bread with non-gluten proteins. Effect on dough rheological properties and bread characteristic. *Food Hydrocolloids, 32*(2), 213–220.

Chapter 6
Tortillas

Alma Haydee Astorga-Gaxiola, Saraid Mora-Rochín, Eslim Sugey Sandoval-Sicairos, and Cuauhtémoc Reyes-Moreno

1 Introduction

The popularity of tortillas has spread worldwide in the last years, reaching Asia and Europe due to population migration. In the United States, tortillas are the second most consumed baked product, just after white bread (Serna-Saldívar & Chuck-Hernández, 2016; Báez-Aguilar et al., 2022). Central America and Mexico have a great dependence on maize products, of which, in Mexico, around 75% are consumed in the form of tortillas. Tortillas from maize (*Zea mays* L.) and wheat (*Triticum aestivum* L.) flour were manufactured by Mesoamerican cultures in Mexico, and since then they are generally enjoyed with vegetables, eggs, meats, and beans. Around 1400 million tortillas are consumed per day (Serna-Saldívar & Chuck-Hernández, 2016; León-López et al., 2019). However, only one of every 10 tortillas consumed is from wheat flour (Waniska, 2015).

Maize and wheat tortillas are excellent sources of carbohydrates, A, D, E and B12 vitamins, and minerals such as Ca, Fe and Zn (Mora-Rochín et al., 2019). Maize tortillas have gained popularity among celiac or gluten intolerant people (Serna-Saldívar, 2015). On the other hand, most wheat tortillas come from refined flours (dry milling) that have been enriched with Fe, Zn, and vitamins B1, B2, niacin, B6, and folic acid (Rooney & Serna-Saldívar, 2015).

Nixtamalization is the traditional process used to manufactured masa and their products (mainly tortillas), where the lime-cooking process produces chemical and nutritional changes in the structure of the kernel, the starch is gelatinized, proteins are denaturalized, and the pericarp tissue is partially hydrolyzed. Currently, new trends and technologies in tortilla production are centric to human health improving

A. H. Astorga-Gaxiola · S. Mora-Rochín (✉) · E. S. Sandoval-Sicairos · C. Reyes-Moreno
Facultad de Ciencias Químico-Biológicas, Universidad Autónoma de Sinaloa,
Blvd. de las Américas y Josefa Ortiz de Domínguez, S/N, Culiacán, Sinaloa, Mexico
e-mail: smora@uas.edu.mx

M. A. Shah et al. (eds.), *Cereal-Based Food Products*,
https://doi.org/10.1007/978-3-031-40308-8_6

the contents of dietary fiber, phenolics and carotenoids (with antioxidants properties), and phytosterols (Corrales-Bañuelos et al., 2016;Iuga et al., 2019; Mora-Rochín et al., 2021).

2 Ingredients and Their Functions

The main ingredients to produce maize flour tortillas are mature maize kernels, water, and lime; and for wheat flour tortillas they are wheat flour, water, shortening, and salt. Other ingredients (gums, emulsifiers, and preservatives) can be added to the flours to prevent microbial growth and increase the textural shelf life of tortillas (Serna-Saldívar & Chuck-Hernández, 2016). Table 6.1 shows the main ingredients for the formulation of tortillas.

Nixtamalized Maize Flour

During the lime cooking process, the maize kernel starch swells and pregelatinizes, which produces a cohesive and adhesive masa that allows the tortillas to be molded. A masa without cohesiveness does not have the consistency to form the tortilla, and a masa with high adhesiveness sticks, breaks, and does not allow the tortilla to be molded to be placed on the griddle for baking (Arámbula-Villa et al., 2001; Serna-Saldívar & Chuck-Hernández, 2016).

Wheat Flour

Proteins (glutenin and gliadin) are some of the most important components of wheat flour because, when mixed with water, they form gluten that traps carbon dioxide during baking, making tortillas with a fluffy crumb, flexible with good foldability,

Table 6.1 Typical formulation for the elaboration of flour tortillas

Ingredients	Flour percent
Flour	100.0
Water (35–40 °C)	45.0–55.0
Salt	1.0–2.0
Shortening (fat)	5.0–15.0
Baking powder	0.6–2.0
Gums	0.1–0.5
Emulsifiers	0.1–0.5
Preservatives	0.2–0.5
Reducing agent	0.003

Adapted from Bejosano and Alviola (2015); Serna-Saldívar & Chuck-Hernández (2016)

and with a longer shelf life (Alviola et al., 2008; Cauvain, 2017). Wheat flour also influences other factors such as texture, shelf life, tortilla shape, baking properties, flavor, as well as translucency (Whitney et al., 2011; Waniska, 2015).

Water

This is the second most abundant ingredient (45–55% of flour weight) in tortilla formulation, whose main function is the hydration to form the gluten. Water is usually warmed to 35–40 °C before being added to the flour. An excess of water can make the masa stickier and more difficult to process (Serna-Saldívar & Chuck-Hernández, 2016; Cauvain, 2017).

Shortening (Fat)

Liquid or solid fats (5–15% of flour weight) make the tortilla adhere less to other surfaces acting like a lubricant. The masa becomes softer and the tortilla less firm. Also, it gives a prolonged shelf life, reheating capacity, and improved extensibility. An excessive amount of oil can make sticky tortillas that feel satiny, fatty, or lubricated (Young & O'sullivan, 2011).

Salt

Common salt (1–2%) enhances the flavor of the ingredients present in the masa and strengthen the gluten. It also produces a reduction in water activity, prolonging the shelf life of the product (Cauvain, 2017).

Lime or Calcium Salts for Nixtamalization Process

The pericarp of maize kernels is hydrolyzed by calcium hydroxide [$Ca(OH)_2$] or lime (1% based on kernel weight), which penetrates the grain mainly through the germ causing the softening of the cell walls of the endosperm, favoring water and heat penetration, which causes swelling and partial gelatinization of the starch granules. Lime cooking increases grain calcium and magnesium content, and niacin bioavailability, as well as decreases contamination (aflatoxins and other mycotoxins) of the maize kernels (Serna-Saldívar & Chuck-Hernández, 2016; Kamau et al., 2020).

Calcium chloride, calcium lactate, and calcium carbonate are salts that have been studied to replace lime, observing that the masa and tortillas produced are similar in texture and color to traditional tortillas(Ruiz-Gutiérrez et al., 2011; Bello-Pérez et al., 2014).

Gums

Gums (0.1–0.5%) decrease the stickiness of baked tortillas and improve masa handling (Serna-Saldívar & Chuck-Hernández, 2016). The texture of tortillas is improved by the addition of xanthan gum because it stabilizes the oil in water emulsion. This gum allows higher water retention but lower water activity and water migration, reducing staling and stickiness extending the shelf life. High amounts of gum could produce more translucent areas on tortillas and a stickier masa, leading to weakened gluten formation (Waniska, 2015).

Emulsifiers

They are used to make the masa softer and more extensible, reduce the rancidity and stickiness of the tortillas, and maintain a uniform outer edge in large tortillas (Bejosano & Alviola, 2015). Sodium stearoyl-2-lactylate, monoglycerides, and diglycerides of fatty acids in combination with amylose form a complex conjugate that decreases starch dispersion during gelatinization, resulting in less dispersed amylose during baking and a weaker amylose gel (Waniska, 2015).

Baking Powder (Leavening Base)

Sodium bicarbonate, or baking powder, produces carbon dioxide when is combined with an acid. Carbon dioxide increases the volume of the tortillas due to the gas bubbles that are trapped in the structure of the masa, making fluffier tortillas, and increasing their diameter (Waniska, 2015).

Preservatives

Antimicrobial agents, like salts (potassium sorbate, calcium propionate, sodium propionate) and acids (citric, fumaric, and phosphoric), act like antifungal agents improving shelf life (Bejosano & Alviola, 2015; Serna-Saldívar & Chuck-Hernández, 2016).

Reducing Agents

Reducing agents weaken the gluten by breaking the bonds between the proteins in the masa reducing the elasticity and mixing time. The hardness of the masa decreases with the use of cysteine and sodium metabisulfite, which are used in the preparation of wheat tortillas (Bejosano & Alviola, 2015).

Enzymes

They are commonly used to prolong the shelf life of the texture of maize and wheat tortillas are α-amylases and xylanases. Transglutaminases, oxidoreductases, and proteases are used to modify the rheological properties of wheat flour masa (Austin, 2015).

3 Preparation Methodology

Maize Tortilla

Ancestral Process

The origin of maize cooking is an ancestral heritage or magical process.The main ingredients for the ancestral nixtamalization process of cooking maize kernels were volcanic or wood ashes and water, generating by-products (nixtamal, masa, and tortillas) with important nutritional characteristics such as protein and mineral contents, dietary fiber, and resistant starch. Archaeological findings suggested the procedure was common practice among Mesoamerican cultures, the Mayans who cooked the maize with wood ashes, and the Aztecs who added lime [$Ca(OH)_2$] instead of ashes, because it allowed the remotion of the pericarp of maize kernel more efficiently and improved the consistency of the masa and tortillas texture (Katz et al., 1974; Mariscal-Moreno et al., 2017; Iuga et al., 2019; Serna-Saldívar & Chuck-Hernández, 2019) (Fig. 6.1). The use of lime, proposed by the Aztec culture, to cook maize is currently used. Maize nixtamalized products (fresh masa, flours, fresh tortillas, tortilla chips and corn chips) are manufactured mainly by the traditional nixtamalization process (Fig. 6.2). In Mexico and several Central American countries, tortillas are the main food. Currently, lime-cooking processes are similar to those used by Aztec civilizations (maize cooking with $Ca(OH)_2$)(Iuga et al., 2019; Serna-Saldívar & Chuck-Hernández, 2019).

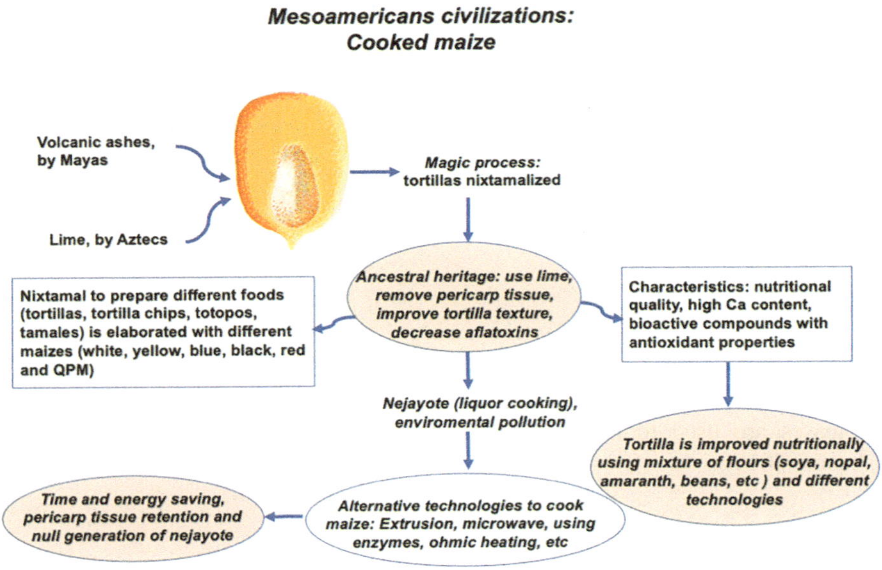

Fig. 6.1 Historical evolution of nixtamalization

Nowadays, nixtamalized products, such as tortillas, are obtained mainly from two main manufacturing processes: *traditional (fresh masa)* and *industrial (dry masa flour)* nixtamalization (Serna-Saldívar, 2010). Both processes use lime solution to cook the maize kernels. In the traditional approach, cooked maize or nixtamal is ground into masa, while the commercial procedure uses industrially processed maize flours made from dried masa (Mora-Rochín et al., 2019).

Nixtamalization Process

Lime Cooking

Maize kernels are cooked in a calcium hydroxide solution (85–95 °C) (Fig. 6.2) (Bello-Pérez et al., 2014). This part of the process is considered the most important stage due to the biochemical and physicochemical changes of the maize kernel. The mineral calcium generates a concentration gradient through the pericarp tissue of the grain. This diffusion of calcium promotes physicochemical changes in the internal anatomical parts such as germ and endosperm. As a result, the rheological properties of the maize kernel are modified allowing the heat and mass transference phenomenon (Gutiérrez-Cortez et al., 2016; García-Armenta et al., 2020).

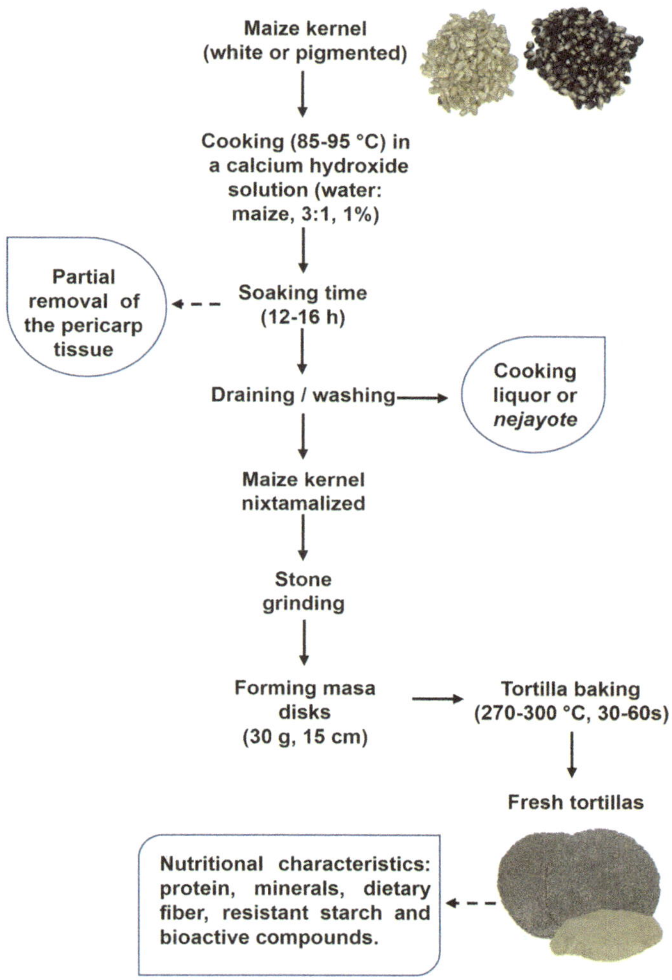

Fig. 6.2 Transformation of maize kernel into tortilla by nixtamalization process

Soaking Time

A particular characteristic of the traditional process is the long *soaking* time (12–16 h), where the total or partial removal of the pericarp tissue happens, which makes possible the getting of nixtamalized products (Gutiérrez-Cortez et al., 2016; García-Armenta et al., 2020). The pericarp tissue is the anatomical part of the maize considered as the main source of dietary fiber and phenolic compounds, whose consumption has benefits in human health (Mora-Rochín et al., 2019). At the end of the soaking time, the cooked maize is washed and hand-rubbed with enough water to facilitate the removal of the pericarp tissue to obtain the nixtamal. It is important to mention that a highly polluting by-product of the nixtamalization process is the *nejayote* or cooking liquor (3–10 L of contaminated effluents/kg of maize), which is considered an environmental problem due to the high solid and alkaline pH contents (Fig. 6.2) (Valderrama-Bravo et al., 2018; Ramirez et al., 2020).

Stone-Grinding

This is a key factor to obtain the masa because the grinding disrupts the swollen starch granules and distributes the hydrated starch and protein around the ungelatinized portions of the endosperm, forming masa with a given granulation (Valderrama-Bravo et al., 2018). Nowadays, the stone-grinders used are built of two matched carved stones (volcanic or synthetic and aluminum oxide), one stationary and the other rotating at 500–700 rpm. The typical stone is 10 cm thick and 40 cm in diameter and is machine- or hand-carved with radial grooves (Serna-Saldívar, 2010).

Tortilla Baking

The typical method to make tortillas is to take small portions (30 g) of fresh masa and pressed and shaped them into flat disks (15 cm) using a manual machine (Casa Herrera, México D.F., México). The masa disks are baked on a hot clay griddle at 270 ± 10 °C for 30–60 s on each side, and then again on the first side until puffing of the tortilla occurs(Fig. 6.2) (Mora-Rochín et al., 2010). This thermal process inactivates microbial load, improves sensorial characteristics (color and flavor) via Maillard reactions, and further gelatinizes the starch for a structured tortilla (Serna-Saldívar & Rooney, 2015).

Currently, two continuous tortilla-forming machines, sheeter/former and extruder, are used. The concept and basic design of these continuous oven machines, of which some are capable of baking 30,000 tortillas per hour, are used by most processors today (Serna-Saldívar, 2010).

Today, the masa obtained by the industrial fresh-masa process is commercialized to small tortilla factories where it is transformed into tortillas using automated methods (Iuga et al., 2019).

The industrial dry masa flours are the most commonly used because by adding enough water and possibly other ingredients a masa suitable to form tortillas can be produced (Serna-Saldívar, 2010). The main advantage to use dry flour is the speed to get products nixtamalized.

Alternative Technologies to Produce Instant Maize Flours

Recent reports about instant maize flours production have suggested other technologies with different advantages such as energy efficiency, low cost, high productivity, little or null generation of polluting waste for the environment, and pericarp tissue retention (products with more protein content, dietary fiber and phenolic compounds), rendering products with higher nutritional qualities (Mora-Rochín et al., 2019) (Fig. 6.1).

Extrusion lime-cooking process uses high temperature in short periods of time. Maize kernels are ground and mixed with calcium hydroxide (0–5% w/w), conditioned (14–45% moisture content) and allowed to rest in refrigeration or at room temperature (for up to 12 h). The conditioned mix is passed through the extruder (100 to 150 °C, by short times and different rpm). The cooked samples are milled to produce the nixtamalized extruded flours with similar characteristics to traditional flours. However, due to the high pressure used during the extrusion process, a high disintegration of starch structures occurred producing a wide retrogradation of amylose fragments, giving a stiffed product as a result (Aguayo-Rojas et al., 2012; Platt-Lucero et al., 2013).

Figueroa-Cárdenas et al. (2011) proposed an *ecological nixtamalization,* whose main characteristic is the reduction of the pH in the residue solution (pH = 4.19 to 7.38) which is different from traditional nixtamalization (pH = 12 to 14). The difference in pH values is attributed mainly to the calcium salts used, carbonate, chloride, and particularly calcium sulfate. Finally, masa and tortillas produced with this technique show excellent organoleptic characteristics (Maya-Cortés et al., 2010; Mariscal-Moreno et al., 2017).

Nixtamalization by microwaves showed physicochemical differences, such as water absorption index, particle size and maximum viscosity, when compared to those obtained by the traditional process. Gums contained in the pericarp tissue are released by the effect of the microwave heating, causing undesirable reactions, the saponification of lipids and total swelling of starch granules (Martínez-Bustos et al., 2000; Escalante-Aburto et al., 2019).

During the *Nixtamalization with malting/extrusion,* maize kernels are germinated (1 day minimum), then they are dried, ground, and conditioned for extrusion at different conditions to obtain suitable flours to make tortillas. However, processing times and equipment requirements increase the costs of production (Rooney & Serna-Saldívar, 2015).

In the *enzymatic nixtamalization* process, maize kernels are hydrated (25–80%) and digested with an enzymatic solution (mainly protease). Subsequently, an alkali is added [$Ca(OH)_2$, NaOH or KOH]. These conditions favor an increase in pH levels and soften pericarp tissue by the protease activity allowing the degradation of protein and facilitating the milling, although the use of enzymes could be expensive (Escalante-Aburto et al., 2019).

Ohmic heating consists in an internal production of heat by passing an electric alternating current through the ground maize. A mix of hydrated ground maize and lime are used to feed the continuous ohmic heater. The masa obtained is dried and milled to produce nixtamalized flours (Gaytán-Martínez et al., 2012; Menera-López et al., 2013).

The major advantage of these all-alternative technologies mentioned is the obtention of products that share similarities with those of traditional nixtamalization. Despite all the nixtamalization alternatives, the industry of masa flours has not included these procedures to cook maize because doing so might hurt its profits by lack of acceptance or by consumer dissatisfaction (Escalante-Aburto et al., 2019).

Wheat Tortillas

History says that, when Cortés conquered Mexico, he ordered his slaves to plant wheat seeds, resulting in its disseminating through out the land (Long, 1996). Wheat tortilla was accepted by Spaniards and mestizos due to the different flavor and characteristics when compare with maize tortilla. First, wheat tortillas were elaborated with flours and maybe a pinch of salt. Later, a magical improvement occurred when animal fat was added in the mixture. The masa was cut and rolled out with a rolling pin into this circle and baked using similar conditions to those used in maize tortillas elaboration. The tortilla obtained was less susceptible to staling, which was an advantage over maize tortillas. Wheat tortillas soon started to be more popular, and their ingredients and quality improved by the addition of vegetable shortening and milk. Nowadays, wheat flour tortillas are consumed in South, Central, and North America. In general, one of every 10 tortillas consumed in Mexico is made with wheat flours, particularly in Sonora where they are known for their large size (diameter: 50–60 cm) and minor thickness (1 mm) (Ramírez-Wong et al., 2007; Waniska, 2015).

Dry Milling Process

Wheat flours are produced by the dry milling of the wheat kernel. This consists in the separation of the bran (the outer covering of the grain), the germ (the embryo contained inside the kernel), and the endosperm (fraction that makes white flour), the three anatomical parts of the wheat kernel. During milling, these parts are separated and recombined to achieve different types of flours (to elaborate cake, breads, pasta, and flour tortilla), which showed different quantity and quality of protein. The minimum protein content is around 11% to elaborate wheat flour tortillas. In addition, the flours used must be enriched, lightly bleached, and produced from a hard wheat (Bejosano & Alviola, 2015).

Elaboration Methods

Die-cut, hand stretch, and hot-press are the methods used in the elaboration of wheat tortillas, the latter being the most common in the industry (>90%) for home consumption. In the hot-press method, small uniform-sized balls are cut from the masa, then heated plattens press the balls into disks, which are finally baked into tortillas. *In* the *Die-cut* procedure, rollers flatten the masa into sheets to be cut into disks and then baked. This procedure fits well in mass production. In the *hand-stretch* process, thinner, larger, and stronger tortillas are obtained; however, the slow production process and the necessary extra work prevent its use in a bigger scale (Bejosano & Alviola, 2015).

4 Quality Aspects

Commercial tortilla manufacturing involves the modification of the ancestral process, causing changes in the quality parameters of the product (texture, symmetry, color, odor, puffing, flexibility, and shelf life). Commercial tortillas tend to be much thinner, have a sweeter taste, and harden faster than traditional tortillas because of the type of raw material used in their manufacturing as the industry seeks to generate large quantities of this product using cheaper materials and methods (Iuga et al., 2019).

Texture

The rollability test, the most common technique for evaluating tortilla texture, monitors textural changes in maize and wheat tortillas during storage or in response to the addition of food additives and consists in rolling a tortilla around a dowel to measure levels of breakage (Suhendro et al., 1998). Results are expressed using a score given by a panel conformed of regular consumers (Brooker, 2015). The scale used influences these results since scales differ between research teams. Reports from Suhendro et al. (1998) used a 1 to 5 scale (1 = extremely firm to 5 = extremely soft and flexible) while Bejosano et al. (2005) and Iuga et al. (2019) reported results using a scale of 1 to 15 (1 = nothing rollable, 15 = it rolls without breaking).

Color

This factor represents the visible attributes to consider when buying food. It may lead the consumer to accept or reject the product, even without having tried it. Color values, determined using a hand-held colorimeter, are expressed as L* (lightness), +a* (redness and greenness), and + b* (yellowness and blueness) (Brooker, 2015; Rodríguez-Noriega et al., 2021).

Table 6.2 Chemical composition of common maize flour and fortified tortillas

Nutritional component (%)	WFT	CMF	EBMF + EAF	CMF + BF	CMF + SF
Ash	2.2	1.20	1.57	1.81	3.6
Protein	8.8	7.85	11.69	10.89	16.4
Fats	9.6	4.71	6.21	4.13	5.3
Carbohydrates	94.4	89.38	69.03	84.25	75.3
Total dietary fiber	6.9	5.23	11.92	12.76	NM

Data from Gámez-Valdez et al., 2021, Treviño-Mejía et al., 2016, Serna-Saldívar et al., 2015, and Obatolu et al., 2007. *WFT:* Wheat flour tortilla, *CMF:* Common maize flour, *EBMF:* Extruded Blue maize flour, *EAF:* Extruded Amaranth flour, *BF:* Bean flour, and *SF:* Soybean flour, *NM:* Not mentioned

Sensory Evaluation

This tool allows measuring, analyzing, and interpreting reactions of human beings to the quality of tortillas, ingredients, or the processing technique employed. The jury must have previous training to produce replicable and reliable results since human perception, our analytical instrument, could handicap the interpretation of the data collected (Rodríguez-Noriega et al., 2021).

Shelf Stability

It represents the period in which food products retain their characteristic "freshness". To produce tortillas that do not mold during extended storage times, manufacturing techniques and/or packaging materials must be improved or changed. Extrinsic and intrinsic factors compromise shelf life as well. Extrinsic factors are those to which the food is exposed during the production chain (e.g., temperature, moisture, or light exposure). Intrinsic factors are those specific to the composition of the food (water content, pH, and added preservatives or additives) (Brooker, 2015; Kong & Singh, 2016).

The pH of Tortillas

A low pH in both masa and tortillas prevents mold development and extends shelf life (<5.5 for propionates and < 6.1 for sorbates) although a pH under 5.2 causes negative changes in the physicochemical and sensorial properties of tortillas. Regarding wheat flour tortillas, sodium bicarbonate neutralizes most acids, forming carbon dioxide, thus lowering the pH (0.2–1.5 units) of the masa during the cooking process and of the tortillas (0.1–0.3 units) in the first 24 h of production (Brooker, 2015).

Moisture Content and Water Activity (a_w)

Moisture average contents in tortillas are 42% to 46%, which favor the growth of microorganisms. For this reason, it has been concluded that the microbiological stability of maize tortillas is 6–12 h at room temperature. Maize and wheat tortillas have an a_w of 0.94 to 0.98%, which makes them susceptible to mold and yeast damage after 48 h of storage at room temperature. The use of polyols (glycerol and propylene glycol) reduces the a_w and increases shelf life and sensory acceptability (Martínez-Flores et al., 2004; Báez-Aguilar et al., 2022).

Nutritional Value

The protein content of maize and wheat flour tortillas is considered low because they are deficient in lysine and tryptophan, essential aminoacids for the optimal development of infants (Table 6.2) (Mora-Rochín et al., 2019). These deficiencies have prompted the development of fortified tortillas using different types of flours: bean, soybean, and amaranth (Obatolu et al., 2007; Treviño-Mejía et al., 2016; Gámez-Valdez et al., 2021).

Adding wheat bran in whole-wheat tortillas production increases contents of both the protein and lysine (Serna-Saldívar et al., 2015).Total carbohydrate content can increase, be constant, or decrease as a consequence of losses in pericarp and starch in the cooking, soaking, and washing processes (Table 6.2) (Ramírez-Wong et al., 2007; Escalante-Aburto et al., 2019). Maize and wheat tortillas have a total fiber content of 5.4–16.45 and 1.8%, respectively, while in whole-wheat tortillas it represents 6.9%. The lipid content in wheat tortillas (Table 6.2) makes them high in energy when shortening or lard are utilized. The degermination happening during milling and nixtamalization processes can reduce the lipid content (Serna-Saldívar et al., 2015).

Bioactive Compounds and Antioxidant Activity

The nutraceutical properties of maize are closely related to its bioactive compounds (Table 6.3). In recent decades, consumer interest in tortillas made with pigmented maize (yellow, red, and blue) has been on the rise because of the health benefits they provide (Corrales-Bañuelos et al., 2016; Mora-Rochín et al., 2016). The processing method is important in the preservation of bioactive properties. Both wheat and maize contain bioactive compounds partially degraded or lost in storage, milling, and manufacturing of tortillas (Serna-Saldívar et al., 2015).

De la Parra et al. (2007) studied the effect of nixtamalization on the phytochemical profiles (total phenols, anthocyanins, ferulic acid and carotenoids) and antioxidant capacity of five types of corn processed into masa, tortillas, and tortilla chips,

Table 6.3 Nutraceuticals associated with maize and wheat

Nutraceutical compound	Location in kernel	Preventive or therapeutic effect
Insoluble dietary fiber	Pericarp, aleurone and cell walls	Increases intestinal transit and volume and weight of feces, reducing constipation. Reduces the glycemic index and thus the risk of diabetes and hypercholesterolemia. Binds bile acids and carcinogens, preventing diverticulosis, polyps, and colorectal cancer.
Soluble dietary fiber		Improves gastrointestinal function and increases the viscosity of bolus. Reduces lipid levels, lower blood pressure, improves blood glucose control and immune function, and reduces inflammation.
Resistant starch	Endosperm	Reduces obesity. Prevents constipation and improves colon health. Decreases insulin resistance in type-2 diabetes. Improves iron and calcium absorption. Presents a prebiotic effect. In the colon is readily fermented into short-chain fatty acids.
Phenolics	Pericarp	Antioxidant effect, prevents cancer, atherosclerosis, and aging.
Anthocyanins and flavonoids	Aleurone layer (mainly found in pigmented landraces)	

Modified from Serna-Saldívar et al., 2015

observing that the nixtamalization significantly reduced the content of total phenols and antioxidant capacity, compared to raw grains. Mora-Rochín et al. (2010) observed a decrement in total phenol content (23.6–12.5%), and anthocyanin content (53.3%) in four maize varieties (white, blue, yellow, and red) during traditional nixtamalization.

The antioxidant activity of cereals is largely due to phenolic compounds. Ferulic acid is the most relevant and the one found in the greatest quantity in maize and wheat (Table 6.4), occurring in bound, free, and condensed forms (Gaxiola-Cuevas et al., 2017). However, different authors have mentioned that maize has a higher antioxidant capacity than wheat (Adom & Liu, 2002; Verma et al., 2018). Disregarding the form in which it can be found, ferulic acid has a broad spectrum of therapeutic applications. Primarily, it has a significant antioxidant activity since not only does it scavenge for free radicals but also inhibits enzymes that catalyze free radical generation (Zduńska et al., 2018).

5 Conclusions and Future Trends

The readily available ingredients that conform tortillas have allowed them to be a staple food in Mexico and become popular the world over. Traditional nixtamalization produces changes in the chemical and nutritional content of tortillas. Thus, other tendencies and technologies derived from this process have been developed

Table 6.4 Bioactive compounds and antioxidant activity of different types of maize tortillas

Maize	Sample	TPC[a]	TAC[b]	TFAC[c]	Antioxidant activity	
					Hidrophilic[d,e]	Lipophilic[f]
White	Raw	159.0	0.27	474.49	19,312[d]	~23
	Dough	165.2	0.28	42.22	590 [e]	~7
	NMT	166.2	0.19	101.66	~9300[d]	~9
	EMT	121.6	0.25	314.52	~17,500[d]	–
Yellow	Raw	126.3	0.38	451.38	19,180[d]	~24
	Dough	182.2	0.31	77.87	600 [e]	~8
	NMT	183.3	0.25	157.65	~11,800[d]	~10
	EMT	110.5	0.31	407.29	~17,700[d]	–
Blue	Raw	147.0	30.80	336.49	12,286[d]	~21
	Dough	158.5	2.63	53.33	590 [e]	~7
	NMT	161.8	13.80	187.79	~10,300[d]	~9
	EMT	128.0	13.11	325.59	~11,600[d]	–
Red	Raw	152.4	0.99	532.16	19,180[d]	~23
	Dough	125.3	2.21	49.83	570 [e]	~7
	NMT	136.5	0.60	208.12	~11,600[d]	~10
	EMT	125.8	0.70	326.45	~14,500[d]	–

Source: Aguayo-Rojas et al., 2012; Mora-Rochín et al., 2010; De la Parra et al., 2007. *NMT:* Nixtamalized maize tortillas, *EMT*: Extruded maize tortilla, *TPC:* Total phenolic content, *TAC:* Total anthocyanin content, *TFAC:* Total ferulic acid content. [a] mg Galic Acid Equivalents (GAE)/100 g, dw; [b] mg Equivalents of Cyanidin-3-glucoside/100 g, dw; [c] mg/100 g, dw; [d] μmol Trolox equivalents (TE)/100 g, dw; [e] Vitamin C eq/100 g, dw; [f] Vitamin E eq/100 g, dw

showing advantages such as polluting waste reduction, energy efficiency, but most importantly, the preservation of the integrity of nutrients and bioactive compounds in the final product.

In further studies, tortillas fortified by the addition of other flours, such as chickpea, beans, soy, nopal, chia, and amaranth, can be evaluated to generate exceptional vehicles for combating malnutrition and chronic degenerative diseases because of the multiple health benefits related to the presence of phenolic compounds with high antioxidant activity.

References

Adom, K. K., & Liu, R. H. (2002). Antioxidant activity of grains. *Journal of Agricultural and Food Chemistry, 50*, 6182–6187. https://doi.org/10.1021/jf0205099

Aguayo-Rojas, J., Mora-Rochín, S., Cuevas-Rodríguez, E. O., Serna-Saldívar, S. O., Gutiérrez, J. A., Reyes-Moreno, C., & Milán-Carrillo, J. (2012). Phytochemicals and antioxidant capacity of tortillas obtained after lime-cooking extrusion process of whole pigmented Mexican maize. *Plant Foods for Human Nutrition, 67*, 178–185. https://doi.org/10.1007/s11130-012-0288-y

Alviola, J. N., Waniska, R. D., & Rooney, L. W. (2008). Role of gluten in flour tortilla staling. *Cereal Chemistry, 85*, 295–300. https://doi.org/10.1094/CCHEM-85-3-0295

Arámbula-Villa, G., Barrón-Ávila, L., González-Hernández, J., Moreno-Martínez, E., & Luna-Bárcenas, G. (2001). Efecto del tiempo de cocimiento y reposo del grano de maíz (*Zea mayz* L.) nixtamalizado, sobre las características fisicoquímicas, reológicas, estructurales y texturales del grano, masa y tortillas de maíz. *Archivos Latinoamericanos de Nutrición, 51*, 187–193.

Austin, D. L. (2015). Enzymes: Extending shelf life and eating quality of tortillas. In L. W. Rooney & S. O. Serna-Saldívar (Eds.), *Tortillas: Wheat flour and corn products* (pp. 201–214). AACC International Press. https://doi.org/10.1016/B978-1-891127-88-5.50010-4

Báez-Aguilar, A. M., Arámbula-Villa, G., Prinyawiwatkul, W., López-Espíndola, M., Ramírez-Rivera, E. J., Contreras-Oliva, A., & Herrera-Corredor, J. A. (2022). Effect of calcium hydroxide mixed with preservatives on physicochemical characteristics and sensory shelf-life of corn tortilla. *Journal of the Science of Food and Agriculture, 102*, 688–695. https://doi.org/10.1002/jsfa.11399

Bejosano, F., & Alviola, J. N. (2015). Dough conditioners in flour tortilla processing. In L. W. Rooney & S. O. Serna-Saldívar (Eds.), *Tortillas: Wheat flour and corn products* (pp. 185–194). AACC International Press. https://doi.org/10.1016/B978-1-891127-88-5.50008-6

Bejosano, F. P., Joseph, S., Lopez, R. M., Kelekci, N. N., & Waniska, R. D. (2005). Rheological and sensory evaluation of wheat flour tortillas during storage. *Cereal Chemistry, 82*, 256–263. https://doi.org/10.1094/CC-82-0256

Bello-Pérez, L. A., Flores-Silva, P. C., Agama-Acevedo, E., de Dios Figueroa-Cárdenas, J., López-Valenzuela, J. A., & Campanella, O. H. (2014). Effect of the nixtamalization with calcium carbonate on the indigestible carbohydrate content and starch digestibility of corn tortilla. *Journal of Cereal Science, 60*, 421–425. https://doi.org/10.1016/j.jcs.2014.05.001

Brooker, D. J. (2015). Quality assurance for corn and wheat flour tortilla manufacturing. In L. W. Rooney & S. O. Serna-Saldívar (Eds.), *Tortillas: Wheat flour and corn products* (pp. 97–123). AACC International Press. https://doi.org/10.1016/B978-1-891127-88-5.50004-9

Cauvain, S. P. (2017). Raw materials. In S. P. Cauvain (Ed.), *Baking Problems Solved* (pp. 33–144). Woodhead Publishing. https://books.google.com.mx/books?id=Wv5PCwAAQBAJ

Corrales-Bañuelos, A. B., Cuevas-Rodríguez, E. O., Gutiérrez-Uribe, J. A., Milán-Noris, E. M., Reyes-Moreno, C., Milán-Carrillo, J., & Mora-Rochín, S. (2016). Carotenoid composition and antioxidant activity of tortillas elaborated from pigmented maize landrace by traditional nixtamalization or lime cooking extrusion process. *Journal of Cereal Science, 69*, 64–70. https://doi.org/10.1016/j.jcs.2016.02.009

De la Parra, C., Serna Saldívar, S. O., & Liu, R. H. (2007). Effect of processing on the phytochemical profiles and antioxidant activity of corn for production of masa, tortillas, and tortilla chips. *Journal of Agricultural and Food Chemistry, 55*, 4177–4183. https://doi.org/10.1021/jf063487p

Escalante-Aburto, A., Mariscal-Moreno, R. M., Santiago-Ramos, D., & Ponce-García, N. (2019). An update of different nixtamalization technologies, and its effects on chemical composition and nutritional value of corn tortillas. *Food Reviews International, 36*, 456–498. https://doi.org/10.1080/87559129.2019.1649693

Figueroa-Cárdenas, C. J. D., Rodríguez-Chong, A., & Veles-Medina, J. J. (2011, August 15). Proceso ecológico de nixtamalización para la producción de harinas, masa y tortillas integrales. *Mexican Patent* 289, 339.

Gámez-Valdez, L. C., Gutiérrez-Dorado, R., Gómez-Aldapa, C. A., Perales-Sánchez, J. X. K., Milán-Carrillo, J., Cuevas-Rodríguez, E. O., & Reyes-Moreno, C. (2021). Effect of the extruded amaranth flour addition on the nutritional, nutraceutical and sensory quality of tortillas produced from extruded creole blue maize flour. *Biotecnia, 23*, 103–112.

García-Armenta, E., Caro-Corrales, J., Cronin, K., Reyes-Moreno, C., & Gutiérrez-Dorado, R. (2020). Heat transfer simulation in corn kernel during nixtamalization process. *Revista Mexicana de Ingeniería Química, 19*, 745–754. https://doi.org/10.24275/rmiq/Alim792

Gaxiola-Cuevas, N., Mora-Rochín, S., Cuevas-Rodríguez, E. O., León-López, L., Reyes-Moreno, C., Montoya-Rodríguez, A., & Milán-Carrillo, J. (2017). Phenolic acids profiles and cellular antioxidant activity in tortillas produced from Mexican maize landrace processed by nixtamalization and lime extrusion cooking. *Plant Foods for Human Nutrition, 72*, 314–320. https://doi.org/10.1007/s11130-017-0624-3

Gaytán-Martínez, M., Figueroa, C. J. D., Vázquez, L. P. A., Morales, S. E., Martínez, F. H. E., & Reyes, V. M. L. (2012). Physiochemical, functional and chemical characterization of nixtamalized corn flour obtained by ohmic heating and traditional process. *CYTA – Journal of Food, 10*, 182–195. https://doi.org/10.1080/19476337.2011.606478

Gutiérrez-Cortez, E., Rojas-Molina, I., Zambrano-Zaragoza, M. L., Espinosa-Arbeláez, D. G., Rojas, A., & García, J. C. (2016). The mass transport phenomenon through pericarp during nixtamalization process. *Food and Bioproducts Processing, 100*, 477–486. https://doi.org/10.1016/j.fbp.2016.09.008

Iuga, M., Ávila-Akerberg, V. D., González-Martínez, T. M., & Mironeasa, S. (2019). Consumer preferences and sensory profile related to the physico-chemical properties and texture of different maize tortilla types. *Food, 8*, 533. https://doi.org/10.3390/foods8110533

Kamau, E. H., Nkhata, S. G., & Ayua, E. O. (2020). Extrusion and nixtamalization conditions influence the magnitude of change in the nutrients and bioactive components of cereals and legumes. *Food Science & Nutrition, 8*, 1753–1765. https://doi.org/10.1002/fsn3.1473

Katz, S. H., Hediger, M. L., & Valleroy, L. A. (1974). Traditional maize processing techniques in the new world. *Science, 184*, 765–773. https://science.org/doi/10.1126/science.184.4138.765

Kong, F., & Singh, R. P. (2016). Chemical deterioration and physical inestability of foods and beverages. In H. Skibted, J. Risbo, & L. A. Monges (Eds.), *The stability and shelf life of food* (pp. 43–76). Woodhead Publishing Limited. https://doi.org/10.1016/B978-0-08-100435-7.00002-2

León-López, L., Reyes-Moreno, C., Ley-Osuna, A. H., Perales-Sánchez, J. X. C., Milán-Carrillo, J., Cuevas-Rodríguez, E. O., & Gutiérrez-Dorado, R. (2019). Improvement of nutritional and nutraceutical value of nixtamalized maize tortillas by addition of extruded chia flour. *Biotecnia, 21*, 55–66.

Long, J. (1996). *Conquista y comida. Consecuencias del Encuentro de Dos Mundos* (1st ed.). Universidad Nacional Autónoma e México.

Mariscal-Moreno, R. M., Figueroa, J. D. C., Santiago-Ramos, D., Rayas-Duarte, P., Véles-Medina, J. J., & Martínez-Flores, H. E. (2017). Nixtamalization process affects resistant starch formation and glycemic index of tamales. *Journal of Food Science, 82*, 1110–1115. https://doi.org/10.1111/1750-3841.13703

Martínez-Bustos, F., García, M. N., Chang, Y. K., Sánchez-Sinecio, F., & Figueroa-Cárdenas, J. D. (2000). Characteristics of nixtamalized maize flours produced with the use of microwave heating during alkaline cooking. *Journal of the Science of Food and Agriculture, 80*, 651–656. https://doi.org/10.1002/(SICI)1097-0010(199904)79:5<713::AID-JSFA229>3.0.CO;2-F

Martínez-Flores, H. E., Gaytán-Martínez, M., Figueroa-Cárdenas, J. D., Martínez-Bustos, F., Reyes-Vega, M. L., & Rodríguez-Vidal, A. (2004). Effect of some preservatives on shelf-life of corn tortillas obtained from extruded masa. *Agrociencia, 38*, 285–292.

Maya-Cortés, D. C., Figueroa, C. J., Garnica-Romo, M. G., Cuevas-Villanueva, R. A., Cortés-Martínez, R. A., Véles-Medina, J. J., & Martínez-Flores, H. E. (2010). Whole-grain corn tortilla prepared using an ecological nixtamalization process and its impact on the nutritional value. *International Journal of Food Science & Technology, 45*, 23–28. https://doi.org/10.1111/j.1365-2621.2009.02095.x

Menera-López, I., Gaytán-Martínez, M., Reyes-Vega, M. L., Morales-Sánchez, E., & Figueroa-Cárdenas, J. D. (2013). Physico-chemical properties and quality assessment of corn flour processed by a continuous ohmic heating system and traditional nixtamalization. *CyTA-Journal of Food, 11*, 8–14. https://doi.org/10.1080/19476337.2012.762692

Mora-Rochín, S., Gutiérrez-Uribe, J. A., Serna-Saldívar, S. O., Sánchez-Peña, P., Reyes-Moreno, C., & Milán-Carrillo, J. (2010). Phenolic content and antioxidant activity of tortillas produced from pigmented maize processed by conventional nixtamalization or extrusion cooking. *Journal of Cereal Science, 52*, 410–416. https://doi.org/10.1016/j.jcs.2010.08.010

Mora-Rochín, S., Gaxiola-Cuevas, N., Gutiérrez-Uribe, J. A., Milán-Carrillo, J., Milán-Noris, E. M., Reyes-Moreno, C., Serna-Saldívar, S. O., & Cuevas-Rodríguez, E. O. (2016). Effect of traditional nixtamalization on anthocyanin content and profile in Mexican blue maize (Zea mays L.) landraces. *LWT-Food Science and Technology, 68*, 563–569. https://doi.org/10.1016/j.lwt.2016.01.009

Mora-Rochín, S., Milán-Noris, A. K., & Milán-Carrillo, J. (2019). Maize. In S. Ahmad Mir, A. Manickavasagan, & M. Ahmad-Shah (Eds.), *Whole grains: Processing, product development, and nutritional aspects* (pp. 87–102). CRC Press.

Mora-Rochín, S., Menchaca-Armenta, M., Milán-Noris, A. K., Gutiérrez-Urive, J. A., Cuevas-Rodríguez, E. O., Reyes-Moreno, C., & Milán-Carrillo, J. (2021). Optimization of lipophilic compounds in tortillas from native pigmented maize obtained from flours by the lime cooking extrusion process. *Biotecnia, 23*, 65–72.

Obatolu, V. A., Augustine, O., & Iken, J. E. (2007). Improvement of home-made maize tortilla with soybean. *International Journal of Food Science &Technology, 42*, 420–426. https://doi.org/10.1111/j.1365-2621.2006.01242.x

Platt-Lucero, L. C., Ramírez-Wong, B., & Torres-Chavéz, P. I. (2013). Effect of xylanase on extruded nixtamalized corn flour and tortilla: Physicochemical and rheological characteristics. *Journal of Food Process Engineering, 36*, 179–7186. https://doi.org/10.1111/j.1745-4530.2011.00667.x

Ramirez, K., Quintero-Soto, M. F., & Rochín-Medina, J. (2020). Enhancement of the antioxidant and antimicrobial activities of maize wastewater by a eco-friendly process. *Journal of Food Measurement and Characterization, 14*, 1682–1689. https://doi.org/10.1007/s11694-020-00416-1

Ramírez-Wong, B., Walker, C. E., Ledesma-Osuna, A. I., Torres, P. I., Medina-Rodríguez, C. L., López-Ahumada, G. A., Salazar-García, M. G., Ortega-Ramírez, R., Johnson, A. M., & Flores, R. A. (2007). Effect of flour extraction rate on white and red winter wheat flour compositions and tortilla texture. *Cereal Chemestry, 84*, 207–2013. https://doi.org/10.1094/CCHEM-84-3-0207

Rodríguez-Noriega, S., Buenrostro-Figueroa, J. J., Rebolloso-Padilla, O. N., Corona-Flores, J., Camposeco-Montejo, N., Flores-Naveda, A., & Ruelas-Chacón, X. (2021). Developing a descriptive sensory characterization of flour tortilla applying flash profile. *Food, 10*, 1473. https://doi.org/10.3390/foods10071473

Rooney, L. W., & Serna-Saldívar, S. O. (2015). Food-grade corn quality for lime-cooked tortillas and snacks. In L. W. Rooney & S. O. Serna-Saldívar (Eds.), *Tortillas: Wheat flour and corn products* (pp. 227–246). AACC International Press. https://doi.org/10.1016/B978-1-891127-88-5.50012-8

Ruiz-Gutiérrez, M., Quintero-Ramos, A., Meléndez-Pizarro, C., Talamás-Abbud, R., Barnard, J., Marquez- Melendez, R., & Lardizabal, D. (2011). Nixtamalization in two steps with different calcium salts and the relationship with chemical, texture and thermal properties in masa and tortilla. *Journal of Food Process Engineering, 35*, 772–783. https://doi.org/10.1111/j.1745-4530.2010.00627.x

Serna-Saldívar, S. O. (2010). *Cereal grains: Properties, processing and nutritional attributes* (p. 794). CRC Press.

Serna-Saldívar, S. O. (2015). Nutrition and fortification of corn and wheat tortillas. In L. W. Rooney & S. O. Serna-Saldivar (Eds.), *Tortillas: Wheat flour and corn products* (pp. 29–63). https://doi.org/10.1016/B978-1-891127-88-5.50002-5

Serna-Saldívar, S. O., & Chuck-Hernández, C. (2016). Tortillas. In B. Caballero, P. Finglas, & F. Toldrá (Eds.), *The encyclopedia of food and health* (Vol. 5, pp. 319–325). Academic.

Serna-Saldívar, S. O., & Chuck-Hernández, C. (2019). Food uses of lime-cooked corn with emphasis in tortillas and snacks. In S. O. Serna-Saldívar (Ed.), *Corn: Chemistry and technology* (pp. 469–500). AACC International Press. https://doi.org/10.1016/B978-0-12-811971-6.00017-6

Serna-Saldívar, S. O., & Rooney, L. W. (2015). Industrial production of maize tortillas and snacks. In L. W. Rooney & S. O. Serna-Saldívar (Eds.), *Tortillas: Wheat flour and corn products* (pp. 247–281). AACC International Press. https://doi.org/10.1016/B978-1-891127-88-5.50013-X

Serna-Saldívar, S. O., Gutiérrez-Uribe, J. A., & García-Lara, S. (2015). Phytochemical profiles and nutraceutical properties of corn and wheat tortillas. In L. W. Rooney & S. O. Serna-Saldívar (Eds.), *Tortillas: Wheat flour and corn products* (pp. 65–96). AACC International Press. https://doi.org/10.1016/B978-1-891127-88-5.50013-X

Suhendro, E. L., Almeida-Domínguez, H. D., Rooney, L. W., & Waniska, R. D. (1998). Objective rollability method for corn tortilla texture measurement. *Cereal Chemistry, 75*, 320–324. https://doi.org/10.1094/CCHEM.1998.75.3.320

Treviño-Mejía, D., Luna-Vital, D. A., Gaytán-Martínez, M., Mendoza, S., & Loarca-Piña, G. (2016). Fortification of commercial nixtamalized maize (*Zea mays* L.) with common bean (*Phaseolus vulgaris* L.) increased the nutritional and nutraceutical content of tortillas without modifying sensory properties. *Journal of Food Quality, 39*, 569–579. https://doi.org/10.1111/jfq.12251

Valderrama-Bravo, M. D. C., Domínguez-Pacheco, A., Aguilar, C., Sánchez-Hernández, G., Pérez-Reyes, C., Contreras-Padilla, M., Rojas-González, T., & Oaxaca-Luna, A. (2018). Changes in masa and tortillas obtained from maize irradiated and nixtamalized with nejayote. *International Agrophysics, 32*, 387–394. https://doi.org/10.1515/intag-2017-0030

Verma, B., Hucl, P., & Chibbar, R. N. (2018). Phenolic content and antioxidant properties of bran in 51 wheat cultivars. *Cereal Chemistry, 85*, 544–549. https://doi.org/10.1094/CCHEM-85-4-0544

Waniska, R. D. (2015). Flour Tortilla Problems. In L. W. Rooney & S. O. Serna-Saldívar (Eds.), *Tortillas: Wheat flour and corn products* (pp. 215–226). AACC International Press. https://doi.org/10.1016/B978-1-891127-88-5.50011-6

Whitney, K., Simsek, S., Berzonsky, W., Ohm, J. B., & Sorenson, B. (2011). The effect of spring wheat starch properties on flour tortilla quality. *Journal of Food Process Engineering, 34*, 697–715. https://doi.org/10.1111/j.1745-4530.2009.00428.x

Young, N. W. G., & O'sullivan, G. R. (2011). The influence of ingredients on product stability and shelf life. In D. Kilcast & P. Subramaniam (Eds.), *Woodhead publishing series in food science, technology and nutrition, food and beverage stability and shelf life* (pp. 132–183). Woodhead Publishing Limited. https://doi.org/10.1533/9780857092540.1.132

Zduńska, K., Dana, A., Kolodziejczak, A., & Rotsztejn, H. (2018). Antioxidant properties of ferulic acid and its possible application. *Skin Pharmacology and Physiology, 31*, 332–336. https://doi.org/10.1159/000491755

Chapter 7
Crackers

Manish Tiwari, Akansha Tiwari, Ankita Kataria, and Komal Chauhan

1 Introduction

The term bakery deals with the production and selling of flour based baked products under the action of dry heat. Nowadays, degree of dryness of baked products can be modified by the amount of steam produced from the baked items (Cappelli et al., 2021).

The concept of a bakery revolves around the creation and marketing of baked goods made primarily from flour, sugar and fat which are subjected to dry heat during the baking process. In contemporary baking, the level of moisture in baked products can be adjusted by controlling the amount of steam generated during the baking operation, as highlighted in a study by Cappelli et al., in 2021. Bakery products are typically categorized based on a range of characteristics, including their shape, size, color, taste, texture, flavor, sugar content, pH level, water activity (a_w), moisture content, and the types of leavening agents utilized. This classification framework is illustrated in Figure 7.1, as described by Manley in 2011.

Unsweetened bakery products contain minimal or less sugar such as rolls, bread, pizza base, crackers (Biljwan et al., 2019; Borczak et al., 2018). Sweetened products namely cakes, cookies, biscuits, pancakes, muffins etc. contain sufficient amounts of sugar. Filled products contains filling such as fruit and meat pies,

M. Tiwari · K. Chauhan (✉)
Department of Food Science and Technology, National Institute of Food Technology Entrepreneurship and Management, Sonipat, Haryana, India

A. Tiwari
Department of Food Technology, Parul University, Vadodara, Gujarat, India

A. Kataria
Department of Food Science and Technology, Maharaja Ranjit Singh Punjab Technical University, Bathinda, Punjab, India

M. A. Shah et al. (eds.), *Cereal-Based Food Products*,
https://doi.org/10.1007/978-3-031-40308-8_7

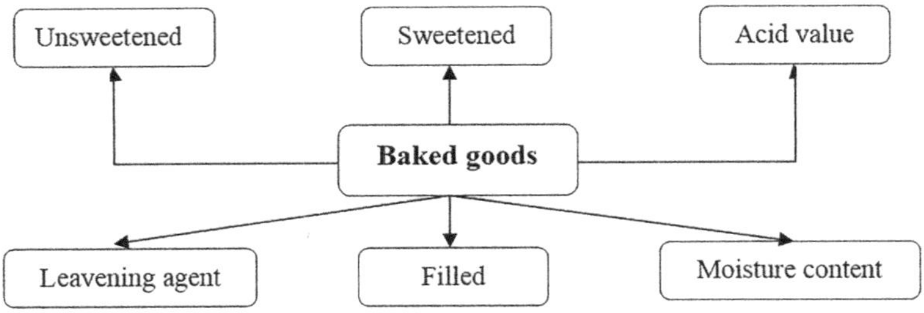

Fig. 7.1 Classification of baked products

sandwiches, pastries, cream cakes, pizza (Smith et al., 2004). Bakery goods are also categorised on the basis of leavening agents used to raise or leaven the volume of the dough (Neeharika et al., 2020; Venturi et al., 2021). Microbial leaveners (*Saccharomyces cerevisiae*) are used to prepare bread and rolls while chemically leavened baking powders, baking soda, acid salts, and inert fillers are used to develop muffins and pancakes (Parapouli et al., 2020; Asyikeen et al., 2013). Unleavened bakery products such as tortillas, pitas, gorditas, lavash and crepes are developed without leavening agents (Wieser et al., 2020). Baked goods are also categorized on the basis of pH and further classified as high acid e.g. sourdough bread, apple pie (pH: < 4.6), low acid e.g. white bread, whole wheat bread, chocolate nut bread (pH: 4.6–7.0) and non-acid e.g. crumpets, banana nut bread, carrot muffin (pH: 7.0).

On the basis of available moisture i.e., water activity (a_w), bakery products are classified as high moisture ($a_w > 0.85$) (cake, pies, breads); intermediate moisture ($a_w > 0.6$ and $a_w < 0.85$) (pastries, soft cookies, doughnuts) and low moisture ($a_w < 0.6$) products (crackers, cookies) (Smith & Simpson, 1995).

Baked products are generally taken as energy dense food products rich in sugar, fat, however, the increased awareness and advanced technology switched the narrow aspects of bakery index into broader one by including healthy ingredients with low sugar and fat content such as crackers (Romano et al., 2018).

Crackers are thin, crisp, dry and leavened or unleavened bakery products with 2–3% moisture content and are available in various shapes (Reibner et al., 2021). Baked crackers makes a cracking or snapping sound when crushed (Meriles et al., 2021).

They are similar to non-sweetened, fermented, laminated, non-sweet type to semi-sweet cookies. Nevertheless, cracker dough is hard with little or no sugar and moderate amount of fat (10–20%) respectively and low moisture content and bakes quickly. Cookie dough can range from soft to firm with moderate sugar and fat content (20–45%) respectively and less than 5% moisture content. Crackers are commonly used due to their great practicality, convenience, and long shelf life (Xu et al., 2020; Faccioli et al., 2021).

Several types of crackers are available such as flavoured, soda, cream and snack crackers that are widely consumed as snacks and provide easy access to nutritious staples (Manley, 2011; Akonor et al., 2017) (Fig. 7.2).

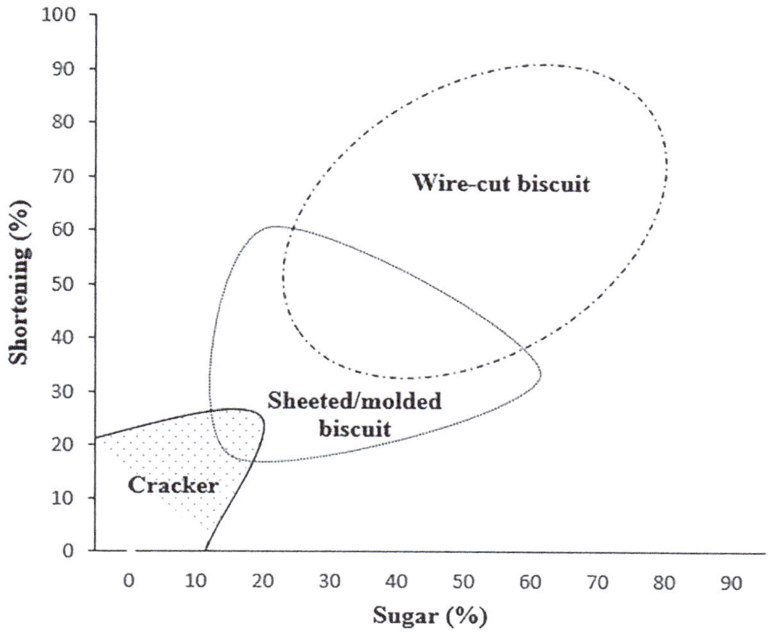

Fig. 7.2 Classification of biscuit in relation to sugar and shortening in 100 parts flour. (Source: Manley, 2011)

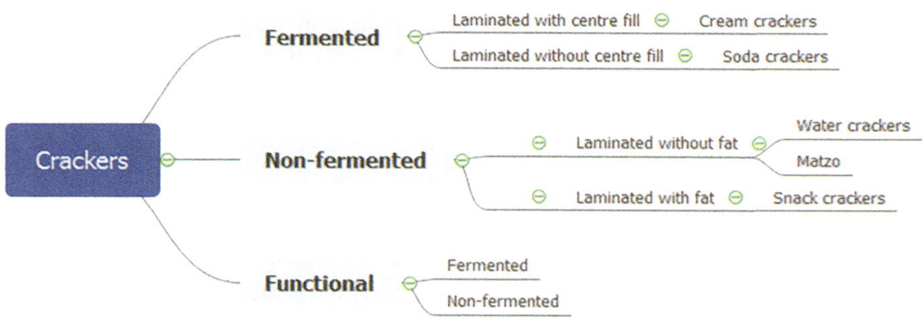

Fig. 7.3 Classification of crackers

Classification of Crackers

Crackers contains low fat and sugar and are functionally classified as fermented and unfermented on the basis of manufacturing process (Mihiranie et al., 2021). They are also classified on the basis of composition or formulation, dough rheology and texture. Further they are categorized into cream, snack, unsweetened soda crackers, semi-sweet biscuits, hard sweet biscuits (Manley, 2011). The shape and size of crackers can be varied (Gonçalves & Salas-Mellado, 2018). During manufacturing, when they are heated they are generally layered or sprayed with oil and thus apparent fat is usually found to be high (Fig. 7.3).

Fermented Crackers

Fermented crackers are produced from the sponge dough which is fermented up to 19 h. During sponge fermentation, both yeast and beneficial bacteria grow, causing the consistency of the sponge to change drastically. After fermentation, the sponge is mixed with other ingredients. Sodium bicarbonate is added to neutralize the acidic dough (pH 7.0). The dough is then allowed to relax for 4–6 h. Thereafter, it is laminated by multiple pairs of heavy steel sheeting rolls. The purpose of laminating the sheeted dough (six to eight layers) is to repair the dough sheet with holes or tear and to turn the folded dough for more uniform stress in all directions. Furthermore, fat is added between layers of dough to create characteristic flaky structure after baking.

Cream Crackers

Cream crackers were first prepared by Irish firm in 1880s, known as Jacobs. The nomenclature indicates that it contains cream which is not true. The term "cream crackers"refers to the technique in which the blend is creamed during processing (Neo et al., 2007). They are rectangular in shape and are comparatively large (65 × 75 mm). Its surface is uneven with light brown blisters at the bottom and top surface. Hence, they have flaky layers with evenly distributed structure throughout the inner surface. The manufacturing process of cream crackers is simple with similar ingredients as in soda crackers. They contain 12–18% shortening, 0.9–1.5% salt, 1.0–2.4% yeast and water (Khatkar, 2013). The dough is made in one step and allowed to ferment for 4–16 h. Similar to saltine crackers, the product is dusted with fine crackers and lamination is applied between layers before baking and cutting. The dust filling in the crackers consists of flour, shortening and salt that separates the layers and provide flaky texture during processing (Khatkar, 2013). There is no use of chemical leaveners, thus they have a bland and somewhat nutty flavour. The moisture content in the finished product is slightly higher (3–4%) than saltines crackers (Caballero et al., 2003).

Soda Crackers

Soda crackers have been commonly consumed in United States for more than 150 years. They are also known as saltine crackers because of the presence of salt. These are long fermented laminated dough products containing high amount of shortening. The dough consists of 0–5% yeast, 8–10% shortening, salt, and optionally, malt or malt syrup on flour weight basis (Li et al.,2013a). They are neutralized with 1% baking soda (sodium bicarbonate) before sponge mixing and fermentation in order to unfold the protein subunits network and make the crackers lighter (Li et al., 2014). The dough ferments without significant decrease in product's pH, resulting in a slightly alkaline end product and therefore termed as 'soda' cracker.

Individual saltines are not formed, baked and pricked by scrap less cutters prior to baking. These perforations create weak points during baking that make it possible to separate the sheet into separate pieces. The soda crackers have docking holes, which can be scattered randomly or in rows. During the saltine cracker manufacturing process, the internal structure shows a successive layer within each docking hole due to lamination (Li et al., 2013b). The dough used for the preparation of soda crackers is comparatively thinner than other products under this category.

Normally, the dimensions of soda crackers are 50 × 50 mm with 4 mm thickness and weight varying between 3 to 3.5 g. The final moisture content of the product is around 2.5% (Saberi et al., 2021). This crispy and cracking texture is due to low moisture content and laminated structure. They generally possess bland flavour and the unique crisp texture makes it suitable to be consumed with soups.

Non-fermented Crackers

Non fermented crackers are chemically leavened unfermented product with final pH of about 6.5. The ingredients are mixed to form adough which is left for 2–4 h at room temperature. Thereafter, the dough is sheeted to form a continuous ribbon that is laminated with light dusting of flour between the layers. The remaining process is the same for both fermented and chemically leavened crackers. The sheeted dough is rolled with docking pins and cut into desired shape and size. The docking pins press the dough together to facilitate steam release from the crackers and help keep the layers from separating. The high temperature, short time baking, puffs the thin dough sheet into crackers by vaporizing internal water. Salting is done before baking. The baked crackers are also buttered or seasoned with seasonings. The low moisture content of the finished product is important for maintaining the crispy, crumbly yet soft texture of the crackers. Most crackers contain about 2% moisture. (Manley, 2011).

Water Crackers

Water crackers are identical to matzos with a ratio of 100: 29: 6.5: 1 as flour: water: fat: salt (Manley, 2011). The dough is not fermented but conditioned before sheeting in presence of proteolytic enzymes that softens the gluten network and make it more elastic. The dough shape is maintained by resting the dough before cutting. The sheet is then laminated, gauged and cut before baking. They are baked at 400 °C for 1 min (Manley, 2011). Water crackers are generally round in shape and possess

70 mm diameter. The cutters must be oval owing to longitudinal shrinkage that takes place in the oven.

The water crackers have strongly blistered surface. Baked water crackers are fairly hard with bland flavour, crisp texture and darker blisters (Davidson, 2018). These crackers can be made crisp when spread with butter, savouries or cheese before serving.

Matzos

Matzos are Jewish crackers that are widely consumed in Israel (Zakupszki et al., 2018). The shape varies from round to large rectangular sheets (water biscuits). Normally, the matzo preparation uses about 38 g of water in 100 g of wheat flour for mix preparation with no fermentation (Yoneya & Nip, 2006). The dough mix is sheeted to 2–6 layers of strong and clear sheet followed by docking (small holes) and cutting. It is then baked at a temperature of 400 °C for 4–5 min (Manley, 2011). Higher baking temperatureresults in the formation of blisters which remove the colour and produces very pale crackers. The final product has 3% moisture content (Miller, 2016).

Snack Crackers

Snack crackers are modified form of fermented crackers and are classified under non-fermented crackers. They do not require long fermentation for its preparation (Ahmed & Abozed, 2015). The dough is matured by means of proteolytic enzymes or sulphites of sodium salt with minimal usage of yeast and malt hence they are termed as chemically leavened crackers (Manley, 2011; Raveendran et al., 2018). The enzymatically matured dough is not deformed during sheeting and cutting. The crackers are baked and thereafter, sprayed with heated oil and sprinkled with seasoning or dusted with sugar (4–10%), herbs, cheese, sesame seeds etc. for flavouring, enhanced mouth feel and appearance (Hussien et al., 2020; Sykes & Davidson, 2020; Beltrami et al., 2018). The final product has a firm texture but is crunchy with good taste, flavour and aroma (Qadri et al., 2018).

They can be of varied shapes and sizes. Generally, round shaped are available at commercial level. They have docking holes which allows even lifting during baking (Manley, 2011). The final product has a moisture content of about 2% with denser structure and soft texture as compared to fermented crackers such as cream and soda crackers (Ahmed & Abozed, 2015; Miller, 2016).

Table 7.1 Value added functional crackers

Cracker types	Value addition	Method of preparation	Baking conditions	Observations	References
Cream crackers	Pumpkin (*Cucurbita pepo L.*) seed flour (PSF) viz. Nevsehir, Edine and Urgup variety	One-step fermentation method	Convection type, 180 °C, 6–7 min.	• 30% PSF of Nevsehir variety found to have maximum ash, protein and total dietary fiber content. • Ratio of tartaric acid: Organic acid varied from 1292 to 2557 mg/kg and fatty acid: Linoleic acid (41.34–43.69%).	Altiner et al., (2021)
	Germinated green lentil extract (GLE)	–	Convection type, 175 °C, 13 min.	• The addition of GLE in crackers was found to contain antioxidant activity (0.84 μmoltrolox/mg) and total phenolic content (3.33 mg GAE/g).	Polat et al. (2020)
	Buckwheat sourdough	Sponge dough method	Convection type, 200 °C, 15 min.	• Crackers with sourdough had higher antioxidant activity (16.84 and 15.84 mmol Fe).	Alibasic et al., (2020)
Soda crackers	Microalgae strains- *chlorella vulgaris*Allma,*Arthrospira platensis* (F&M-C256),*Phaeodactylumtricornutum* (F&M-M40) and *Tetraselmissuecica* (F&M-M33)	Sponge and dough method	Convection type, 180 °C, 10 min.	• Highest antioxidant activity was found in crackers having strains of *A. platensis*. • *P. tricornutum*and*T. Suecica* crackers showed maximum antioxidant activity and phenolic content.	Batista et al., (2019)
	Roselle (*Hibiscus sabdariffa*) calyxes	Straight dough method	Convection type, 210 °C, 15 min	• Acceptable crackers obtained with high in fiber content, phenols, flavonoid and antioxidant activity by incorporating up to 3.75% Roselle calyxes into the cracker's formula.	Ahmed and Abozed (2015)

Functional Crackers

Crackers are functionally classified on the basis of ingredients, phytochemicals and the health or therapeutic benefits rendered (Mir et al., 2017; Polat et al., 2020). They are further categorized as fermented and unfermented on the basis of manufacturing process. A variety of functional crackers are available such as gluten-free, low salt and low-fat to be used in several disease conditions (Hussien et al., 2020). Studies have been conducted to prepare value added functional crackers (Table 7.1).

2 Ingredients

Ingredients are the base of any food formulation. The use of right ingredients not only determines the sensory acceptability but also the quality of the final product. The common ingredients used for developing crackers are wheat flour, shortening, leavening agents, sweeteners, enzymes or malt flour, emulsifiers, whey, eggs, flavourings, acidulants for chemical leavening, yeast, or a mixture thereof salt and dough conditioners (Hui, 2006). Nowadays, several bioactive components or phytocomponents are also added to crackers to enhance functional attributes. Table 7.2 depicts major and minor ingredients used for making crackers along with the commercial brands available.

Major Ingredients

Flour

Wheat flour is considered as the most important and primary ingredient in bakery product. The processing of the crackers requires higher gluten content as compared to other biscuits in order to retain the dough integrity which gives uplifted finished product (Cauvain & Clark, 2019).

Soft wheat flour of 8 to 9% protein value is generally used for developing crackers. The dough is extensible and possesses weak gluten network due to the presence of low protein content (Li et al., 2013b). Conversely the low starch flours results in lower swelling, gelatinization, little gluten formation, resulting into soft dough that break on stretching (Mamat & Hill, 2014). The puffs and fermented crackers are exceptions that requires medium strength flour with protein content greater than 10.5%. A high protein content flour results in hard texture and rough surface. The flour needs to be sieved and sifted to aerate the dough during mixing operations.

Table 7.2 Major and minor ingredients used in the preparation of different types of commercially available crackers

Crackers type	Variety	Ingredients		Brand
		Major	Minor	
Unsweetened Unsweetened	Gluten free	Chickpea flour, oats flour	Rosemary, chia seeds, olive oil, salt	MO'S
	Nutrichoice sugar free	Refined wheat flour, edible vegetable oil, sugar, yeast	Milk solids, emulsifier	Britannia
	Almond flour crackers	Almond flour, egg white, cheddar cheese, butter, coconut flour, flaxseed, parmesan cheese cultured milk	Enzymes, salt, baking powder, xanthan gum, sea salt, annatto, paprika, garlic, rosemary, onion	Fat snax
Low carb	Keto cheese crackers	Watermelon seed, dehydrated cheese (low carb), water, sunflower seed, vigna mungo, flax seed, whey protein, sesame seed, vachellia	Nilotica gum, himalayan pink salt, guar gum, pumpkin seed	Ketofy
Sweetened	Rovira export sodas classic soda crackers	Enriched wheat flour (niacin, reduced iron, thiamine mononitrate, riboflavin, folic acid)	Palm oil, salt, sodium bicarbonate, malt syrup, corn syrup, yeast, enzyme	Rovira
	O'Rice crackers	Japonica rice, palm oil, sugar	Potato starch	ORION
Slightly sweetened	Krack jack crackers	Wheat flour, edible vegetable oils, sugar	Raising agents, invert sugar syrup, iodised salt, yeast, acidity regulators, emulsifier, dough conditioner and improver	Parle
	Cheese crackers	Wheat flour, edible vegetable oils, sugar	Liquid glucose, iodized salt, butter, malt, processed cheese, permitted emulsifiers, leavening agents, acid salt, dough conditioner, dough improvers, contains added permitted artificial flavour (cheese)	Priyagold
	Cheddar cauliflower crackers	Cauliflower, whole grain rice, rice flour,potato starch, cassava starch	Potato flakes, palm oil,maltodextrin, natural flavors, salt, baking powder,sugar, paprika, onion powder,sunflower lecithin, lentil flour, vegetable blend (spinach, carrots,broccoli, tomato, shiitake mushroom, beet), lactic acid,garlic powder, rosemary extract	From the ground up

Sweetener

Sweeteners are mainly used as flavour enhancer in bakery industry. It provides sweet taste, improves crust color, texture and extends shelf-life (Cauvain, 2017). Sucrose, high fructose corn syrup, corn syrup, glucose syrups, honey, invert sugar and molasses are the common sweeteners used in bakery industry. Among these, the principal sweetener used is granulated sugar, also known as sucrose (Zhou et al., 2014). Granulation sugar is important in developing crackers as it cause more spread due to the incorporation of air within the dough thereby affecting its textural, physicochemical properties and eating quality, Very fine sugar results into insufficient aeration, toughness, dense texture with poor eating quality (Mamat & Hill, 2018). Additionally, sugar act as a fuel for yeast activity and regulate the rate of fermentation processes. Yeast fermented dough imparts a characteristic colour, and crisp texture to the crackers.

Fat or Shortening

Fat or shortenings are used as lubricants rendering tenderness to crackers thereby enhancing dough stability and the eating quality (Devi & Khatkar, 2016). Smooth, plastic hydrogenated shortenings are generally utilised for developing crackers.

Some crackers needs higher fat content with little or no water added to achieve final consistency. Granular shortenings are not suitable as they do not aerate the dough well and result in uneven distribution of fat (Jacob & Leelavathi, 2007). The fat used for making crackers should not melt but should be able to cream and integrate aeration. Thus, it is desirable if the fats employed as surface coatings, applied as a spray of warm oil and thus have minimal absorption into the crackers with glossy surface. Butter augmentation enhances the crackers'flavour and taste without degrading the creaming process. Before blending butter with hydrogenated fat, it should be softened to prevent it from breaking into lumps that are difficult to homogenise. (Srikanlaya et al., 2017).

Water

The formation of dough is an essential process for the preparation of bakery products. Wheat flour comprises the albumins, globulins, glutenins and the gliadins which are present in inactive form (Cauvain & Young, 2009). The hydration of these proteins results in gluten formation that is responsible for rheological properties of dough and the finished products (Faridi, 1990). The formation of gluten network resists the spread ratio of the crackers. Enzymatic or chemical leavening agents require less water at the initial stage of dough making.

The addition of water varies with the method used for dough preparation and comparatively less percentage of water is required than for bread dough (Khatkar, 2013). There are three methods for dough mixing for crackers: straight dough with fermentation time; straight dough with no fermentation time; sponge and dough method. The water to flour ratio may differ for each method, hence the rheological properties that occurs post-mixing or during dough processing. In straight-dough method, the mixed dough is sheeted with little or no fermentation, laminated (if flakiness is desired), cut, docked, and baked. Often, crackers from this method are cut into circles or other shapes that create a large amount of scrap dough that must be recycled to the mixer. Unlike the sponge and dough system, where a bulk fermentation period is used, this process does not include a fermentation step after mixing. Sponge and dough process change the dough properties such as elasticity and sponginess of the dough mix due to fermentation (Manley, 2011).

Minor Ingredients

Ingredients including leavening or raising agents and emulsifiers play an important role in textural properties of the crackers (López-Tenorio et al., 2015; Nguyen et al., 2014). Baking powder, bicarbonates of soda and ammonia and some enzymes are commonly used as raising agents (Nguyen et al., 2014). Some enzymes invertase, maltase and zymase are also used as dough modifiers and play a vital role in dough rheological properties. The former softens the texture and maintains the product volume during baking by releasing CO_2. Lecithin as an emulsifier is utilised to enhance mixing, lessen dough stickiness, and enhance dough handling and machining qualities. In top dressings, it can also be used with oil. The dispersion of spices in salad dressing oils have been aided by lecithin. Based on the weight of the flour, a use level of 0.2–0.5% is recommended. (Manley, 2011).

Leavening Agents

Chemical Leavening Agents

Baking Powder

Baking powder or sodium bicarbonate is an acid salt (phosphates, tartarates, and sulphates). During baking, they release CO_2, resulting in leavening thereby improving texture besides making digestion easier. Hazen (2013) reported that sodium bicarbonate must yield at least 12% CO_2 of the total commercial acid salts such as tartaric, fumaric, cream of tartar, glucono delta lactone. The neutralization of baking powder denotes the reactivity power of baking powder. Neutralization is defined as the amount of soda (g) that neutralizes 100 g of acidic salt (Cioroianu et al., 2011).

Baking powders are available as slow acting, fast acting and double acting (Otero-Guzmán et al., 2020). In fast acting powders, most of the gaseous part in the form of CO_2 releases at room temperature while slow acting powders release a part of CO_2 during mixing but liberates maximum CO_2 during baking. Double acting baking powders are widely used in bakery industry as it reacts and release sufficient amount of gas during mixing as well as baking.

Baking Soda
Baking soda is also known as sodium carbonate. It is a widely used as leavening agent in bakery industry and liberates CO_2 at higher temperature. The produced gases within the cookie dough expands the cell wall making them tender (Hazen, 2013). Baking soda increases the pH of baked goods that improvise the color, flavor, crumb texture, and gluten strength of the dough (Ahmad et al., 2022).

Ammonium Bicarbonate
Ammonium carbonate is another chemical leavening agent and that produces CO_2 and NH_3 during baking with no solid residue and detectable odour. Ammonium bicarbonate can be used in low moisture baked products. The reaction is rapid at 60 °C (Vetter, 2003). The products allow full escape of the ammonia due to porosity developed within the cell structure during baking.

Microbial Leavening Agents – Yeast

Yeast is a single-cell microbe and is widely used in bakery sector. *Saccharomyces cerevisiae* is commonly used for the dough fermentation, hence called "*Baker's Yeast*" (Ali et al., 2012). It can grow under anaerobic conditions and produces simple sugars and CO_2. During development of baked products yeast act upon starch using natural α and β amylases present in cereal flour resulting in the production of simple sugars that promotes its growth. The yeast fermented dough for crackers develops fluffy texture due to aeration. Baker's yeast is available in several forms such as active dry, fresh (or cake), liquid, and instant (Naeem et al., 2013).

Fresh yeast is susceptible to change in conditions like temperature, available water, added salt and sugar. The liberation of CO_2 is three times at 30 °C and the yeast inactivates rapidly at 54 °C. They are generally activated in water prior to mixing. Fresh yeast is usually added to water in the ratio of 1:5. Frozen yeast is generally added to lukewarm water for 15 min before using. Dried yeast is twice as potent on wet weight basis as compared to fresh yeast and can be reactivated by mixing it with water at 40 °C with little sugar for 10–15 min (Hidalgo & Brandolini, 2014).Yeast should not be activated in salt water as 2% salt solution will hamper the growth of yeast. Likewise, sugar concentration should be less than 5% (Ali et al., 2012) for optimum growth.

Enzymes

Enzymes are usually added to modify dough rheology in making crackers (Panghal et al., 2011). In laminated dough crackers, bacterial proteases have been reported to be helpful in preventing cracks, curling, and the formation of gas bubbles (Heredia-Sandoval et al., 2016). Yeast mechanism is principally controlled by enzymes such as invertase, maltase and zymase complex. Invertase hydrolyss the sucrose into glucose and fructose whereas maltase breakdown the maltose to two units of glucose. Zymase works on simple sugars that are produced by the action of invertase and maltase. The simple sugars are converted by zymase to alcohol and CO_2 that causes the dough to rise and swell. The enzymatic reactions are as under:

$$\underset{\text{Sucrose}}{C_{12}H_{22}O_{11}} \xrightarrow{\text{Inveratase}} \underset{\text{Glucose}}{C_6H_{12}O_6} + \underset{\text{Fructose}}{C_6H_{12}O_6}$$

$$\underset{\text{Maltose}}{C_{12}H_{22}O_{11}} \xrightarrow{\text{Maltase}} \underset{\text{Glucose}}{C_6H_{12}O_6} + \underset{\text{Glucose}}{C_6H_{12}O_6}$$

$$\underset{\text{Dextrose}}{C_6H_{12}O_6} \xrightarrow{\text{Zymase}} \underset{\text{Carbon dioxide}}{2\,CO_2} + \underset{\text{Ethanol}}{C_2H_5OH}$$

3 Preparation of Crackers

The manufacturing process used to produce crackers consist of mixing, shaping, cutting, baking, cooling and packaging as shown in Fig. 7.4. The dough is fermented and leavened using ingredients like yeast, sodium bicarbonate, and ammonium bicarbonate. The final dough has a moisture content of 15–25%. It depends on the initial moisture content of the flour (Manley, 2011). During sheeting, the dough is often laminated into multiple thin layers to form sheets. During baking, the spring in dough shows open and flaky texture in finished good (Khatkar, 2013).

Mixing of ingredients to make dough followed by fermentation is generally carried out under batch processes while sheeting or forming, cutting, baking, oil spraying, cooling and packaging are continuous in-line processes. Addition of water during cracker preparation in sponge dough process and direct process is kept low to minimize gluten development and shorten the baking time (Khatkar, 2013). All types of crackers follow similar mode of preparation with little difference. Some crackers require colour contrast between light background colour and dark blisters while others can be cut into strips or entire sheets and baked before being broken into individual biscuits. Traditional American soda crackers, are typically cooked on heavy wire mesh oven bands while traditional English crackers are baked on light wire-mesh bands such as cream crackers and water biscuits. The bands are quickly warmed and heat transferred within pieces is carried out via conduction. In order to bake crackers to low moisture content (1.5–2.5%), a lot of energy is required (Li et al., 2013b). The main difference is that they are developed using reducing agents to facilitate optimal processing and better machining. Common reducing agents used are sodium bisulfites and l-cysteine (Fort, 2016).

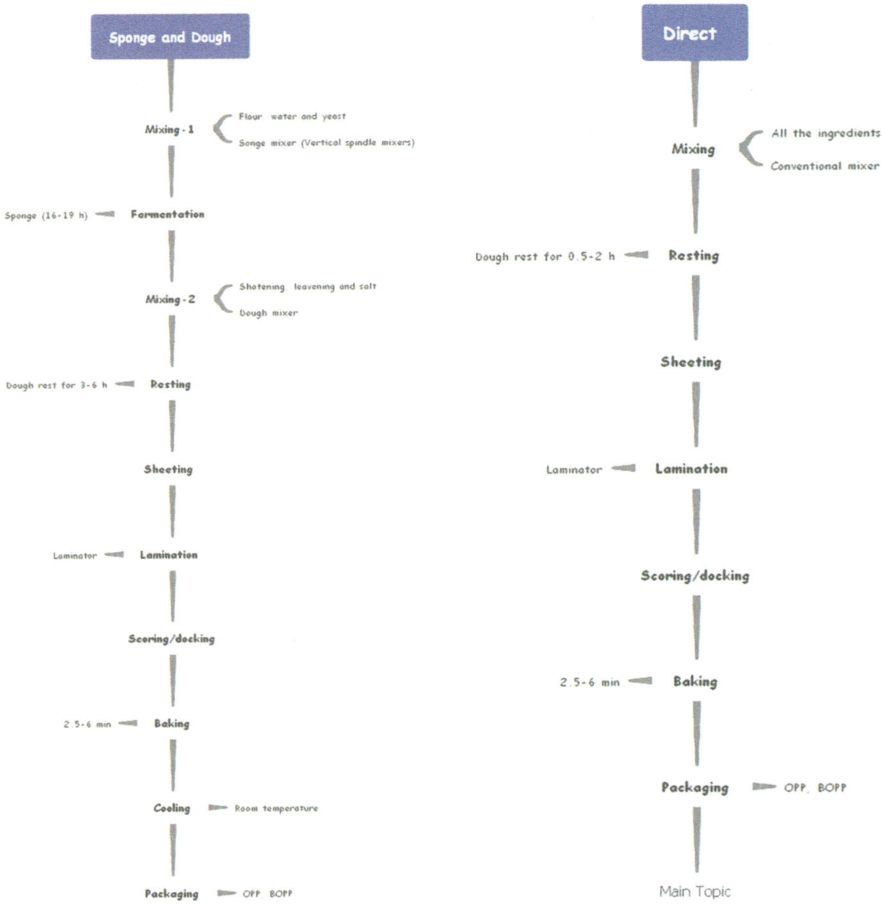

Fig. 7.4 Scematic representation of developing crackers using different methods. (Source: Kweon et al., 2011)

Sponge Dough Method

The major ingredients such as flour (60–70%), water and yeast are weighed and made to a sponge dough at a low speed for 2–3 min in vertical spindle type mixer (Khatkar, 2013). The dough is allowed to ferment for 16–19 h at 27 °C and 75% relative humidity (Fig. 7.4). The pH of dough decreases from 6.0 to 4.0 during fermentation resulting in desirable end flavor. Flour also contains proteolytic enzymes with an optimum pH of about 4.1 that modifies the dough handling properties such as its extensibility and elasticity (Espinoza-Herrera et al., 2021). The resulting dough is more extensible and exhibits less elastic behaviour, therefore, easily sheeted. Thereafter, the sponge dough is mixed with other ingredients including addition of base (sodium bicarbonate) that produces carbon dioxide which raises the pH of the dough and makes it fluffy (Hazen, 2013). The final dough is kept for 3–6 h under similar conditions. The dough is then sheeted into the thin sheets in three-roll type sheeters for final forming (Khatkar, 2013). It is then laminated for 2–4 mm final thickness with 7–8 layers of a preset thickness in a laminator. The docking is done to put the pattern holes onto the crackers using pins and rotary cutters or stamping cutter to cut the crackers into desired shapes and sizes. Baking is done in a quadrupled zone in direct gas fired oven on a heavy mesh oven band for about 2.5–6 min at 230–315 °C (Manley, 2011).

The high temperature produces steam that expands the cracker volume and allows the moisture to escape from the dough. Final zone of the baking oven sets the final color of the crackers that ranges from light yellow to golden brown. The final crackers have about 2–2.5% moisture content. Oil spraying is done in some of the crackers to make the appearance glossy and acceptable followed by cooling and packaging in biaxially oriented polypropylene or oriented polypropylene films as it has good moisture barrier properties, high clarity and graphic reproduction, good seal ability and shelf appeal (Davidson, 2018).

Fermented soda crackers are generally manufactured using a sponge and dough method. These crackers have a flaky layered texture and are relatively dry. They are prepared using chemical leavening agents, however, enzyme crackers are made using proteolytic enzymes (Li et al., 2013b). Enzymes enhance the dough sheetability and decrease its elasticity. The crackers are immediately coated with oil after baking. In terms of size and structure, chemical crackers and enzyme crackers have similar appearance.

Direct Method

Direct method of crackers preparation is also known as straight dough process and do not fermentation (Manley, 2011). The ingredients (Table 7.3) are added to the horizontal high-speed mixer to form a homogeneous mass. During mixing, flour hydration, fat and water emulsification, gluten formation and activation of leavening agents occurs. The dough is allowed to rest for 0.5 to 2 h (Fig. 7.4). Reducing agents such as L-cysteine and sodium metabisulphite are generally used to reduce the gluten network. Soda crackers are generally sheeted, laminated and baked. Snack crackers are prepared using straight dough method. The shape and size of these crackers are different from soda crackers thus cutting also varies. Cutting process yields more residual dough for recycling (Davidson, 2018).

Table 7.3 Formulation for cracker preparation using sponge-dough and direct method

Ingredients	Sponge and Dough method		Direct method
	Sponge	Dough	
Flour (%)	65	35	100
Water (%)	30	1	25
Shortening (%)	6	6	12
Yeast (%)	0.2	0	0.5
Starter (%)	0.1	0	0
Dough conditioner (%)	0.01	0	0.15
Salt (%)	0	1.4	1.4
Liquid sour (%)	0	0	20
Soda (%)	0	0.7	0.7
Malt (%)	1	0	0
Fermentation (h)	18–20	3–4	0.5–2

(Source: Van Wakeren & Popper, 2002; Yoneya & Nip, 2006)

4 Quality Aspects

Quality refers to the degree of acceptance of the finished product among consumers. The quality measures for crackers include color, weight, shape, size, spread ratio, volume index, density and percentage puffiness. The foremost quality parameter in the industry is to measure the weight of the cracker after cooling at room temperature. It denotes the fare availability of the crackers as mentioned on the packaging to the consumers. The dimensional properties of crackers such as length, breadth (for non-spherical crackers), diameter (for spherical crackers) and thickness are important criteria for calculating spread ratio of the crackers. Lower thickness of crackers is generally preferred by the consumers.

Spread Ratio

Spread ratio indicates the suitability of ingredients with finished goods. Spread ratio of crackers deals with the grain fineness, texture, bite and overall mouthfeel (Nagi et al., 2012). High spread ratio is desirable for the crackers (5.0–7.1) as compared to cookies ($\leq$1.0) (AACC, 2000; Hussein et al., 2011).

$$\text{Spread factor} = \frac{\text{Diameter or width} \times \text{Correction factor} \times 10}{\text{Thickness}}$$

Volume Index

Cracker volume can be assessed by using quality indicator called volume index. It is an important parameter that determines the quality of crackers and is affected by the ingredients used. Higher volume index of crackers shows better acceptability among consumers (Qadri et al., 2018).

Density

Density is an important quality parameter in the selectivity of crackers among consumers. Less dense and light weight crackers are largely acceptable by consumers. The density of the developed crackers ranges from 0.651 to 0.718 g/cm^3 (Nammakuna et al., 2009; Qadri et al., 2018).

Percentage Puffiness

Puffiness is a very important quality parameter in crackers as per consumer's point of view. Percentage puffiness in crackers varies from 15 to 60% to its original volume (Qadri et al., 2018).

5 Conclusions and Future Aspects

An increase in the baked products over the shelf has shifted the research areas and public interest towards making them more palatable and healthy so as to improve the nutritional status and acceptability by the consumers. Crackers are nutritionally rich and are consumed by a large diaspora. They are categorised under various categories including fermented and non-fermented which can be further classified as soda, cream and snack crackers. However, the unit operations during manufacturing are similar such as uniform mixing, fermentation (for fermented crackers), dough preparation, lamination, sheeting, baking, cooling and packaging. Flour is the major ingredient since the protein and starch content influence the fermentation process and the texture of the final product. The leavening agent also varies and the crackers may be leavened biologically (yeast) or chemically. Alternative flours apart from wheat may be used individually or as replacement to develop functional crackers. However, elaborate research and planned approach is necessary to comprehend the manufacturing process as well as the imbalance of sensorial properties. The understanding of the ingredients and manufacturing steps, and variations caused by them in the quality of the final product is highly mandatory for the preparation of functional crackers which can be provided for regular consumption especially for the health conscious consumers.

References

AACC. (2000). *Approved methods of the AACC* (10th ed.). American Association of Cereal Chemists.

Ahmad, S., Naz, A., Usman, M., Amjad, A., Pasha, I., & Farooq, U. (2022). Impediment effect of chemical agents (additives) on gluten development in cookie dough. *Journal of Food Science and Technology, 59*(4), 1396–1406.

Ahmed, Z. S., & Abozed, S. S. (2015). Functional and antioxidant properties of novel snack crackers incorporated with Hibiscus sabdariffa by-product. *Journal of Advanced Research, 6*(1), 79–87.

Akonor, P. T., Dziedzoave, N. T., Buckman, E. S., Mireku Essel, E., Lavoe, F., & Tomlins, K. I. (2017). Sensory optimization of crackers developed from high-quality cassava flour, starch, and prawn powder. *Food Science & Nutrition, 5*(3), 564–569.

Ali, A., Shehzad, A., Khan, M. R., Shabbir, M. A., & Amjid, M. R. (2012). Yeast, its types and role in fermentation during bread making process-A. *Pakistan Journal of Food Sciences, 22*(3), 171–179.

Alibašić, H., Junuzović, H., Selimović, A., Selimović, A., & Brčina, Tijana. (2020). Chemical Composition and Sensory Properties of Gluten-Free Crackers with Buckwheat Sourdough. *International Journal for Research in Applied Sciences and Biotechnology, 7*, 4. https://ssrn.com/abstract=3722868

Altıner, D. D., Sabuncu, M., & Şahan, Y. (2021). Chemical and nutritional characteristics of crackers substituted with Cucurbita pepo L. seed flour. *Emirates Journal of Food and Agriculture*. 33. https://doi.org/10.9755/ejfa.2021.v33.i5.2700

Asyikeen, Z. N., Ma'aruf, A. G., Sahilah, A. M., Khan, A. M., & Aida, W. W. (2013). A new source of Saccharomyces cerevisiae as a leavening agent in bread making. *International Food Research Journal, 20*(2), 967.

Batista, A. P., Niccolai, A., Bursic, I., Sousa, I., Raymundo, A., Rodolfi, L., Biondi, N., & Tredici, M. R. (2019). Microalgae as functional ingredients in savory food products: *Application to wheat crackers. Foods, 8*(12). https://doi.org/10.3390/foods8120611

Beltrami, M. C., Döring, T., & De Dea Lindner, J. (2018). Sweeteners and sweet taste enhancers in the food industry. *Food Science and Technology, 38*, 181–187.

Biljwan, M., Naik, B., Sharma, D., Singh, A., & Kumar, V. (2019). Recent development in dough based bakery products: A mini review. *The Pharma Innovation Journal, 8*(5), 654–658.

Borczak, B., Sikora, E., Sikora, M., Kutyła-Kupidura, E. M., & Kapusta-Duch, J. (2018). Nutritional properties of sugar-free wheat-flour cookies. *Zeszyty Problemowe Postepow Nauk Rolniczych, 595*, 21–28.

Caballero, B., Trugo, L. C., & Finglas, P. M. (2003). *Encyclopedia of food sciences and nutrition*. Academic.

Cappelli, A., Lupori, L., & Cini, E. (2021). Baking technology: A systematic review of machines and plants and their effect on final products, including improvement strategies. *Trends in Food Science & Technology, 115*, 275–284.

Cauvain, S. P. (2017). *Baking problems solved*. Woodhead Publishing.

Cauvain, S. P., & Clark, R. H. (2019). *Baking technology and nutrition: Towards a healthier world*. Wiley.

Cauvain, S. P., & Young, L. S. (2009). *Bakery food manufacture and quality: Water control and effects*. Wiley.
Cioroianu, G., Ognean, C. F., & Ognean, M. (2011). The effects of different formulation of baking powder on short biscuits characteristics. Bulletin of University of Agricultural Sciences and veterinary medicine Cluj-Napoca. *Agriculture, 68*(2).
Davidson, I. (2018). *Biscuit, cookie and cracker production: Process, production and packaging equipment*. Academic.
Devi, A., & Khatkar, B. S. (2016). Physicochemical, rheological and functional properties of fats and oils in relation to cookie quality: A review. *Journal of Food Science and Technology, 53*(10), 3633–3641.
Espinoza-Herrera, J., Martínez, L. M., Serna-Saldívar, S. O., & Chuck-Hernández, C. (2021). Methods for the modification and evaluation of cereal proteins for the substitution of wheat gluten in dough systems. *Food, 10*(1), 118.
Faccioli, L. S., Klein, M. P., Borges, G. R., Dalanhol, C. S., Machado, I. C. K., Garavaglia, J., & Dal Bosco, S. M. (2021). Development of crackers with the addition of olive leaf flour (Olea europaea L.): Chemical and sensory characterization. *LWT, 141*, 110848.
Faridi, H. (1990). Application of rheology in the cookie and cracker industry. In *Dough rheology and baked product texture* (pp. 363–384). Springer.
Fort, E. L. (2016). *Effect of reducing agents on batter consistency and physical characteristics of bread from sorghum flour*. (Doctoral dissertation,. Kansas State University.
Gonçalves, L. S., & Salas-Mellado, M. M. (2018). Evaluation of technology and sensory quality of cream cracker enriched with minced cobia (Rachycentroncanadum). *International Food Research Journal, 25*(3), 1197–1203.
Hazen, C. (2013). Cookies and crackers. *Food Product Design, 23*(3), 1–5.
Heredia-Sandoval, N. G., Valencia-Tapia, M. Y., Calderón de la Barca, A. M., & Islas-Rubio, A. R. (2016). Microbial proteases in baked goods: Modification of gluten and effects on immunogenicity and product quality. *Food, 5*(3), 59.
Hidalgo, A., & Brandolini, A. (2014). Bread from wheat flour. In R. K. Robinson (Ed.), *Encyclopedia of food microbiology*. Academic Press.
Hui, Y. H. (Ed.). (2006). *Handbook of food science, technology, and engineering* (Vol. 149). CRC Press.
Hussein, A. M. S., Amal, S., Amany, M. H., Abeer, A. A., & Gamal, H. R. (2011). Physiochemical, sensory and nutritional properties of corn-fenugreek flour composite biscuits. *Australian Journal of Basic and Applied Sciences, 5*(4), 84–95.
Hussien, H., El-Adly, N. A., Shams, O. S., & Mohamed, E. (2020). Production of high nutritional value gluten free crackers with sesame and turmeric powder. *Egyptian Journal of Food Science, 48*(2), 291–302.
Jacob, J., & Leelavathi, K. (2007). Effect of fat-type on cookie dough and cookie quality. *Journal of Food Engineering, 79*(1), 299–305.
Khatkar, B. S. (2013). A manual on soft wheat products and processes. *Post graduate diploma in bakery science and technology (PGDBST-06)*. Guru Jambheshwar University of Science and Technology, Hisar. 125, 3–31.
Kweon, M., Slade, L., & Levine, H. (2011). Development of a benchtop baking method for chemically leavened crackers. I. Identification of a diagnostic formula and procedure. *Cereal Chemistry, 88*(1), 19–24.
Li, J., Hou, G. G., & Chen, Z. (2013a). Whole grain saltine crackers: Formulation, processing, and quality improvements. *Cereals Food World, 58*, 180–185.
Li, J., Hou, G. G., Chen, Z., & Gehring, K. (2013b). Effects of endoxylanases, vital wheat gluten, and gum Arabic on the rheological properties, water mobility, and baking quality of whole-wheat saltine cracker dough. *Journal of Cereal Science, 58*(3), 437–445.
Li, J., Hou, G. G., Chen, Z., Chung, A. L., & Gehring, K. (2014). Studying the effects of whole-wheat flour on the rheological properties and the quality attributes of whole-wheat saltine cracker using SRC, alveograph, rheometer, and NMR technique. *LWT-Food Science and Technology, 55*(1), 43–50.

López-Tenorio, J. A., Rodríguez-Sandoval, E., & Sepúlveda-Valencia, J. U. (2015). The influence of different emulsifiers on the physical and textural characteristics of gluten-free cheese bread. *Journal of Texture Studies, 46*(4), 227–239.

Mamat, H., & Hill, S. E. (2014). Effect of fat types on the structural and textural properties of dough and semi-sweet biscuit. *Journal of Food Science and Technology, 51*(9), 1998–2005.

Mamat, H., & Hill, S. E. (2018). Structural and functional properties of major ingredients of biscuit. *International Food Research Journal, 25*(2).

Manley, D. (2011). Classification of biscuits. In *Manley's Technology of Biscuits, crackers and cookies* (pp. 271–278). Woodhead Publishing.

Meriles, S. P., Penci, M. C., Steffolani, M. E., & Ribotta, P. D. (2021). Effect of heat-treated wheat germ on dough properties and crackers quality. *International Journal of Food Science & Technology, 56*(4), 1837–1843.

Mihiranie, S., Jayasundera, M., & Perera, N. (2021). Development of snack crackers incorporated with defatted coconut flour. *Journal of Microbiology, Biotechnology and Food Sciences, 2021*, 153–159.

Miller, R. (2016). Biscuits, cookies and crackers: Nature of the products.

Mir, S. A., Bosco, S. J. D., Shah, M. A., Santhalakshmy, S., & Mir, M. M. (2017). Effect of apple pomace on quality characteristics of brown rice based cracker. *Journal of the Saudi Society of Agricultural Sciences, 16*(1), 25–32.

Naeem, A., Abbas, T., Hasnain, A., & Ali, T. M. (2013). 19 comparative study of commercially available Baker's yeasts to evaluate their gas production ability. *Journal of Asian Medical Student Association, 2*(S3), 8–10.

Nagi, H. P. S., Kaur, J., Dar, B. N., & Sharma, S. (2012). Effect of storage period and packaging on the shelf life of cereal bran incorporated biscuits. *American Journal of Food Technology, 7*(5), 301–310.

Nammakuna, N., Suwansri, S., Thanasukan, P., & Ratanatriwong, P. (2009). Effects of hydrocolloids on quality of rice crackers made with mixed-flour blend. *Asian Journal of Food and Agro-Industry, 2*(4), 780–787.

Neeharika, B., Suneetha, W. J., Kumari, B. A., & Tejashree, M. (2020). Leavening agents for food industry. *International Journal of Current Microbiology and Applied Sciences, 9*(9), 1812–1817.

Neo, Y. P., Tan, C. H., & Ariffin, A. (2007). Fatty acid composition of five Malaysian biscuits (cream crackers) with special reference to trans-fatty acids. *ASEAN Food Journal, 14*(3), 197.

Nguyen, T. T., Le, T. Q., Songsermpong, S., Le, T. T., & Truong, K. T. P. (2014). Effects of baking power concentrations on the texture and sensory evaluation of shrimp cassava cracker-contained oil puffed by microwave technique. In *The 16th food innovation Asia conference* (pp. 12–13).

Otero-Guzmán, N. C., Rodríguez-Sandoval, E., & Tabares-Londoño, J. A. (2020). Influence of different types of baking powder on quality properties of muffins. *Dyna, 87*(214), 9–16.

Panghal, A., Navnidhi, S., & Khatkar, B. S. (2011). Effect of enzymes on cookies quality. *Annals of Agri-Bio Research, 16*(1), 75–78.

Parapouli, M., Vasileiadis, A., Afendra, A. S., & Hatziloukas, E. (2020). Saccharomyces cerevisiae and its industrial applications. *AIMS Microbiology, 6*(1), 1.

Polat, H., Capar, T. D., Inanir, C., Ekici, L., & Yalcin, H. (2020). Formulation of functional crackers enriched with germinated lentil extract: A response surface methodology box-Behnken design. *LWT, 123*, 109065.

Qadri, T., Hussain, S. Z., Rather, A. H., Amin, T., & Naseer, B. (2018). Nutritional and storage stability of wheat-based crackers incorporated with brown rice flour and carboxymethyl cellulose (CMC). *International Journal of Food Properties, 21*(1), 1117–1128.

Raveendran, S., Parameswaran, B., Ummalyma, S. B., Abraham, A., Mathew, A. K., Madhavan, A., et al. (2018). Applications of microbial enzymes in food industry. *Food Technology and Biotechnology, 56*(1), 16.

Reibner, A. M., Struck, S., Alba, K., Proserpio, C., Foschino, R., Turner, C., et al. (2021). Cross-national differences in consumer responses to savoury crackers containing blackcurrant pomace. *International Journal of Food Science & Technology, 56*(10), 5007–5016.
Romano, A., Masi, P., Bracciale, A., Aiello, A., Nicolai, M. A., & Ferranti, P. (2018). Effect of added enzymes and quinoa flour on dough characteristics and sensory quality of a gluten-free bakery product. *European Food Research and Technology, 244*(9), 1595–1604.
Saberi, F., Kouhsari, F., Abbasi, S., Rosell, C. M., & Amini, M. (2021). Effect of baking in different ovens on the quality and structural characteristics of saltine crackers. *International Journal of Food Science & Technology, 56*(12), 6559–6571.
Smith, J. P., & Simpson, B. K. (1995). Modified atmosphere packaging of bakery and pasta products. In J. M. Farber & K. L. Dodds (Eds.), *Principles of modified atmosphere and sous vide product packaging* (pp. 207–242). Academic.
Smith, J. P., Daifas, D. P., El-Khoury, W., Koukoutsis, J., & El-Khoury, A. (2004). Shelf life and safety concerns of bakery products—A review. *Critical Reviews in Food Science and Nutrition, 44*(1), 19–55.
Srikanlaya, C., Therdthai, N., Ritthiruangdej, P., & Zhou, W. (2017). Effect of butter content and baking condition on characteristics of the gluten-free dough and bread. *International Journal of Food Science & Technology, 52*(8), 1904–1913.
Sykes, G., & Davidson, I. (2020). *Biscuit, cookie and cracker process and recipes*. Academic.
Van Wakeren, J., & Popper, L. (2002). Enzymes in soda cracker production. *Cereal Foods World, 47*(8), 382.
Venturi, M., Galli, V., Pini, N., Guerrini, S., Sodi, C., & Granchi, L. (2021). Influence of different leavening agents on technological and nutritional characteristics of whole grain breads obtained from ancient and modern flour varieties. *European Food Research and Technology, 247*(7), 1701–1710.
Vetter, J. L. (2003). Leavening agents. In B. Caballero, L. C. Trugo, & P. M. Finglas (Eds.), *Encyclopedia of food sciences and nutrition*. Academic.
Wieser, H., Koehler, P., & Scherf, K. A. (2020). *Wheat-an exceptional crop: Botanical features, chemistry, utilization, nutritional and health aspects*. Elsevier.
Xu, J., Zhang, Y., Wang, W., & Li, Y. (2020). Advanced properties of gluten-free cookies, cakes, and crackers: A review. *Trends in Food Science & Technology, 103*, 200–213.
Yoneya, T., & Nip, W. K. (2006). Cracker manufacture. In F. Yuksel & A. Kayacier (Eds.), *Bakery products: science and technology* (Vol. 411). Blackwell Publishing.
Zakupszki, Z. B., Véha, A., & Szabó, B. P. (2018). Development of matzo made of wholemeal rye flour and its implementation in the industry. *Analecta Technica Szegedinensia, 12*(1), 8–14.
Zhou, W., Therdthai, N., & Hui, Y. H. (2014). Introduction to baking and bakery products. In W. Zhou, Y. H. Hui, I. De Leyn, et al. (Eds.), *Bakery products science and technology* (pp. 1–16). Wiley.

Chapter 8
Popped and Puffed Cereal Products

Sabah Mounir, Atef Ghandour, Eman Farid, and Adel Shatta

1 Introduction

Nowadays, snack foods represent an important sector in the food industry and are produced by popping, puffing, Instant controlled pressure drop (DIC), and extrusion-cooking. The design of snack foods is a complicated operation to satisfy the different tastes of consumers. Ready-to-eat cereal snacks have great popularity because they are easy to digest, light, and crispy. The increased demand for these snacks motivates food scientists and technologists to produce new, distinctive, and unique products. Food snacks include a huge number of products, such as puffed or popped products, which are usually produced from whole grains of some cereals like rice, corn, wheat, oat, and barley (Hoke et al., 2007). Due to the presence of bran layer, which contains antioxidants, proteins, minerals, and fiber (Erba et al., 2011; Hidalgo & Brandolini, 2008, 2014), popped and puffed cereals are considered to be nutritive and healthy products. Popping and puffing of cereals have been used for hundreds of years as an easy, cheap, and rapid conventional process of dry heating for the production of weaning foods and ready-to-eat expanded products. Most snack producers apply some of the existing techniques for designing expanded cereals and incorporating differences to improve the health image of the final products (Mishra et al., 2014). Traditional snacks, including popcorn, popped and puffed rice, popped sorghum, popped and roasted wheat, expanded finger millet, and popped millets, are widely available worldwide (Jaybhaye et al., 2014).

S. Mounir (✉) · E. Farid
Department of Food Science, Faculty of Agriculture, Zagazig University, Zagazig, Egypt
e-mail: smounir@zu.edu.eg

A. Ghandour
Agricultural Engineering Research Institute, Agriculture Research Center (ARC), Giza, Egypt

A. Shatta
Food Technology Department, Faculty of Agriculture, Suez Canal University, Ismailia, Egypt

M. A. Shah et al. (eds.), *Cereal-Based Food Products*,
https://doi.org/10.1007/978-3-031-40308-8_8

2 Cereals Suitable for Popping and Puffing

Cereals are the main source of carbohydrates, providing the energy necessary for different activities. Some of these cereals are rich in protein, fiber, minerals, and vitamins (Oghbaei & Prakash, 2016). Rice, wheat, and corn are the main cereals of human nutrition, accounting for 85, 72, and 19%, respectively, of human consumption (Maclean et al., 2002).

Maize (corn) is used to produce the most common popcorn (Gundboudi, 2006), which is commercialized today all over the world. Besides, wheat grains are used in the manufacturing of puffed and roasted wheat, which are also commercialized by manufacturing industries (Murthy et al., 2008). Rice grains are used to produce popped or puffed rice, which is consumed in Southeast Asian countries. In India and some African countries, crispy popped sorghum with increased digestibility has been obtained from sorghum grains (Saravanabavan et al., 2013). Millet grains are used to produce an expanded finger millet (Ushakumari et al., 2007), which is considered a new-generation snack product commercialized in India (Sharma et al., 2014; Ushakumari et al., 2007). Other grains may also be used to produce expanded products such as oat, barley, rye, and pseudo-cereals like amaranth (Inoue et al., 2009), buckwheat, and quinoa (Haldkar et al., 2022; Zapana et al., 2020; Mandhare et al., 2020; Subramani et al., 2020a, b). Some chemical and physical characteristics must be taken into consideration while selecting grains to be popped or puffed, such as grain hardness, degree of husk interlocking, and amylose content. These characteristics will be discussed later in the section dealing with the factors affecting the quality of the final product.

3 Principles of Popping and Puffing

Both popping and puffing of cereal grains are based on the cooking of the grains at high temperatures by different heating means, generating internal superheated steam within the grains. During cooking, the matrix of the grains undergoes a phase transition from an amorphous to a rubbery state, wherein the internal superheated steam creates and develops locally high pressure inside the molten matrix. In a rubbery state, the molten matrix has a viscoelastic character, which allows this steam to expand within the matrix, expanding the original structure of the grain or creating a new internal structure, producing crispy products with an open porous structure. Once the superheated steam pressure is released from the expanded grain, a rapid cooling of the matrix takes place, and the matrix returns again to an amorphous state, which allows for maintaining the obtained expanded structure.

Popping Process

Popping is defined as a process of starch cooking using dry heat; moisture-conditioned grains are subjected to a high temperature for a short time (Mishra et al., 2014), expanding the endosperm of grains. Starch is cooked by superheated steam generated by the instant heating of the kernel's moisture. Afterward, a locally high pressure is created within the grain matrix as a result of superheated steam expansion, leading to pericarp rupture and thus forcing out the superheated steam from the grain at the popping time.

Puffing Process

Puffing of cereal grains is defined as a high temperature-short time (HTST) treatment of dehulled parboiled grains. During the parboiling step, the cracks in the grain are sealed because the crystalline starch molecules become amorphous (Gulati & Datta, 2016), hardening the grain's outer layer, which acts afterward as a small pressure chamber, allowing moisture to entrap within the grain matrix (Hoke et al., 2005; Joshi et al., 2014a). As a result of intense heating, the kernel becomes extensible due to the glass transition, which allows the superheated steam to expand within the grain matrix. While releasing the superheated steam, either the original structure of the grain is expanded or a new internal structure is created (Gulati & Datta, 2016).

4 Popping/Puffing Methods

There are different methods to pop and puff the cereal grains, including some traditional and innovative methods. These methods are classified into: (1) high temperature techniques at normal atmospheric pressure, such as hot oil, hot sand or salt, and hot air, (2) high pressure techniques followed by pressure drop, like gun/explosion puffing or popping, extrusion-cooking, and instant controlled pressure drop, and (3) rapid heating techniques using dielectric heating, such as microwave. Recently, infrared puffing has been applied as an innovative technique to produce puffed rice and popcorn. Figure 8.1 summarizes the different methods for popping and puffing cereals.

High-Temperature Techniques at Atmospheric Pressure

Hot Oil

Hot oil is an ancient and common technique used to produce popcorn (Swarnakar et al., 2022). This method is easy, even for less experienced cooks. Expanded products are obtained by heating vegetable oil (which acts as a heating medium) at

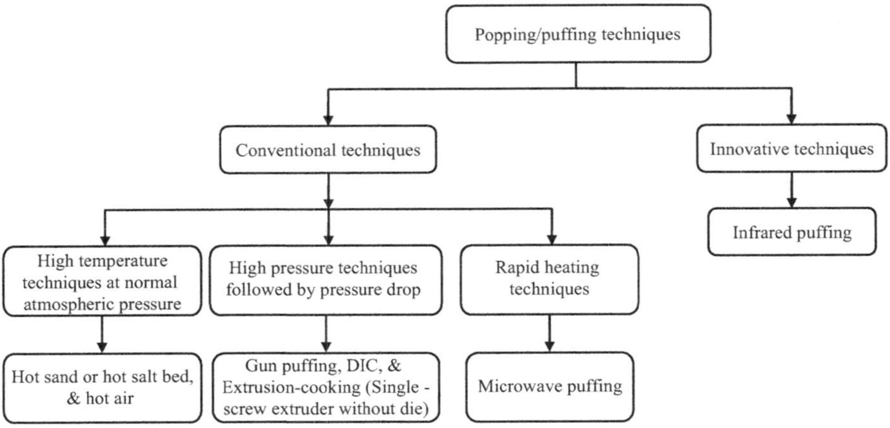

Fig. 8.1 Conventional and innovative popping and puffing techniques

temperatures ranging between 200 and 220 °C for a few seconds. For instance, coconut oil has been used as a heating mediumn and the puffing is performed at 210–220 °C for 4–8 s. As a result of heating, the outer layer of grain is rapidly heated to the temperature of the oil, and the water in the grains gets thus heated and forced out into the surrounding oil in the form of steam (Balaji, 2019). Although the quality of the final product is improved due to the coating with oil, the main disadvantage of this process is the oxidation of the fatty acids constituting the oil during the storage period (Mishra et al., 2014; Joshi et al., 2014a); therefore, the expanded products by this method have a shorter shelf life compared to the other techniques. This method can be used for popping and puffing different cereal grains, such as corn and parboiled rice.

Hot Sand Bed

Popping or puffing using a hot sand bed is a popular technique in Southeast Asia for producing expanded cereal grains like rice, paddy, corn, wheat, sorghum, millet, rye, and amaranth. Briefly, pre-conditioned grains are agitated with the bed of hot sand at temperatures of 220–260 °C for 9–15 s, depending on the grain type (Dutta et al., 2015). The energy required for sand heating is provided by burning wood or coal. Once the required sand temperature is attained, the grains are placed in the vessel containing the sand bed and vigorously agitated with a stirrer. In this method, a uniform heating of grains is ensured by the hot sand as well as the agitation of grains with the hot sand because sand can retain heat, which is transferred to the grains afterward through the agitation. As a result, internal superheated steam is

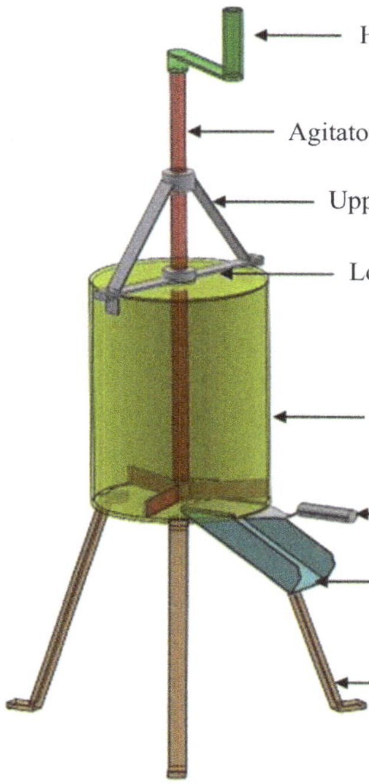

Fig. 8.2 Schematic diagram of manually operated hot sand bed popping/puffing machine. (From Swarnakar et al., 2022)

generated within the grains and forced out of the grain while expanding the kernel. Subsequently, the expanded grains are rapidly removed from the sand bed to avoid burning (Swarnakar et al., 2022). During the puffing, some sand particles may adhere to the expanded grains, which can be removed by sieving the expanded grains (Kora, 2019) and returned again to the sand bed, and the process is repeated. Figure 8.2 shows a schematic diagram of a manual machine for rice puffing with a capacity of 5–6 kg/hr. This machine is mainly composed of a steel vessel with a concave bottom to hold the hot sand and an agitator of three blades in a hollow pipe fixed in the steel vessel, which rotates with the help of a handle. Although this method is easy, it has some limitations, such as: (1) batch operation (Mohapatra et al., 2012), (2) adhesion of some sand particles to the puffed grains, (3) higher storage and transportation costs due to the lower bulk density of puffed grains, and (4) the susceptibility of grains to moisture absorption during the storage period due to their open porous structure; therefore, such these products require proper and effective packaging to avoid moisture absorption and to maintain their crispness.

Hot Air

Contrary to hot oil and hot sand puffing methods that are based on conductive heat transfer, hot air-puffing is mainly based on convective heat transfer and requires a slight pressure gradient. Fluidized bed puffing is a common method of hot air puffing in which grains transform into a fluid-like state as the fluidization of grains ensures uniform heating by hot air. Besides, the surface area of fluidized grains increases during the fluidization, which in turn enhances both heat and mass transfers (Tavanandi et al., 2021); therefore, this technique is more energy-efficient compared to hot oil and hot sand techniques. The crucial parameters affecting the quality of hot air-puffed grains are air temperature, air velocity, and processing time. The air temperature varies from 250 to 300 °C (Zapotoczny et al., 2006), the air velocity has to be kept between the fluidization velocity and the terminal velocity of grains (Das, 2005), and the processing time is mainly dependent upon the air temperature and mode of heat transfer (Swarnakar et al., 2022). Besides the mentioned parameters, the quality of hot air-puffed grains is also affected by the initial moisture content of the grain as well as the relative humidity of the air (Konishi et al., 2004). All these parameters can be optimized to enhance the different quality attributes of hot air-puffed grains (Zapotoczny et al., 2006). The hot air method has been effectively applied for the popping of several cereals, such as wheat (Solís-Morales et al., 2009), corn (Byrd & Perona, 2005), and amaranth (Solanki et al., 2018; Inoue et al., 2009; Markowski et al., 2006; Iyota et al., 2005).

A continuous fluidized bed machine with a capacity of 150 kg/h of wheat has been developed in India by Murthy et al. (2008), using liquefied petroleum gas (LPG) as a fuel. This machine can also be used to puff maize, sorghum, paddy, and rice at temperatures ranging from 280 to 350 °C with a minimum fluidization velocity of 2–3 m s^{-1}. Likewise, Tavanandi et al. (2021) designed and developed another machine for continuous popping and puffing of cereals. This machine mainly consists of (1) a combustion chamber fixed on a centrifugal blower, (2) a screw feeder to introduce the grains into the machine, where the feeding rate is controlled by its rotation, and (3) a wired mesh inside the combustion chamber to hold the grains as well as help in the air distribution, ensuring a good fluidization. During the process, the hot air is generated by burning the LPG at 750–800 °C which is then mixed with the air current generated by the centrifugal blower at adequate ratios to attain a temperature of 250–300 °C in the combustion chamber. Depending on the type of grain, the air temperature can be adjusted by controlling the supply of LPG. Cereals like maize, paddy, rice, and jowar have been successfully puffed at temperatures in the range of 180–250 °C with a fluidization velocity of 2 to 7 m s^{-1}.

High-Pressure Techniques Followed by Pressure Drop

Gun Puffing

Gun puffing is also known as explosion puffing and is mostly applied to puff various cereals such as rice, millet, wheat, barley, and quinoa (Hidalgo et al., 2016; Zapana et al., 2020; Hoke et al., 2007). This technique is based on the heating of grains under pressure for a short time, followed by an instant release of pressure to atmospheric pressure. The temperature ranges from 200 to 210 °C, and the processing time is in the range of 3–7 min (Saha & Roy, 2020). The expansion of grains takes place while moisture is flashing off from the grains in the form of superheated steam due to the release of pressure towards the atmospheric pressure (from the high pressure inside the chamber to the atmospheric pressure). The pressure of internal superheated steam and processing time are crucial factors for achieving the expected texture of the final product because very low pressure may produce expanded grains with low crispness and very high steam pressure may shatter and burst the grain (Mishra et al., 2014). A sufficient processing time helps the internal superheated steam cook the grains, where the grain's state is changes from an amorphous state to a rubbery state. In a rubbery state, the grain has viscoelastic characteristics that enable the grain structure to expand, producing crispy puffed grains.

Instant Controlled Pressure Drop

Instant controlled pressure drop (Détente Instantanée Contrôlée DIC) is defined as a high temperature-short time (HTST) treatment followed by an instant pressure drop towards a vacuum (Mounir et al., 2012). Therefore, the DIC process involves two main stages, where the first stage is a thermal treatment performed with saturated steam having a pressure ranging between 0.1 and 0.6 MPa, corresponding to a temperature ranging from 100 to 160 °C, for a short time varying from 5 to 60 s. The second stage is an instant drop in pressure towards a vacuum (4–5 KPa) with a rate greater than 0.5 MPa/s, which is maintained for 5–20 s. During the first stage, heat is transferred to the product from the saturated steam, and the product is thus heated to produce internal superheated steam within it. The superheated steam expands inside the products, inducing some mechanical stresses that rupture the cell walls and create some pores and vacuoles within the product. During the second stage, the instant dropping of pressure towards a vacuum induces the autovaporization of a part of the product's moisture, which allows a rapid cooling of the product, thus maintaining the final structure.

A DIC reactor generally consists of (1) a stainless steel processing chamber, (2) a vacuum tank havinga volume 100 times greater than that of the processing chamber, and (3) an instant valve that assures a perfect connection between the

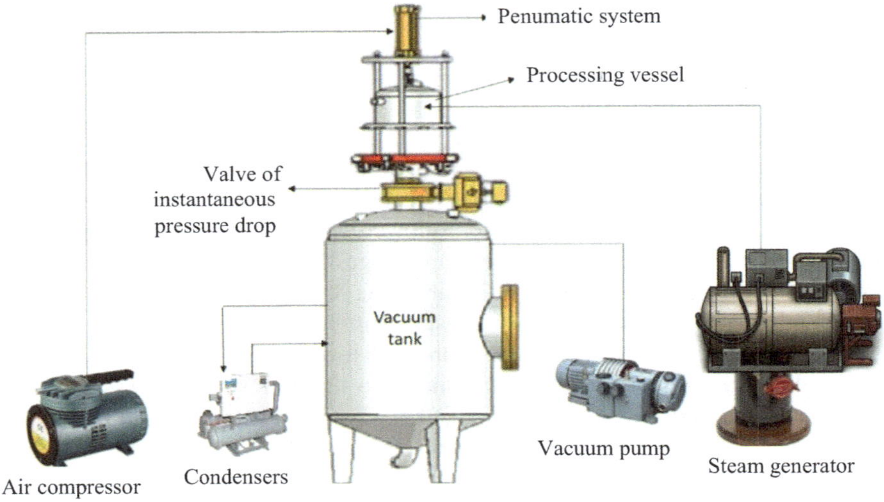

Fig. 8.3 Schematic diagram of lab-scale DIC reactor. (From Mounir et al., 2012)

processing chamber and the vacuum tank in less than 0.1 s (Mounir et al., 2012). Figure 8.3 represents the main parts of lab-scale DIC equipment.

Instant controlled pressure dropwas firstly applied for the puffing of fruits, vegetables, meat, shrimp, and cheese (Mounir et al., 2011a, b; Mounir, 2015; Mounir et al., 2020a, b, 2021a), and the roasting of coffee beans and carob pods (Kamal et al., 2008; Mounir et al., 2021b). Then, DIC has been used for cereal steaming, such as rice, to improve the quality attributes of the final product (Pilatowski et al., 2010; Thai & Allaf, 2014). The crucial parameters affecting the expansion ratio of DIC-puffed products are the pressure of saturated steam, processing time, and the initial moisture content of the raw material. Yağci and Evci (2015) studied the effect of DIC-puffing on some quality attributes of wheat snacks obtained from moisture-conditioned wheat grains. The authors found that the effect of the processing time and pressure of saturated steam on the physicochemical and functional characteristics of wheat snacks was more pronounced compared to the moisture content. Besides, the final product had improved functional and nutritional characteristics.

Extrusion Puffing

Extrusion puffing is used in the production of puffed chips or snacks from cereals with high starch content, such as rice, wheat, and corn (Paznocht et al., 2021). Standard extrusion-cooking involves combining several unit operations such as conveying, mixing, heating, kneading, shearing, homogenization, melting, sterilization, and drying. These unit operations cause some physical and chemical changes in raw materials in order to create different high temperature and pressure zones inside the extruder. These changes have been reported and discussed in many studies (Huang et al., 2022; Ramos Diaz et al., 2016; Mościcki, 2011). During extrusion-cooking, the moisture of the melt is vaporized in the form of steam when the superheated steam is released at atmospheric pressure, creating a porous structure and thus expanding the final extrudate (Li et al., 2019). The chemical composition, especially the amylose content, determines the extent of expansion; Vanier et al. (2016) reported that grains with higher amylose content showed a lower expansion ratio and higher hardness compared to grains with lower amylose.

Tikekar and Karwe (2009) developed a continuous method for puffing amaranth seeds using a single-screw extruder with a tapered screw without a die. The authors found that puffing efficiency depended on barrel temperature and screw speed. About 86% of efficiency was attained under operating conditions of 290 °C and 125 rpm for barrel temperature and screw speed, respectively. While puffed seeds at 250 °C and 75 rpm had the highest sensory characteristics compared to other samples.

Dielectric Heating

Microwave Puffing

Microwave heating has many applications in food processing, such as blanching, cooking, thawing, baking, drying, and decontamination (Chandrasekaran et al., 2013). It can also be used for producing popped or puffed cereals such as corn, sorghum, paddy, and rice (Joshi et al., 2014a; Maisont & Narkrugsa, 2010; Mishra et al., 2015; Mom et al., 2020; Swarnakar et al., 2020). The microwave provides a volumetric heating, and the grains get heated by the vibrational energy. During microwave puffing, the internal superheated steam is generated from the grain moisture, which in turn cooks the grains. Cooked grains undergo a phase transition from an amorphous to a rubbery state, which allows the internal superheated steam to create and develop local high pressure within the grain matrix. Grain expansion takes place when the internal superheated steam vigorously escapes from the grains (Dash & Das, 2019a). After releasing the pressure and stopping microwave heating, the grains cool down to maintain the obtained structure, and the grain matrix returns again to the amorphous state. The moisture content of the grains plays an important role in the expansion of microwave-popped/puffed grains, where neither too low nor too high moisture content allows achieving the expected expansion (Boischot et al., 2003). Therefore, the grains should have an appropriate moisture content to achieve the expected expansion. Microwave parameters that affect the quality of puffed grains include power level, power density, and processing time (Joshi et al., 2014a, b; Swarnakar et al., 2014; Maisont & Narkrugsa, 2010; Hoke et al., 2005). Moreover, different coatings can be applied to the grains, such as NaCl, $NaHCO_3$, vegetable oil, and butter, to attain the maximum amount of popped grains. According to Maisont and Narkrugsa (2010), the high volumetric expansion of microwave-puffed rice and high puffed yield could be achieved by using soaked paddy rice in a 2% salt solution with 13% moisture content and puffing at 700–800 W microwave power.

Innovative Techniques

Infrared Puffing

Infrared radiation has several applications in food processing, such as heating, drying, cooking, roasting, baking, and microbiological decontamination (Aboud et al., 2019; Shavandi et al., 2020). The first study dealing with infrared puffing has been carried out by Shavandi et al. (2022a) to produce infrared-popcorn, followed by another study on white dehulled rice (Shavandi et al., 2022b), using a batch pilot (Fig. 8.4). This machine consists of a stainless steel chamber with dimensions of 65 × 62 × 100 cm, a power supply, and two infrared lamps with a length of 35 cm (1000 Watts). Different processing parameters were studied, such as the infrared power (350, 450, and 550 Watts) and sample distance (10, 20, and 30 cm). The authors reported that the volumetric expansion of rice grains was higher in infrared-puffing compared to traditional puffing. Moreover, the volumetric expansion significantly increased, while the bulk density decreased as the infrared power increased. On the contrary, the sample distance showed a contradictory effect on these characteristics.

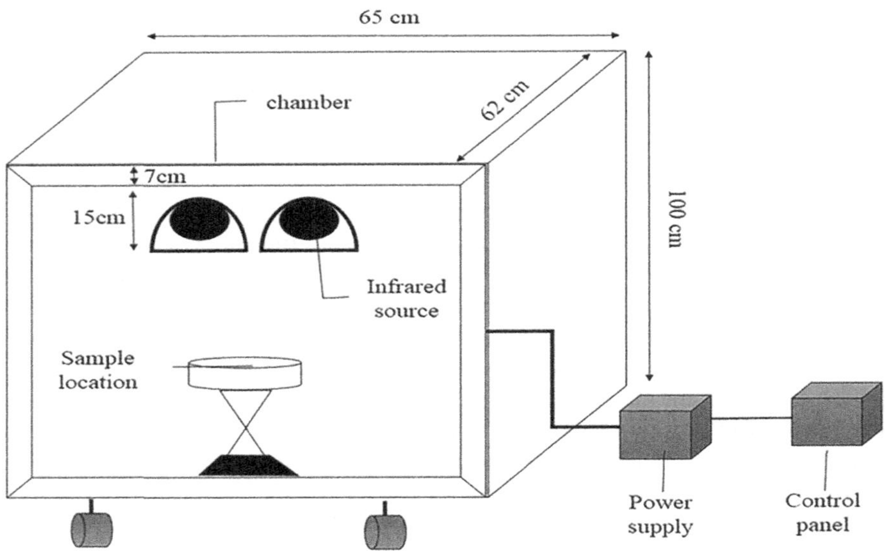

Fig. 8.4 Schematic diagram of infrared puffing. (From Shavandi et al., 2022a, b)

5 Quality Attributes of Popped and Puffed Cereal Grains

Physical Properties

Popping and puffing induce volume expansion within the grain, which increases the original dimensions of the grain, including length, width, and thickness. Besides, a decrease in the mass of grains is also observed after popping/puffing due to the loss of moisture, which is vaporized with the release of superheated steam. These changes cause alterations in some physical properties, such as bulk density, true density, and porosity. The bulk density decreases, while the porosity increases as the volumetric expansion increases (Joshi et al., 2014a, b). On the other hand, these dimensional modifications also cause changes in the sphericity, geometric mean diameter, aspect ratio, and slenderness ratio of grain (Saha & Roy, 2022). For example, popcorn may be obtained in different shapes, such as mushroom-shape or butterfly-shape (Fig. 8.5). Butterfly-shaped popcorn is characterized by having a multilateral expansion and is preferred for home consumption. While mushroom-shaped popcorn is commercially preferred because it is less susceptible to breakage during packaging and transportation than the butterfly-shaped popcorn (Sweley et al., 2013). In popping, the volumetric expansion ranges between 6 and 8 times (Bhatupadya et al., 2008), while it ranges from 4 to 10 times in the puffing process (Saha & Roy, 2022). The dimensional changes have been reported in several studies. For instance, Mir et al. (2016a, b) observed a considerable decrease in the length/breadth ratio after puffing the brown rice.

Nutritive Quality

Heat treatments, including parboiling, popping, and puffing, cause some changes in the nutritional profile of grains with respect to carbohydrates, protein, lipids, and bioactive compounds. Starch loses its native structure and the related crystallinity

Fig. 8.5 Different shapes of popcorn, right: mushroom-shaped popcorn;left: butterfly-shaped popcorn. (From Sweley et al., 2013)

(Briffaz et al., 2012) and may be partially degraded into small molecular components, increasing the sugar content in puffed/popped grains (Jia et al., 2021). It has been reported that the puffing resulted in the breakingdown of intermolecular bonds in starch, such as hydrogen and glycoside bonds (Huang et al., 2018). Additionally, protein content and amino acid profile differ considerably during the rice parboiling (especially in cases of excessive soaking) or during the rice milling, and that may be due to the loss of amino acids found in the bran layer. Moreover, heating during the steaming process induces protein denaturation and hydrolysis (Bhattacharya, 2011), inducing the formation of disulfide bonds that enhance the grain hardness (Derycke et al., 2005). Concerning the fat content, a loss of fat is also observed in puffed/popped grains, and that may be due to 1) the liberation of oil from the lipid granule, which migrates to the bran layer and is lost during the milling operation (Oli et al., 2014; Siriamornpun et al., 2008), and 2) the formation of a starch-lipid complex (Huang et al., 2018). On the contrary, Gugal and Yenagi (2016) found that the proximate chemical composition of sorghum increasedduring popping. Similarly, Dhadke et al. (2022) reported that the contents of protein, crud fiber, and mineral increased after sorghum popping without changing the fat content. While Anjitha et al. (2021) observed that the chemical composition of sorghum was preserved after popping using a developed infrared – assisted hot air-popping machine. A decrease in the total phenolic content (TPC) was also observed after puffing (Mir et al., 2016a; Sharma & Gujral, 2011; Yağci & Evci, 2015), which is attributed to the degradation of phenolic compounds as a result of oxidative and thermal effects during puffing or popping (Randhir et al., 2008). Mir et al. (2016a) also found that the antioxidant activity was increased during parboiling, while decreased in puffed brown rice. This behavior may be related to the Millard reaction that occurs in the parboiling process, resulting in the formation of brown pigments with antioxidant properties (Randhir et al., 2008). Fat-soluble and water-soluble vitamins may also be lost during parboiling as a result of oil liberation and excessive soaking, respectively. Heating and pressure applied during popping and puffing also have significant adverse effects on vitamins because they are heat-sensitive by nature. To improve the nutritive quality of popped and puffed products, some procedures have been suggested, such asfortifying with essential vitamins after popping and puffing orusing whole grains for popping and puffing (Patel et al., 2021; Mir et al., 2016b).

Color Characteristics

Color change can be visually observedafter the popping or puffing of grains, where the extent of this change varies according to the expansion degree, grain type, and the conditions applied during popping or puffing. For instance, in rice parboiling, the Maillard reaction (non-enzymatic browning) and the migration of husk pigments to the endosperm are the main factors affecting the grain color (Mir et al., 2016a, b), where the grain color is changed from white to a darker color, as a result of the formation of brownish pigments. The parboiling temperature and soaking time directly affect the grain color. Soaking brown rice at about 65 °C for 5 h caused an increase in L^*value and a reduction in a^* and b^* values, while this behavior was less obvious when the parboiling was performed at 45 °C, even for 5 h. It may be due to leaching more grain components since their solubility increases at higher temperatures (Lamberts et al., 2006). But after puffing, this color becomes lighter due to the expansion of the grain as well as the exposure of the inner white endosperm, increasing the grain whiteness (Mir et al., 2016a, b). However, severe conditions applied during popping or puffing cause scorching of the final grains, negatively affecting their color. All available literature reported an increase in the color lightness and a decrease in a^* and b^* values after popping or puffing. The decrease in a^* and b^* values may be related to the breakdown of Maillard reaction products and grain pigments. Devi and Das (2017) investigated the impact of moisture content and salt coating on the color change of microwave-popped rice and found that a^* and b^* values increased, while L^* decreased with an increase in the salt level up to 5% (w/w), compared to popped rice without salt. But moisture content showed a contradictory effect on these color parameters.

Microstructural Modifications

Cereal grain consists of the pericarp layer (bran layer), endosperm, and embryo (germ) in a compact and uniform structure (Mariotti et al., 2006). Grain expansion induces some microstructural changes, including (1) the separation of the pericarp layer, (2) changes in the structure of the endosperm, and (3) changes in the native structure of starch and protein. The nature and degree of these microstructural modifications vary according to the type of grains and their chemical composition, the technique used for popping or puffing, and the expansion level (Mariotti et al., 2006). Starch gelatinization plays an important role in the microstructural changes, wherein a very soft molten matrix causes structure collapse during popping or puffing due to the lack of mechanical resistance to withstand the superheated steam pressure (Altan, 2014; Boischot et al., 2003). Popped and puffed grains have a honeycomb-like structure as a result of expanding their structure. Such this open porous structure results from some mechanical stresses created by the expansion of the internal superheated steam within the grain matrix (Mounir et al., 2021a). According to Mir et al. (2015), the compact internal structure of raw grain is transformed to an open porous structure with several vacuoles and pores of different sizes (Fig. 8.6). Furthermore, Tikekar and Karwe (2009) found that the extrusion-puffed amaranth seeds at 290 °C and 125 rpm were thin with a porous structure due to the formation of honeycomb-like structures as a result of the release of the superheated steam (Fig. 8.7). The microstructural modifications may also affect the sensorial characteristics of popped or puffed grains like crispness, brittleness, hardness, and surface smoothness (Roopa et al., 2009).

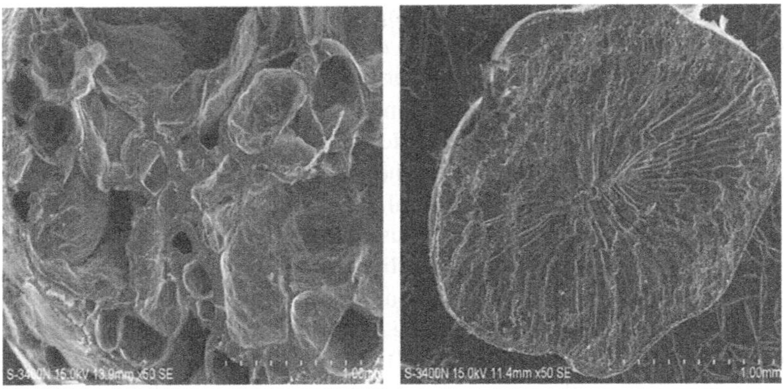

Fig. 8.6 Scanning electron microscopy of rice grains, right: raw brown rice; left: popped brown rice. (From Mir et al., 2015)

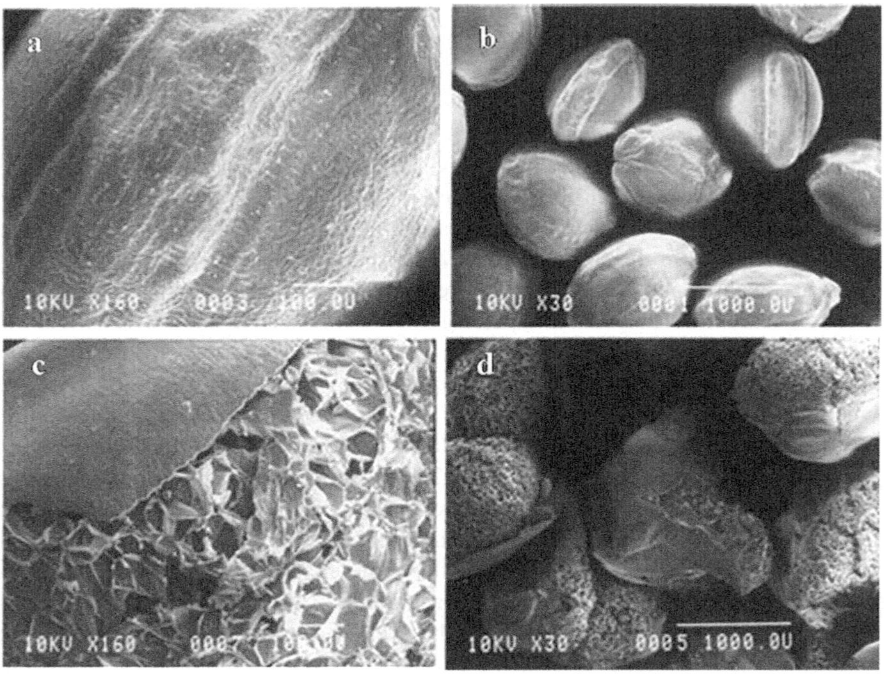

Fig. 8.7 Scanning electron microscope of amaranth seeds, (**a**) raw seeds (unpuffed) at 160×; (**b**) raw seeds at 30×; (**c**) puffed seeds at 160×; and (**d**) puffed seeds at 30× magnification. (From Tikekar & Karwe, 2009)

Sensorial Properties

As previously mentioned, the intense heat treatments induce some macromolecular transformations that are responsible for varying the sensorial characteristics of popped or puffed products (Schlörmann et al., 2020; Sweley et al., 2013). These transformations substantially affect the color, taste, flavor, and mouthfeel of popped or puffed products (Mishra et al., 2014). But the degree of variation in the sensorial characteristics depends on the processing conditions. For instance, the exposure to intensive heat during popping and puffing causes grain charring and discoloration, which in turn alter the color and taste of the final products (Swarnakar et al., 2019). On the other hand, insufficient temperature or inadequate processing time may not allow obtaining the expected expansion, reducing the crispnessof the final products. Mouthfeel and texture characteristics (crispness or hardness) are directly related to the expansion of grains, which is influenced by the initial moisture content of the grain prior to popping or puffing. The effect of moisture content on the expansion of popped or puffed grains will be discussed in the next section. Additives such as salt and oil can also improve the sensorial characteristics (taste, flavor, and texture) of popped or puffed products (Mishra et al., 2015). Salt is not necessary for the popping process, but it affects the color and texture of popped products (Devi & Das, 2017; Maisont & Narkrugsa, 2010).

6 Factors Affecting the Quality of Popped/Puffed Products

Chemical Composition of Grains

Moisture Content

The initial moisture content of grain is a crucial parameter in the popping and puffing processes, as it affects the quality of popped or puffed products, especially the expansion extent. Adequate initial moisture content allows for the generation of a sufficient amount of internal superheated steam within the grain matrix, which in turn develops the pressure necessary to achieve the expected expansion (Swarnakar et al., 2014). In this context, very low moisture content may not allow for creating sufficient internal pressure, which is necessary for the expansion, leading to poor grain expansion (Dutta et al., 2015; Joshi et al., 2014a). Besides, grains with very low moisture dry much more quickly during the popping or puffing process; therefore, the grain matrix becomes unable to be converted to the rubbery state that is necessary for the internal superheated steam expansion within the molten matrix. While too a high moisture content decreases the glass transition temperature (T_g) of starch (plasticization effect: absence of the viscoelastic character), in other words, the grain always remains in a rubbery state even after releasing the internal superheated steam and becomes unable to revert again to the amorphous state. In both cases, the expected expansion and, thus, the crispness of the grains could not be achieved (Gulati & Datta, 2016). The optimum moisture varies according to the grain type or variety and the popping/puffing method, but it has been reported that a moisture content of 3–4% (wb) or below is optimum for producing crispy expanded products (Mazumder et al., 2007).

Starch and Gelatinization

The starch content in the grains may either increase or decrease the expansion of popped or puffed products (Mahanta & Bhattacharya, 2010). Besides, the physicochemical properties of starch also have a significant effect on the expected expansion during popping or puffing because the expansion extent is directly related to the degree of starch gelatinization. Amylose and amylopectin are the main components of starch granules. The expansion extent depends on the relative proportions of amylose and amylopectin, which vary according to the type of grain. Amylose content in rice grains determines the suitable variety for puffing. For instance, high-amylose grains showed better expansion than low-amylose grains (Mishra et al., 2015; Kamaraddi & Prakash, 2015), owing to the fact that low-amylose grains may be susceptible to starch degradation and thus amylose leaching from the starch granule (Mahanta & Bhattacharya, 2010). The formation of a more firm gel during parboiling of high-amylose grains is another supporting explanation (Dutta & Mahanta, 2014). Maisont and Narkrugsa (2009) found a linear correlation between amylose content and each of puffed yield, expansion volume, and expansion ratio. Besides, the amylose-lipid complex improves the expansion of parboiled rice because this complex restricts the swelling and degradation of the starch during the parboiling process, which in turn prevents amylose leaching. Therefore, the effect of starch gelatinization as well as the formation of amylose-lipid complexes allow the internal superheated steam to create and develop high pressure within the molten matrix, leading to increased expansion (Saha & Roy, 2022; Joshi et al., 2014b).

Protein Content

The presence of high amounts of proteins may adversely affect the expansion of popped or puffed products (Mishra et al., 2015; Kamaraddi & Prakash, 2015; Joshi et al., 2014b) by lowering the complexation between lipid and amylose because proteins couldform complexes with both amylose and lipid. The effect of protein on lowering the expansion of popped or puffed products can be concluded as follows: 1) the formation of protein-amylose complexes that reduce the hydration of starch, lowering thus the gelatinization degree (Saeleaw & Schleining, 2011; Ibáñez et al., 2007; Hagenimana et al., 2006), 2) the formation of protein-amylose complexes also contributes to lowering the gel strength of gelatinized starch by hindering the formation of amylose networks (Ibáñez et al., 2007), and 3) protein-protein interactions may also contribute to lowering the expansion owing to the fact that the protein-protein complexes have stronger and more stable networks than those of starch-starch complexes. Therefore, grains with low contents of protein and lipids are suitable for popping or puffing (Ibáñez et al., 2007); however, Saha and Roy (2022) found that rice with a higher protein content had a higher expansion ratio, which is attributed to the protein denaturation during puffing, leading to increased binding between protein and starch. This increases the structural strength of the grain matrix, leading to higher internal pressure and thus increasing the expansion.

Temperature and Processing Time

The expansion of grains is also influenced by the conditions applied during popping or puffing, such as temperature and processing time. Temperature is an important parameter in the process of puffing or popping because it helps the grain matrix pass from an amorphous to a rubbery state as well as generate internal superheated steam within the molten matrix, which then escapes from the grain matrix during the pressure release while the expansion. For example, too low a temperature does not allow the phase transition to be achieved successfully, while too high a temperature may cause charring of the grain as well as increase the rate of moisture loss, leading to excessive drying of the grain. Since popping or puffing is a temperature-time combination, the processing time can be reduced by increasing the temperature applied. These parameters differ depending on the method of popping or puffing and can be changed in specific ranges to obtain expanded grains with high-quality attributes. Mandhare et al. (2020) reported that the expansion ratio varied according to the puffing method; the expansion ratio of quinoa was 4.36 at 260 °C for 16 s using a traditional puffing method (cooking pan), while it was 3.05 at a power level of 100% for 135 s. Moreover, Chandrasekhar and Chattopadhyay (1989) reported a time-temperature combination of 240–270 °C and 7–9.7 s for optimal expansion by fluidized air puffing. On the other hand, the processing time depends on the popping/puffing method used. For instance, Villareal et al. (1987) found that the processing time in hot oil puffing ranged between 4 and 8 s at temperatures ranging between 210 and 220 °C, while in gun puffing, the processing time was longer (3–7 min) at the same temperature applied. Therefore, the popping/ puffing temperature and the processing time may be variedin specific ranges to obtain expanded grains with high-quality attributes.

Physical Characteristics of Grains

Physical characteristics of grains (size, shape, length/width ratio, and hardness) may affect the popping/puffing quality (Kamaraddi & Prakash, 2015; Joshi et al., 2014b). It has been reported that the husk content, degree of husk interlocking, and grain hardness had substantial positive impacts on the puffing quality using the hot sand bed method (Murugesan & Bhattacharya, 1991). In contrast, Maisont and Narkrugsa (2009) showed that the effect of these physical properties was insignificant in the microwave puffing of rice. This contradictory behavior may be explained by the fact that the microwave provides volumetric heating where heat penetrates the husk layer. On the other hand, since the heating is superficial in the hot sand bed, these physical characteristics help heat transfer to the grain core. Tian et al. (2001) reported that the closely packed structure of starch endosperm in hard grains has a significant positive impact on the expansion of popcorn. Grains must have a suitable hardness for expansion because very hard grains produce lower expansion than grains with medium hardness (Mishra et al., 2015; Joshi et al., 2014b). In this context, a sorghum variety with a hardness of 103 N showed higher expansion than the varieties with a hardness of 95 or 107 N.

Aging and Storage of Grains

The chemical composition and physical characteristics of grains are affected by storage conditions and storage duration (McDonough et al., 2004). The chemical composition of grains is changed during their storage, which includes changes inthe contents of moisture, protein, and amylose (Zhou et al., 2002). It was reported that the moisture-conditioned rice grains may absorb or lose moisture during their storage, which affects the popping/puffing quality (Kumar & Das, 2015). Additionally, the formation of starch-lipid and starch-protein complexes may also take place during the storage period of the grains, which in turn affects the quality of popping/puffing, as previously discussed. Moreover, the gradual loss of moisture during the long storage period results in grains drying as well as the formation of starch-protein complexes at the cell walls of the grains, making them much harder (McDonough et al., 2004).

Parboiling of Grains

Parboiling is defined as a hydrothermal treatment of grains(togelatinize the starch) followed by a drying step to form compact starch. This process is mainly used as a pre-conditioning step for the popping and puffing of rice and wheat. The Parboiling process induces some modifications in the starch, including physical, structural, and rheological changes (Dutta & Mahanta, 2014). Parboiling causes starch gelatinization in order to ensure the expected expansion of the grains (Mahanta & Bhattacharya, 2010) and the stability of the final structure of popped and puffed grains. Parboiling also develops the grain's hardness as a result of the filling of internal cracks inside the grain kernel (Bhattacharya, 2011). The hard and dense structure of starch (the amorphous state) is then changed to a molten structure (the rubbery state) during the popping or puffing process and then to a porous structure after releasing the superheated steam. All these changes are necessary to create superheated steam pressure within the grain kernel during popping and puffing. Several parboiling techniques have been applied to prepare rice grains for popping and puffing. These techniques include: 1) normal parboiling; steaming under atmospheric pressure conditions, 2) high pressure steaming; at pressure higher than the atmospheric pressure, and 3) dry-heat parboiling at 250 °C for 3 to 5 min (Swarnakar et al. (2022). It has been reported that rice grains parboiled at high pressure were more suitable for popping or puffing and exhibited a higher expansion ratio than those parboiled at atmospheric pressure (Mahanta & Bhattacharya, 2010), because pressure parboiling causes a higher extent of starch gelatinization, which in turn improves the popping and puffing quality (Bhattacharya, 2011). In addition, Swarnakar et al. (2020) reported that the optimum parboiling conditions to obtain puffed product grains from parboiled unpolished rice were a steam pressure ranging from 200 to 300 kPa and a steaming time of 15 min.

It is worthwhile mentioning that too high steam pressure during rice parboiling could cause starch to break down, which in turn decreases the expansion of the grain (Mahanta & Bhattacharya, 2010).

Extent of Milling

The presence of bran (the outer layer) causes certain resistance in the expansion of puffed grains, affecting their appearance, wherein the color of puffed grains becomesdull. For these reasons, polished parboiled rice is more suitable for puffing. However, paddy rice(unpolished) is used in the popping process, wherein the rice kernel is expanded more while maintaining the bran layer with the popped grains. According to the available literature, about 6–7% amilling extent of rice is recommended to obtain the expected expansion, but a significant loss of nutrients and dietary fibers was observed in the puffed polished rice. Therefore, the puffing of brown rice has become more interesting (Swarnakar et al., 2019, 2020; Mir et al., 2016a, b).

Salting

Adding salt to grains before heating in the puffing process has a considerable impact on the expansion of grains, where it enhances heat transfer into the kernel towards the core of grains during the popping and puffing processes as well as the diffusion of superheated steam generated (Hoke et al., 2005). But the addition of salt is not necessary for grain popping. Devi and Das (2017) found that the salt addition during the microwave popping of paddy grains increases the dielectric characteristics of the paddy, resulting in more energy absorption, accelerating the heating process, and thus enhancing the quality of the popped grains. Besides, salt is added to improve the palatability of the expanded grains (Mohapatra et al., 2012; Swarnakar et al., 2019). Salts such as NaCl, $CaCl_2$, $NaHCO_3$, $NaHSO_3$, and $K2SO_4$ can be used in the production of expanded grains. Among these salts, Dash and Das (2019b) found that NaCl exhibiteda significant satisfying effect on the puffing quality. For rice puffing, the optimum salt level is found to be in the range of 2–4% (w/w) (Mohapatra et al., 2012).

7 Conclusion and Future Trends

Popped and puffed products are the most popularly consumed snacks around the world, and they are commercialized with various designs and names. Their manufacturing techniques vary from ancient and simple to modern and costly techniques, including hot sand bed, hot oil, hot air, gun puffing, microwave puffing, extrusion, and instant controlled pressure drop. Recently, infrared puffing has been adopted to produce these products. The quality features of these products are affected by some interrelated parameters, including the physical and chemical characteristics of the grain, parboiling conditions, mode of heat transfer, and popping and puffing conditions. Therefore, the optimization of these parameters allows for achieving the expected expansion and enhancing the different characteristics of the expanded product. In addition to expansion, quality parameters such as nutritive quality, bulk density, color, hardness, microstructural changes, and sensory characteristics are necessary to characterize and define the popping and puffing process. Finally, it could be said that (1) the puffing and popping processes need more deep studies to enhance the quality of expanded products, in particular the nutritive value, (2) in-depth studies on the glass transition and its relationship with processing parameters and the quality attributes of the expanded products are still required, and (3) infrared and DIC puffing also need further deep studies to increase the diversity of infrared- and DIC-expanded products.

References

Aboud, S. A., Altemimi, A. B., Al-Hilphy, A. R. S., Yi-Chen, L., & Cacciola, F. (2019). A comprehensive review on infrared heating applications in food processing. *Molecules, 24*(22), 4125. https://doi.org/10.3390/molecules24224125

Altan, A. (2014). Effects of pre-treatments and moisture content on microstructure and physical properties of microwave expanded hull-less barley. *Food Research International, 56*, 126–135. https://doi.org/10.1016/j.foodres.2013.12.029

Anjitha, P. K., Baskaran, N., Venkatachalapathy, N., & Tito Anand, M. (2021). Nutritional changes of sorghum after popping by a developed infrared assisted hot air popping machine. *International Journal of Current Microbiology and Applied Science, 10*(01), 3620–3627. https://doi.org/10.20546/ijcmas.2021.1001.427

Balaji, B. H. (2019). *Development of semi-automated multigrain continuous hot air puffing system*. PhD thesis submitted to department of Agricultural Process Engineering, Post Graduate Institute.

Bhattacharya, K. R. (2011). *Rice quality: A guide to rice properties and analysis*. Elsevier. A volume in Woodheadpuplishing series in Food Science, Technology and Nutrition. p. 578.

Bhatupadya, V. G., Bhat, R. S., Shenoy, V. V., & Salimath, P. M. (2008). Physico chemical characterization of popping-Special rice accessions. *Karnataka Journal Agricultural Science, 21*(2), 184–186.

Boischot, C., Moraru, C. I., & Kokini, J. L. (2003). Expansion of glassy amylopectin extrudates by microwave heating. *Cereal Chemistry, 80*, 56–61. https://doi.org/10.1094/CCHEM.2003.80.1.56

Briffaz, A., Mestres, C., Escoute, J., Lartaud, M., & Dornier, M. (2012). Starch gelatinization distribution and peripheral cell disruption in cooking rice grains monitored by microscopy. *Journal of Cereal Science, 56*(3), 699–705. https://doi.org/10.1016/j.jcs.2012.09.001

Byrd, J. E., & Perona, M. J. (2005). Kinetics of popping of popcorn. *Cereal Chemistry, 82*(1), 53–59. https://doi.org/10.1094/CC-82-0053

Chandrasekaran, S., Ramanathan, S., & Basak, T. (2013). Microwave food Processing – A review. *Food Research International, 52*(1), 243–261. https://doi.org/10.1016/j.foodres.2013.02.033

Chandrasekhar, P. R., & Chattopadhyay, P. K. (1989). Heat transfer during fluidized bed puffing of rice grain. *Journal of Food Process Engineering, 11*(2), 147–157. https://doi.org/10.1111/j.1745-4530.1989.tb00026.x

Das, H. (2005). *Food processing operations analysis (1st ed.)* (pp. 339–340). Asian Books Private Limited.

Dash, K. K., & Das, S. K. (2019a). Optimization of fluidized bed preconditioning for microwave puffed rice using integrated artificial neural network and genetic algorithm approach. *Journal of Food Process Engineering, 42*(6), e13158. https://doi.org/10.1111/jfpe.13158

Dash, K. K., & Das, S. K. (2019b). Puffing quality and sensory evaluation of reduced sodium puffed rice. *Quality Assurance and Safety of Crops and Foods, 11*(8), 757–768. https://doi.org/10.3920/QAS2018.1414

Derycke, V., Veraverbeke, W. S., Vandeputte, G. E., De Man, W., Hoseney, R. C., & Delcour, J. A. (2005). Impact of proteins on pasting and cooking properties of nonparboiled and parboiled rice. *Cereal Chemistry Journal, 82*(4), 468–474. https://doi.org/10.1094/CC-82-0468

Devi, M. K., & Das, S. K. (2017). Kinetics of color changes of popped rice during microwave popping: Effect of salt and moisture content. *Journal of Food Process Engineering, 40*(6), e12560. https://doi.org/10.1111/jfpe.12560

Dhadke, S. G., Pawar, V. S., & Wanole, P. D. (2022). Effect of popping on nutritional composition of sorghum. *Biological Forum-An International Journal, 14*(4), 1199–1202.

Dutta, H., & Mahanta, C. L. (2014). Traditional parboiled rice-based products revisited: Current status and future research challenges. *Rice Science, 21*(4), 187–200. https://doi.org/10.1016/S1672-6308(13)60191-2

Dutta, A., Mukherjee, R., Gupta, A., Ledda, A., & Chakraborty, R. (2015). Ultrastructural and physicochemical characteristics of rice under various conditions of puffing. *Journal of Food Science and Technology, 52*(11), 7037–7047. https://doi.org/10.1007/s13197-015-1808-1

Erba, D., Hidalgo, A., Bresciani, J., & Brandolini, A. (2011). Environmental and genotypic influences on trace element and mineral concentrations in whole meal flour of einkorn (*Triticummonococcum* L. subsp. *monococcum*). *Journal of Cereal Science, 54*, 250–254. https://doi.org/10.1016/j.jcs.2011.06.011

Gugal, R. H., & Yenagi, N. B. (2016). Nutrient composition and nutraceutical properties of processed pop sorghum cultivars. *Journal of Farm Sciences, 29*(1), 70–75.

Gulati, T., & Datta, A. K. (2016). Coupled multiphase transport, large deformation and phase transition during rice puffing. *Chemical Engineering Science, 139*, 75–98. https://doi.org/10.1016/j.ces.2015.08.057

Gundboudi, Z. A. (2006). Nutritional and processing qualities of pop sorghum cultivars and value addition. In *Master of home science (foods and nutrition) thesis, submitted to department of food science and Nutrition College of Rural Home Science*. University of Agricultural Sciences. UAS Dharwad.

Hagenimana, A., Ding, X., & Fang, T. (2006). Evaluation of rice flour modified by extrusion cooking. *Journal of Cereal Science, 43*(1), 38–46. https://doi.org/10.1016/j.jcs.2005.09.003

Haldkar, P., Tiwari, V. K., Rumandla, S. K., & Behera, K. (2022). Effect of puffing and cooking on phytochemical profile of psoudo-cereals: A review. *The Pharma Innovation Journal, SP-11*(12), 511–514.

Hidalgo, A., & Brandolini, A. (2008). Protein, ash, lutein and tocols distribution in einkorn (*Triticummonococcum* L. subsp. *monococcum*) seed fractions. *Food Chemistry, 107*, 444–448. https://doi.org/10.1016/j.foodchem.2007.08.009

Hidalgo, A., & Brandolini, A. (2014). Nutritional properties of einkorn wheat. *Journal of the Science of Food and Agriculture, 94*, 601–612. https://doi.org/10.1002/jsfa.6382

Hidalgo, A., Scuppa, S., & Brandolini, A. (2016). Technological quality and chemical composition of puffed grains from einkorn (*Triticummonococcum* L. subsp. *monococcum*) and bread wheat (*Triticumaestivum* L. subsp. *aestivum*). *LWT-Food Science and Technology, 68*, 541–548. https://doi.org/10.1016/j.lwt.2015.12.068

Hoke, K., Housova, J., & Houska, M. (2005). Optimum conditions of rice puffing. *Czech Journal of Food Science, 23*(1), 1–11. https://doi.org/10.17221/3365-CJFS

Hoke, K., Houška, M., Průchová, J., Gabrovská, D., Vaculová, K., & Paulíčková, I. (2007). Optimisation of puffing snaked barley. *Journal of Food Engineering, 80*, 1016–1022. https://doi.org/10.1016/j.jfoodeng.2006.08.012

Huang, R., Pan, X., Lv, J., Zhong, W., Yan, F., Duan, F., & Jia, L. (2018). Effects of explosion puffing on the nutritional composition and digestibility of grains. *International Journal of Food Properties, 21*(1), 2193–2204. https://doi.org/10.1080/10942912.2018.1514634

Huang, X., Liu, H., Ma, Y., Mai, S., & Li, C. (2022). Effects of extrusion on starch molecular degradation, order–Disorder structural transition and digestibility—A review. *Foods, 11*, 2538. https://doi.org/10.3390/foods11162538

Ibáñez, A. M., Wood, D. F., Yokoyama, W. H., Park, I. M., Tinoco, M. A., Hudson, C. A., Mckenzie, K. S., & Shoemaker, C. F. (2007). Viscoelastic properties of waxy and nonwaxy rice flours, their fat and protein-free starch, and the microstructure of their cooked kernels. *Journal of Agricultural and Food Chemistry, 55*(16), 6761–6771. https://doi.org/10.1021/jf070416x

Inoue, T., Iyota, H., Uemura, T., Yamagata, J., Konishi, Y., & Tatemoto, Y. (2009). Optimal conditions for popping amaranth seeds. *Drying Technology, 27*(7–8), 918–926. https://doi.org/10.1080/07373930902988254

Iyota, H., Konishi, Y., Inoue, T., Yoshida, K., Nishimura, N., & Nomura, T. (2005). Popping of amaranth seeds in hot air and superheated steam. *Drying Technology, 23*(6), 1273–1287. https://doi.org/10.1081/DRT-200059502

Jaybhaye, R. V., Pardeshi, I. L., Vengaiah, P. C., & Srivastav, P. P. (2014). Processing and technology for millet based food products: a review. *Journal of Ready to Eat Food, 1*(2), 32–48.

Jia, L., Huang, R., Wang, S., Dong, Y., Jingang, L. V., Zhong, W., & Yan, F. (2021). Effects of Explosion Puffing on the Composition, Structure, and Functional Characteristics of Starch and Protein in Grains. *ACS Food Science and Technology, 1*(10), 1869–1879. https://doi.org/10.1021/acsfoodscitech.1c00232

Joshi, N. D., Mohapatra, D., Joshi, D. C., & Sutar, R. F. (2014a). Puffing characteristics of parboiled milled rice in a domestic convective–microwave oven and process optimization. *Food and Bioprocess Technology, 7*(6), 1678–1688. https://doi.org/10.1007/s11947-013-1220-7

Joshi, N. D., Mohapatra, D., & Joshi, D. C. (2014b). Varietal selection of some Indica rice for production of puffed rice. *Food Bioprocess Technology, 7*, 299–305. https://doi.org/10.1007/s11947-013-1056-1

Kamal, I. M., Sobolik, V., Kristiawan, M., Mounir, S., & Allaf, K. (2008). Structure expansion of green coffee beans using instantaneous controlled pressure drop process. *Innovative Food Science and Emerging Technologies, 9*(4), 534–541. https://doi.org/10.1016/j.ifset.2008.01.004

Kamaraddi, V., & Prakash, J. (2015). Assessment of suitability of selected rice varieties for production of expanded rice. *Cogent Food and Agriculture, 1*(1), 1112675. https://doi.org/10.1080/23311932.2015.1112675

Konishi, Y., Iyota, H., Yoshida, K. M., Moritani, J., Inoue, T., Nishimura, N., & Nomura, T. (2004). Effect of moisture content on the expansion volume of popped amaranth seeds by hot air and superheated steam using a fluidized bed system. *Bioscience Biotechnology and Biochemistry, 68*(10), 2186–2189. https://doi.org/10.1271/bbb.68.2186

Kora, A. J. (2019). Applications of sand roasting and baking in the preparation of traditional Indian snacks: Nutritional and antioxidant status. *Bulletin of the National Research Centre, 43*(158), 1–11. https://doi.org/10.1186/s42269-019-0199-2

Kumar, N. M., & Das, S. K. (2015). Moisture sorption isotherm of preconditioned pressure parboiled brown rice. *Journal of Food Processing and Technology, 6*(12), 1. https://doi.org/10.4172/2157-7110.1000519

Lamberts, L., Brijs, K., Mohamed, R., Verhelst, N., & Delcour, J. A. (2006). Impact of browning reactions and bran pigments on color of parboiled rice. *Journal of Agricultural and Food Chemistry, 54*(26), 9924–9929. https://doi.org/10.1021/jf062140j

Li, X., Masatcioglu, M. T., & Koksel, F. (2019). Physical and Functional Properties of Wheat Flour Extrudates Produced by Nitrogen Injection Assisted Extrusion Cooking. *Journal of Cereal Science, 89*(102), 811. https://doi.org/10.1016/j.jcs.2019.102811

Maclean, J. L., Dawe, D. C., Hardy, B., & Hettel, G. P. (2002). *Rice Almanac: Source book for the most important economic activity on earth* (3rd ed.). IRRI.

Mahanta, C. L., & Bhattacharya, K. R. (2010). Relationship of starch changes to puffing expansion of parboiled rice. *Journal of Food Science and Technology, 47*(2), 182–187. https://doi.org/10.1007/s13197-010-0038-9

Maisont, S., & Narkrugsa, W. (2009). Effect of some physicochemical properties of paddy rice varieties on puffing qualities by microwave "ORIGINAL". *Kasetsart Journal Natural Science, 43*, 566–575.

Maisont, S., & Narkrugsa, W. (2010). Effect of salt, moisture content and microwave power on puffing qualities of puffed rice. *Kasetsart Journal Natural Science, 44*(2), 251–261.

Mandhare, L. L., More, D. R., & Nagulwar, M. M. (2020). Effect of popping methods on popping characteristics of quinoa seed. *Journal of Pharmacognosy and Phytochemistry, 9*(5), 1943–1945.

Mariotti, M., Alamprese, C., Pagani, M. A., & Lucisano, M. (2006). Effect of puffing on ultrastructure and physical characteristics of cereal grains and flours. *Journal of cereal science, 43*(1), 47–56. https://doi.org/10.1016/j.jcs.2005.06.007

Markowski, M., Ratajski, A., Konopko, H., Zapotoczny, P., & Majewska, K. (2006). Rheological behavior of hot-air- puffed amaranth seeds. *International Journal of Food Properties, 9*(2), 195–203. https://doi.org/10.1080/10942910600596076

Mazumder, P., Roopa, B. S., & Bhattacharya, S. (2007). Textural attributes of a model snack food at different moisture contents. *Journal of Food Engineering, 79*(2), 511–516. https://doi.org/10.1016/j.jfoodeng.2006.02.011

McDonough, C. M., Floyd, C. D., Waniska, R. D., & Rooney, L. W. (2004). Effect of accelerated aging on maize, sorghum, and sorghum meal. *Journal of Cereal Science, 39*(3), 351–361. https://doi.org/10.1016/j.jcs.2004.01.001

Mir, S. A., Bosco, S. J. D., Shah, M. A., & Mir, M. M. (2015). Effect of puffing on physical and antioxidant properties of brown rice. *Food Chemistry, 191*, 139–146. https://doi.org/10.1016/j.foodchem.2014.11.025

Mir, S. A., Bosco, S. J. D., Shah, M. A., & Mir, M. M. (2016a). Effect of puffing on physical and antioxidant properties of brown rice. *Food Chemistry, 191*, 139–146. https://doi.org/10.1016/j.foodchem.2014.11.025

Mir, S. A., Bosco, S. J. D., Shah, M. A., Mir, M. M., & Sunooj, K. V. (2016b). Process optimization and characterization of popped brown rice. *International Journal of Food Properties, 19*(9), 2102–2112. https://doi.org/10.1080/10942912.2015.1105254

Mishra, G., Joshib, D. C., & Panda, B. K. (2014). Popping and Puffing of Cereal Grains: A Review. *Journal of Grain Processing and Storage, 1*(2), 34–46.

Mishra, G., Joshi, D. C., Mohapatra, D., & Babu, V. B. (2015). Varietal influence on the microwave popping characteristics of sorghum. *Journal of Cereal Science, 65*, 19–24. https://doi.org/10.1016/j.jcs.2015.06.001

Mohapatra, M., Kumar, A., & Das, S. K. (2012). Mixing characteristics of rice by tracer technique in an agitated cylindrical preconditioner developed for puffed rice production system. *Journal of Food Process Engineering, 35*(5), 784–791. https://doi.org/10.1111/j.1745-4530.2010.00628.x

Mom, V., Chanlat, P., & Songsermpong, S. (2020). Characteristics and process optimization of rice snack (khao-tan) puffing by home microwave oven. *Journal of Food Processing and Preservation, 44*(5), e14413. https://doi.org/10.1111/jfpp.14413

Mościcki, L. (2011). Extrusion, Effect on Physical and Chemical Properties. In J. Gliński, J. Horabik, & J. Lipiec (Eds.), *Encyclopedia of Agrophysics. Encyclopedia of Earth Sciences Series*. Springer. https://doi.org/10.1007/978-90-481-3585-1_56

Mounir, S. (2015). Texturing of chicken breast meat as an innovative way to intensify drying: use of a coupled washing/diffusion cwd phenomenological model to enhance kinetics and functional properties. *Drying Technology, 33*(1), 1369–1381. https://doi.org/10.1080/07373937.2015.1030029

Mounir, S., Besombes, C., Al-Bitar, N., & Allaf, K. (2011a). Study of instant controlled pressure drop DIC-treatment in manufacturing snack and expanded granule powder of apple and onion. *Drying Technology, 29*(3), 331–341. https://doi.org/10.1080/07373937.2010.491585

Mounir, S., Halle, D., & Allaf, K. (2011b). Characterization of pure cheese snacks and expanded granule powders textured by the instant controlled pressure drop (DIC) process. *Dairy Science and Technology, 91*(4), 441–455. https://doi.org/10.1007/s13594-011-0023-8

Mounir, S., Allaf, T., Mujumdar, A. S., & Allaf, K. (2012). Swell drying: Coupling instant controlled pressure drop DIC to standard convection drying processes to intensify transfer phenomena and improve quality-An overview. *Drying Technology, 30*(14), 1508–1531. https://doi.org/10.1080/07373937.2012.693145

Mounir, S., Téllez-Pérez, C., Sunooj, K. V., & Allaf, K. (2020a). Texture and color characteristics of swell-dried ready-to-eat Zaghloul date snacks: Effect of operative parameters of instant controlled pressure drop process. *Journal of Texture Studies, 51*(2), 276–289. https://doi.org/10.1111/jtxs.12468

Mounir, S., Amami, E., Allaf, T., Mujumdar, A., Allaf, K., & k. (2020b). Instant controlled pressure drop (DIC) coupled to intermittent microwave/airflow drying to produce shrimp snacks: Process performance and quality attributes. *Drying Technology, 38*(5–6), 695–711. https://doi.org/10.1080/07373937.2019.1694537

Mounir, S., Ghandour, A., Téllez-Pérez, C., Aly, A. A., Mujumdar, A. S., & Allaf, K. (2021a). Phytochemicals, chlorophyll pigments, antioxidant activity, relative expansion ratio, and microstructure of dried okra pods: swell-drying by instant controlled pressure drop versus conventional shade drying. *Drying Technology, 39*(15), 2145–2159. https://doi.org/10.1080/07373937.2020.1756843

Mounir, S., Ghandour, A., Mustafa, R., & Allaf, K. (2021b). Can hydro-thermo-mechanical treatment by instant controlled pressure-drop (DIC) be used as short time roasting process? Effect of processing parameters on sensory, physical, functional, and color attributes of Egyptian carob powder. *Journal of Food Science and Technology, 58*, 451–464. https://doi.org/10.1007/s13197-020-04553-9

Murthy, K. V., Ravi, R., Bhat, K. K., & Raghavarao, K. S. M. S. (2008). Studies on roasting of heat using fluidized bed roaster. *Journal of Food Engineering, 89*(3), 336–342. https://doi.org/10.1016/j.jfoodeng.2008.05.014

Murugesan, G., & Bhattacharya, K. R. (1991). Basis for varietal difference in popping expansion of rice. *Journal of Cereal Science, 13*(1), 71–83. https://doi.org/10.1016/S0733-5210(09)80028-3

Oghbaei, M., & Prakash, J. (2016). Effect of primary processing of cereals and legumes on its nutritional quality: A comprehensive review. *Cogent Food and Agriculture, 2*, Article ID: 1136015. https://doi.org/10.1080/23311932.2015.1136015

Oli, P., Ward, R., Adhikari, B., & Torley, P. (2014). Parboiled rice: Understanding from a materials science approach. *Journal of Food Engineering, 124*, 173–183. https://doi.org/10.1016/J.JFOODENG.2013.09.010

Patel, H., Rajput, L. P. S., Ansari, F., Rana, G. K., & Soni, R. (2021). Effects of physicochemical and processing characteristics on nutritional and sensorial profile of puffed rice. *Journal of Pharmacognosy and Phytochemistry, 10*(1), 635–642.

Paznocht, L., Burešová, B., Kotíková, Z., & Martinek, P. (2021). Carotenoid Content of Extruded and Puffed Products Made of Colored-Grain Wheats. *Food Chemistry, 340*(127), 951. https://doi.org/10.1016/j.foodchem.2020.127951

Pilatowski, I., Mounir, S., Haddad, J., Cong, D. T., & Allaf, K. (2010). The instant controlled pressure drop process as a new post-harvesting treatment of paddy rice: Impacts on drying kinetics and end product attributes. *Food Bioprocess Technology, 3*, 901–907. https://doi.org/10.1007/s11947-010-0332-6

Ramos Diaz, J. M., Sundarrajan, L., Kariluoto, S., Lampi, A.-M., Tenitz, S., & Jouppila, K. (2016). Effect of extrusion cooking on physical properties and chemical composition of corn-based snacks containing amaranth and quinoa: application of partial least squares regression. *Journal of Food Process Engineering, 40*(1). https://doi.org/10.1111/jfpe.12320

Randhir, R., Kwon, Y. I., & Shetty, K. (2008). Effect of thermal processing on phenolics, antioxidant activity and health-relevant functionality of select grain sprouts and seedlings. *Innovative Food Science and Emerging Technologies, 9*(3), 355–364. https://doi.org/10.1016/j.ifset.2007.10.004

Roopa, B. S., Mazumder, P., & Bhattacharya, S. (2009). Fracture behavior and mechanism of puffed cereal during compression. *Journal of Texture Studies, 40*(2), 157–171. https://doi.org/10.1111/j.1745-4603.2009.00174.x

Saeleaw, M., & Schleining, G. (2011). Effect of frying parameters on crispiness and sound emission of cassava crackers. *Journal of Food Engineering, 103*(3), 229–236. https://doi.org/10.1016/j.jfoodeng.2010.10.010

Saha, S., & Roy, A. (2020). Puffed rice: A materialistic understanding of rice puffing and its associated changes in physicochemical and nutritional characteristics. *Journal of Food Process Engineering, 43*(9), e13479. https://doi.org/10.1111/jfpe.13479

Saha, S., & Roy, A. (2022). Selecting high amylose rice variety for puffing: A correlation between physicochemical parameters and sensory preferences. *Measurement: Food, 5*, 100021. https://doi.org/10.1016/j.meafoo.2021.100021

Saravanabavan, S. N., Shivanna, M. M., & Bhattacharya, S. (2013). Effect of popping on sorghum starch digestibility and predicted glycemic index. *Journal of Food Science and Technology, 50*(2), 387–392. https://doi.org/10.1007/s13197-011-0336-x

Schlörmann, W., Zetzmann, S., Wiege, B., Haase, N. U., Greiling, A., Lorkowski, S., Dawczynski, C., & Glei, M. (2020). Impact of different roasting conditions on sensory properties and health-related compounds of oat products. *Food Chemistry, 307*, 125–548. https://doi.org/10.1016/j.foodchem.2019.125548

Sharma, P., & Gujral, H. S. (2011). Effect of Sand Roasting and Microwave Cooking on Antioxidant Activity of Barley. *Food Research International, 44*(1), 235–240. https://doi.org/10.1016/j.foodres.2010.10.030

Sharma, V., Champawat, P. S., & Mudgal, V. D. (2014). Process development for puffing of Sorghum. *International Journal of Current Research and Academic Review, 2*(1), 164–170.

Shavandi, M., Sadeghi, A., & Sarani, A. (2020). Modeling the effect of different infrared treatment on *B. cereus* in cardamom seeds and using genetic algorithm-artificial neural network. *Journal of Food and Bioprocess Engineering, 3*(1), 29–34. https://doi.org/10.22059/JFABE.2020.75644

Shavandi, M., Javanmard, M., & Basiri, A. (2022a). Novel popping through infrared: Effect on some physicochemical properties of popcorn (*Zea Mays* L. var. Everta). *LWT-Food Science and Technology, 155*(1), 112955. https://doi.org/10.1016/j.lwt.2021.112955
Shavandi, M., Javanmard, M., & Basiri, A. (2022b). Novel infrared puffing: Effect on physicochemical attributes ofpuffed rice (*Oryza sativa* L.). *Food Science and Nutrition, 00*, 1–11. https://doi.org/10.1002/fsn3.3022
Siriamornpun, S., Sareepuang, K., Wiset, L., & Meeso, N. (2008). Effect of soaking temperature on physical, chemical and cooking properties of parboiled fragrant Rice. *World Journal of Agricultural Sciences, 4*(4), 409–415.
Solanki, C., Indore, N., Saha, D., & Aleksha Kudos, S. K. (2018). Effect of popping methods on popping characteristics of amaranth grain. *International Journal of Chemical Studies, 6*(2), 2779–2782.
Solís-Morales, D., Sáenz-Hernández, C. M., & Ortega-Rivas, E. (2009). Attrition reduction and quality improvement of coated puffed wheat by fluidised bed technology. *Journal of Food Engineering, 93*(2), 236–241. https://doi.org/10.1016/j.jfoodeng.2009.01.020
Subramani, D., Tamilselvan, S., Murugesan, M., & Shivaswamy, M. S. (2020a). Optimization of variables that influence puffing quality of hot air oven puffed quinoa. *International Journal of Pharmaceutical Research, 12*(3), 736–743. https://doi.org/10.31838/ijpr/2020.12.03.136
Subramani, D., Tamilselvan, S., Murugesan, M., & Shivaswamy, M. S. (2020b). Optimization of sand puffing characteristics of quinoa using surface methodology. *Current Research in Nutrition and Food Science, 8*(2), 504–515. https://doi.org/10.12944/CRNFSJ.8.2.16
Swarnakar, A. K., Devi, M. K., & Das, S. K. (2014). Popping characteristics of paddy using microwave energy and optimization of process parameters. *International Journal of Food Studies, 3*(1), 45–49. https://doi.org/10.7455/ijfs/3.1.2014.a4
Swarnakar, A. K., Srivastav, P. P., & Das, S. K. (2019). Optimization of preconditioning process of pressure parboiled brown rice (unpolished) for microwave puffing and its comparison with hot sand bed puffing. *Journal of Food Process Engineering, 42*(3), e13007. https://doi.org/10.1111/jfpe.13007
Swarnakar, A. K., Srivastav, P. P., & Das, S. K. (2020). Optimization of pressure parboiling conditions and pre-conditioned moisture content of brown rice (unpolished rice) for microwave puffing and its comparison with hot sand bed puffing. *International Journal of Food Studies, 9*, SI1-SI16. https://doi.org/10.7455/ijfs/9.SI.2020.a1
Swarnakar, A. K., Mohapatra, M., & Susanta Kumar Das, S. K. (2022). A review on processes, mechanisms, and quality influencing parameters for puffing and popping of grains. *Journal of Food Processing and Preservation, 46*, e16891. https://doi.org/10.1111/jfpp.16891
Sweley, J. C., Rose, D. J., & Jackson, D. S. (2013). Quality traits and popping performance considerations for popcorn (*Zea mays*Everta). *Food Reviews International, 29*(2), 157–177. https://doi.org/10.1080/87559129.2012.714435
Tavanandi, H. A., Das, A. K., Venkateshmurthy, K., & Raghavarao, K. S. M. S. (2021). Design and development of a machine for continuous popping and puffing of grains. *Journal of Food Science and Technology, 58*(5), 1703–1714. https://doi.org/10.1007/s13197-020-04680-3
Thai, C. D., & Allaf, K. (2014). DIC-Assisted Parboiling of Paddy Rice. In T. Allaf & K. Allaf (Eds.), *Instant Controlled Pressure Drop (D.I.C.) in Food Processing. Food Engineering Series*. Springer. https://doi.org/10.1007/978-1-4614-8669-5_3
Tian, Y., Buriak, P., & Eckhoff, S. R. (2001). Effect of hybrid and physical properties of individual popcorn kernels on expansion volume. *Cereal Chemistry, 78*(5), 578–582. https://doi.org/10.1094/CCHEM.2001.78.5.578
Tikekar, R. V., & Karwe, M. V. (2009). Development of a continuous method for puffingamaranth (*amaranthus* spp.) Seeds. *Journal of Food Process Engineering, 32*(2), 265–277. https://doi.org/10.1111/j.1745-4530.2007.00215.x
Ushakumari, S. R., Rastogi, N. K., & Malleshi, N. G. (2007). Optimization of process variables for the preparation of expanded finger millet using response surface methodology. *Journal of Food Engineering, 82*(1), 35–42. https://doi.org/10.1016/j.jfoodeng.2007.01.012

Vanier, N. L., Vamadevan, V., Bruni, G. P., Ferreira, C. D., Pinto, V. Z., Seetharaman, K., Zavareze, E. R., Elias, M. C., & Berrios, J. D. J. (2016). Extrusion of Rice, Bean and Corn Starches: Extrudate Structure and Molecular Changes in Amylose and Amylopectin. *Food Science, 81*(12), E2932–E2938. https://doi.org/10.1111/1750-3841.13545

Villareal, C. P., Juliano, B., & o. (1987). Varietal differences in quality characteristics of puffed rices. *Cereal Chemistry, 64*(4), 337–342.

Yağci, S., & Evci, T. (2015). Effect of instant controlled pressure drop process on some physico-chemical and nutritional properties of snacks produced from chickpea and wheat. *International Journal of Food Science and Technology, 50*(8), 1901–1910. https://doi.org/10.1111/ijfs.12843

Zapana, F., de Bruijn, J., Vidal, L., Melín, P., González, M. E., Cabrera, G., Williams, P., & Bórquez, R. (2020). Physical, chemical and nutritional characteristics of puffed quinoa. *International Journal of Food Science and Technology, 55*(1), 313–322. https://doi.org/10.1111/ijfs.14290

Zapotoczny, P., Markowski, M., Majewska, K., Ratajski, A., & Konopko, H. (2006). Effect of temperature on the physical, functional, and mechanical characteristics of hot-air- Puffed amaranth seeds. *Journal of Food Engineering, 76*, 469–476. https://doi.org/10.1016/j.jfoodeng.2005.05.045

Zhou, Z., Robards, K., Helliwell, S., & Blanchard, C. (2002). Ageing of stored rice: Changes in chemical and physical attributes. *Journal of Cereal Science, 35*(1), 65–78. https://doi.org/10.1006/jcrs.2001.0418

Chapter 9
Pasta

Antima Gupta and Savita Sharma

1 Introduction

In today's dynamic lifestyle, we require food products that are instant, versatile and ready-to-cook/eat/drink and pasta is one of those products which has gained popularity in the past two decades because of its versatility, nutritive value, low cost, ease of preparation, longer shelf-life and palatability. Pasta is an Italian word for dough and is defined as a type of dough extruded into different shapes for cooking (Adegunwa et al., 2012).

Pasta widely available in the market made of only two ingredients durum wheat (*Triticum durum*) semolina and water. Durum wheat is majorly suitable as it is high in protein and during milling it gives coarse flour due to hard nature of endosperm resulted in pasta with excellent technological attributes. Durum wheat during extrusion forms a continuous 3-D network of gluten and holds the starch in place. In Italy, largest producer and consumer of pasta suggests only durum wheat for the pasta preparation and if any low-grade durum wheat semolina is utilized then package must contain the information as "low grade durum wheat semolina pasta" (Italian Legislation Law N° 5,801,967, Italian Presidential Decree N° 1,872,001) as per the law (Murray et al., 2017). The reason is quality of pasta, the soft wheat flour doesn't form extensive gluten protein network resulted in inferior quality of pasta produced in terms of texture and cooking quality (Resmini & Pagani, 1983). Moreover, use of hexaploid wheat is forbidden in Italy. However, the molecular weight of glutenin also affects the cooking quality and gluten strength of pasta. Among high molecular weight glutenin subunits (HMWGS) 6 + 8 or 7+ 8 gives

A. Gupta (✉) · S. Sharma
Department of Food Science and Technology, Punjab Agricultural University, Ludhiana, Punjab, India
e-mail: antimagupta@pau.edu

M. A. Shah et al. (eds.), *Cereal-Based Food Products*,
https://doi.org/10.1007/978-3-031-40308-8_9

better results while for low molecular weight glutenin subunits (LMWGS) 2 gives better results than HMWGS 20 and LMWGS 1 (Kovacs et al., 1997).

Semolina is rich in complex carbohydrates, protein, fat and vitamin. But it lacks certain essential fatty acids and amino acids, vitamins, dietary fiber, minerals and phytochemicals. Therefore, there is a need to improve its nutritional quality. Both World Health Organization (WHO) as well as Food Agriculture Organization (FAO) have recommended pasta as one of the most effective enrichment media for various nutrients such as minerals, lipids, proteins, vitamins, etc. (Chillo et al., 2008). To meet the demand of consumer for healthy and convenience food products, non-conventional sources are being searched for developing gluten free pasta, protein rich pasta, fiber rich pasta, micronutrient rich pasta, and antioxidant rich pasta. Various raw materials like agro-industrial waste, whole flour, by products from cereal and fruit juice industry, fish powders, meat powder, egg albumin, pseudocereals, millet flour, multigrain flour are being added to improvise the nutritional value of the pasta and to meet the demand of healthy pasta (Nilusha et al., 2019).

Numerous challenges are faced by the researchers to make healthy pasta. Single variable change in pasta can hamper its cooking quality and technological attributes. For instance, incorporation of fiber or protein in durum wheat will disrupt the continuous gluten network which is the most important factor to retain the integrity of pasta structure, it is responsible for the mechanical strength of pasta, firmness and cooking loss. Therefore, addition of non-conventional ingredient needs detailed optimization of the proportion of non-conventional ingredient, amount of additive added, processing parameter, time, temperature, relative humidity, hydration level etc. This chapter highlight the equipment and machinery involved, types of pasta, processing method, quality parameters and future trend in pasta industry.

2 Global Market Scenario: Socio-economic Opportunities and Challenges

Pasta have been around for thousands of years and is an important component of the food and culture of Asian nations like China and Japan. Globally, pasta market is driven on account of convenience, versatility, nutritional composition, low cost, transportation and longer shelf life (Nilusha et al., 2019). World total production of pasta reached 16.9 million tonnes in 2021 with Italy ranked one (3.89 million tonnes) and India ranked 23rd (1.18 million tonnes) in the production worldwide (UNAFPA, 2022). In India, pasta market is growing at a faster rate and the primary factors are rising urbanization, shifting lifestyle and increasing demand of ready-to-eat food. In addition, the market is impacted by rising disposable incomes and a rise in the employment rate of women.

The COVID-19 pandemic is predicted to have a positive impact on the pasta market worldwide than anticipated and the global market exhibited a growth of 7.84% in 2020 as compared to 2019. The increased consumption of pasta globally during pandemic was majorly due to its affordability and convenience. The introduction of the new pasta products creates an opportunity for the incremental growth of pasta market. Additionally, employment is increasing that also led to surge in the demand of ready to cook, cheap, convenience and healthy food products.

There are n number of players in pasta market like Nestle, ITC, Bambino, Del Monte, MTR and many more. The pasta market size was valued USD 14.1 Billion in 2022 and is estimated to reach USD 17.7 Billion by 2030 growing with the CAGR of 3.32% from 2023 to 2030 (Fig. 9.1). The pasta market is primarily split into fresh, instant and dried pasta based on types while on application basis it is categorized as family and commercial. Presently, dried pasta taken charge globally holding the major percentage in the market due to longer shelf life than that of instant and fresh pasta. Additionally, dried pasta absorbs ingredients and sauces better than instant and fresh pasta, resulting in good acceptability of varietal recipe.

The pasta is primarily made from durum wheat semolina, to further improve the market share of pasta manufacturers are not only adding flavour but also working on improvising the nutritional value of pasta by adding more protein, fiber, antioxidants, micronutrients, industrial waste and gluten free flour. Moreover, due to the paradigm shift in the consumption trend of healthier food products due to the change in the lifestyle pattern (Fig. 9.2). This recent trend to manufacture healthier pasta is expected to boost growth substantially in the coming years.

There are n number of challenges for the pasta industries such as the prime ingredient of pasta durum wheat, grown only 5% to the total world wheat production and is generally sold at a higher price than common wheat (Fuad & Prabhasankar, 2010). Consequently, to meet the need of rising pasta consumption and healthy pasta, non-conventionalsources are also need to be searched (Nilusha et al., 2019). Consumer appreciates the new developed product with respect to taste, cooking behaviour and nutritional feature of pasta. However, they ignore the challenges behind producing these healthier options, with change in single variable like type of raw material can affect the entire process and quality of pasta. In this context, process development and product formulation from non-conventional source are need to be studied in detail to obtain high quality end product (Bresciani et al., 2022).

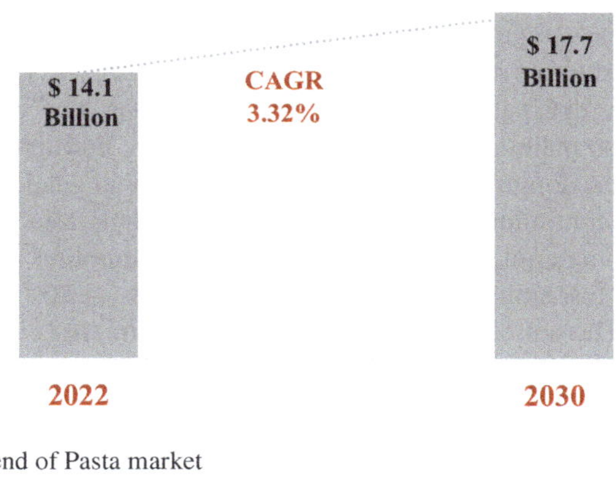

Fig. 9.1 Global trend of Pasta market

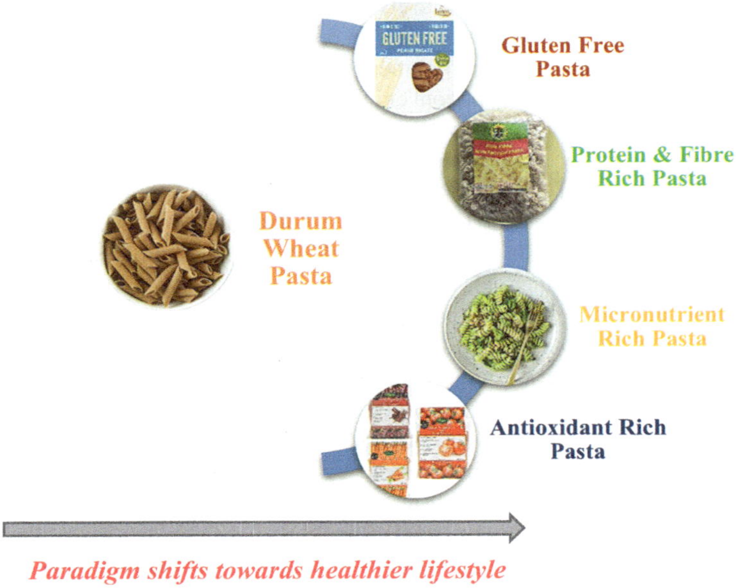

Fig. 9.2 Trends in pasta industry

3 Equipment and Machinery Involved in PASTA Processing

Pasta is an extruded product made from cold extrusion technology with single screw. Cold extrusion is the technology where product process below 50 °C (Manthey & Twombly, 2006). The prime steps in pasta manufacturing are mixing, kneading, extrusion/shaping, drying and packaging. Above all, mixing, kneading and extrusion or shaping steps are done in the extruder while drying and packaging needs additional machinery. In Fig. 9.3 all the important components of pasta machine are labelled. The extrusion screw, which is used to force the dough through the die once it has been mixed and sent to the extruder, also kneads the dough into a homogeneous mass, regulates the production rate, and affects the overall quality of the finished product. In high-capacity extruder, the barrel is equipped with water cooling jacket to dissipate the heat generated during the extrusion process. Additionally, it also maintains the extrusion temperature throughout the process to 51 °C, because if the dough is too hot the gluten protein will get denature and affects its ability to make continuous 3-D network which ultimately affect the pasta quality.

The basic raw material for pasta is durum wheat semolina and water. Durum wheat is milled into semolina with the particle size between 300 and 500 µm into a milling machine. The semolina is then transferred into a mixing chamber of extruder from the inlet and added with moisture @30% to hydrate all the particles of

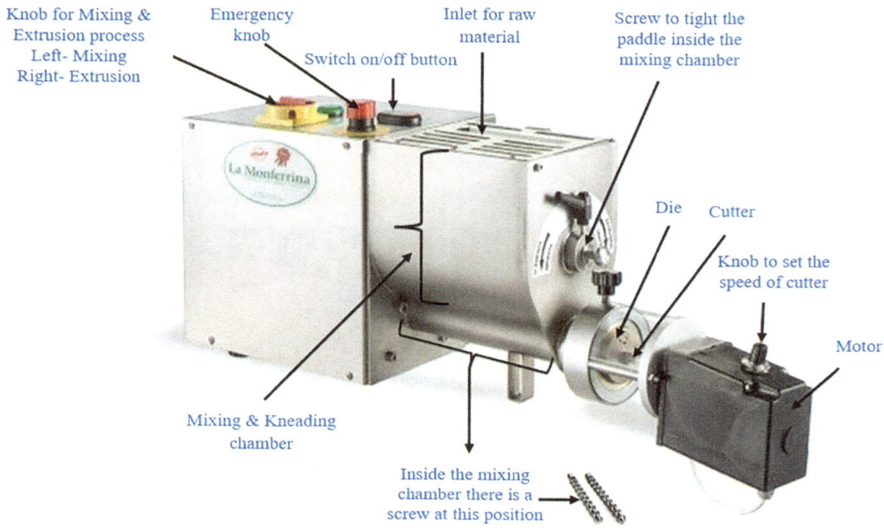

Fig. 9.3 Cold extruder for pasta making

semolina. The particles are being mixed in the mixing chamber for around 5–10 min depending upon the capacity with the paddle/barrel that rotates clockwise during mixing and anticlockwise during extrusion. The temperature of the barrel usually reaches 50 °C, with mixer vacuum of <45 cm Hg. After mixing the dough is passed through the die. The screw present inside the chamber has two functions: (1) It conveys and pushes the material from back to front, towards the die and, (2) It carries the material with pressure to the die, that will result in the continuous 3-D network of gluten in pasta. The speed moving at a speed of 20 rpm. Dies are available in different forms resulting in different shapes of pasta (Fig. 9.4). The two types of dies are widely used including bronze die and modern Teflon™ coated bronze die. It is well observed that bronze dies produce rough surface and a larger final diameter while Teflon™ coated bronze dies produces very smooth surface (Lucisano et al., 2008; Mercier et al., 2011; Pollini et al., 2012). After the extrusion of dough into different shapes the pasta is subjected to drying at 40 °C in a tray dryer with 95% relative humidity (Approved by AACCI in 1961).

Pasta was traditionally either made fresh and consumed right away or air dried for storage and consumed later. However, in air drying chances of microbial growth are much higher as this is less controlled than modern drying and greatly dependent upon the weather conditions. In modern drier, temperature, relative humidity and air flow can be controlled. After drying pasta is packed so packaging machine required. The pasta products are packed in cellophane bags or in boxes.

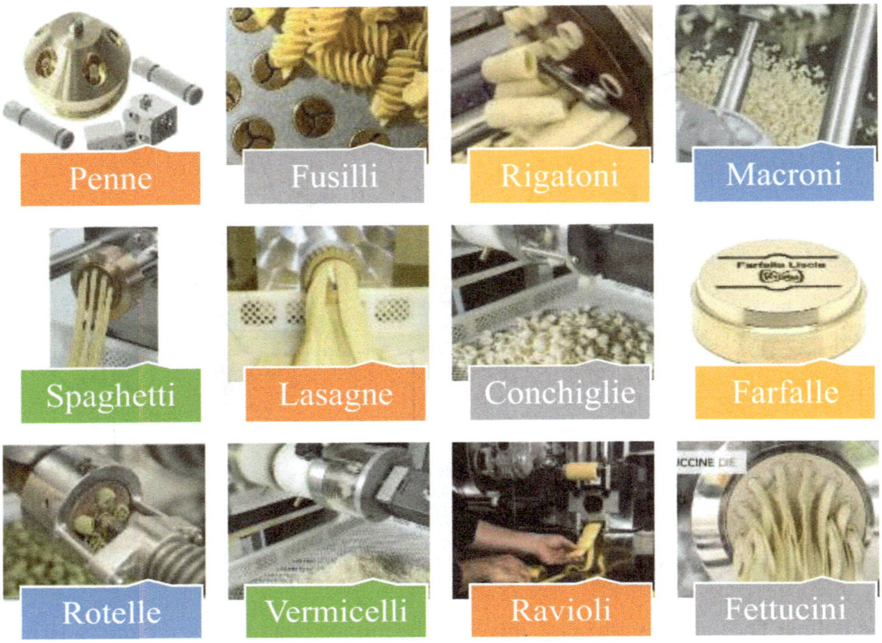

Fig. 9.4 Types of die used for various pasta products

4 Types of Pasta

In the recent century industries and scientist are working on the products that are most suited for the consumer and consumer are demanding minimally processed, healthier and convenience food products. In this pasta is the excellent vehicle to carry all the nutrients in a convenient way. Therefore, to meet the demand of the consumers, incorporation of gluten free flour, proteins, fibers, micronutrients and antioxidants into the pasta products have been done. These newly developed pasta not only improves the nutritional profile of the pasta but also its biofunctionality (Table 9.1).

Gluten FREE Pasta

Pasta globally available in the market are solely prepared from durum wheat which contains gluten. A mixture of protein gliadin and glutenin, the gliadin component of the gluten is responsible for typical digestive related condition where an individual is unable to response to gluten is known as celiac disease. To avoid such problems in the gluten intolerant individual's gluten free diets are recommended and most ideal. As we know gluten is one of the important constituents responsible to give

Table 9.1 Impact of supplementation of non-conventional food to the quality attributes of pasta

S.No.	Type of Pasta	Dietary ingredient	Impact on the Pasta Properties	Reference
1	Gluten free pasta	Soy flour, Channa flour, sorghum flour and whey protein concentrate	• Slightly higher cooking loss than the acceptable limit due to the total devoid of gluten content in the flour. • High shear value was notices for all pasta. • The *in vitro* starch digestibility of gluten free pasta without gums was slightly higher than control and with gums it is lower because of encapsulation of starch with gums. • Low *in vitro* protein digestibility of gluten free pasta.	Susanna and Prabhasankar (2012)
		100% faba, lentil and black gram flour	• Highest protein in black gram flour pasta then lentil and faba bean pasta. • The % histidine, isoleucine, leucine, valine, lysine, cysteine and methionine, tyrosine and threonine content is higher in all the pasta than the RDA levels. • The resistant starch was twice in all legume-based pasta in comparison to durum wheat pasta. • The cooking time of legume-based pasta was lower than the durum wheat pasta. The order is black gram pasta<faba bean pasta< lentil pasta. • Cooking loss was higher for lentil pasta as they are gluten free. • The gluten free pasta was less firm than the durum wheat pasta which represents the absence of strong protein network.	Laleg et al. (2016)
		Sorghum flour, amaranth flour and maize flour	• Firmness was low in comparison to industrial sample which impact the consumer acceptance. • The cooking time of pasta was in the acceptable limit between 7 to 8.5 min. • Cooking loss was less. • Pasta with 60% maize, 30% sorghum flour, 10% amaranth flour 12% egg white. 2.5% guar gum and 36% water was found to have excellent pasta attributes by sensory panellists.	Paux and Rosentrater (2018)
		Rice and lupine bean	• Higher values of protein ranging from 13.64 to 23.31%, as lupine is good source of protein containing essential amino acids. • Maximum average value of cooking loss was 10.85% which well within range. Slight increase was due to the absence of gluten. • The acceptable ratio of ingredients is 20% lupin flour, 82% rice flour, 30% egg, 0.15% guar gum. • The acceptable gluten free pasta contains 18.6% protein, 0.68% fiber, 7% unsaturated fat.	Albuja-Vaca et al. (2019)
		Germinated quinoa flour (0 to 30%) in rice and corn semolina	• Increasing quinoa levels increase the protein from 8.1 to 12.7%. • Antioxidant activity was also increased from 13.4 to 28.8%. • Minerals like Ca, Fe, K, Mg, P, and Zn increased with quinoa addition. • Cooking loss was increased with quinoa levels but it was well within limit of acceptability. • Firmness of the quinoa-based gluten free pasta was increased.	Demir and Bilgicli (2020)

(continued)

Table 9.1 (continued)

S.No.	Type of Pasta	Dietary ingredient	Impact on the Pasta Properties	Reference
2	Protein enriched pasta	Egg albumen, Rice bran protein concentrate, whey protein concentrate and soy protein concentrate	• The order for cooking time was whey protein<egg albumin = soy protein<rice bran protein concentrate. • The order for cooking loss was Egg albumin<Whey protein<Soy protein< rice bran protein concentrate. • The order for firmness was Soy protein< rice bran protein concentrate< whey protein< egg albumin. • Individually with increasing percentage of protein cooking time, cooking loss and firmness increased.	Phongthai et al. (2017)
		Oyster mushroom powder	• Cooking time, water absorption and cooking loss increased with mushroom level. • Extensibility of the noodles increased. • The mushroom powder @4% was acceptable in terms of taste, texture and flavor. • The nutritional composition of 4% enriched pasta was 11.32%, 18.27% fat, dietary fiber 1.96%.	Arora et al. (2018)
		Soy protein isolate	• Cooking time and cooking loss increased. • Hardness of the pasta increased. • Color values are impacted, L*value decreased, a* and b* value increased. • Soy protein inclusion up to 10% was acceptable. • *Invitro* protein digestibility of soy protein fortified noodles increased.	Khatkar and Kaur (2018)
		Pangas protein isolate	• Protein content increased to 21.30%. • Cooking time, water absorption and cooking loss, increased by manifold. • Yellowness of the pasta was prominent in 10% PPI supplemented pasta.	Reddy Surasani et al. (2019)
		Quinoa protein isolate	• The pasta was acceptable with 8% fortification of QPI containing 18.20% crude protein. • The cooking time, cooked weight, swelling index, water absorption and cooking loss increased but to an acceptable limit. • L* and b* value decreased while a* value increased after enrichment. • Firmness increased with protein levels.	Gupta et al. (2021)

3	Fiber enriched pasta	Wheat, rice, barley and oat bran	• Cooking time was reduced due to the disruption of gluten network. • Water absorption increased with levels of bran. • Loss of solids in the cooking water increased to a greater extend with fiber enrichment. • The suggested cereal bran (wheat, rice and oat bran) can be added upto 15% and 10% (barley bran) without much affecting the quality characteristics of pasta.	Kaur et al. (2013)
		Long-chain inulin, short-chain inulin, Glucagel, psyllium and oat	• All combination of fiber enriched pasta increased the cooking time. • Swelling index and cooking loss increased due to the dilution of gluten and starch. • Out of all dietary fiber, oat is suggested in improving the cooking properties of pasta. • For some fibers like oat, glucagel the firmness of the pasta improved while for other it decreased. • Pasta enriched with dietary fiber showed darker colorcolor than control.	Foschia et al. (2015)
		Wheat bran	• Incorporation of wheat bran induced an increase in cooking time, swelling index and cooking loss of the enriched pasta. • The color of the bran enriched pasta was comparatively darker. • Nutritional quality significantly improved.	Amel et al. (2015)
		Oat β-glucan	• Incorporation of oat fiber powder increased the swelling index, water uptake ratio and cooking loss values. • Oat fiber substituted pasta are darker, softer and less sticky. • Pasta enriched with 8% oat fiber is recommended with acceptable pasta quality attributes and can be labelled under "high in fiber" category.	Piwińska et al. (2016)
		Insoluble dietary fiber from wheat bran (0,2,4,6,8,10%)	• Addition of insoluble dietary fiber improves the viscoelastic and mechanical strength of the sheeted dough for noodles by filling the gluten network. • The cooking loss in noodle increased over 4% insoluble dietary fiber and maximum tensile strength was reduced.	Lei et al. (2021)

(continued)

Table 9.1 (continued)

S.No.	Type of Pasta	Dietary ingredient	Impact on the Pasta Properties	Reference
4	Micro-nutrient fortified pasta	Vitamin A, folic acid and iron	• Rice noodle prepared by steaming contains higher amount of protein, fat, ash and gelatinized starch than boiling. • Noodles prepared from steaming showed better cooking and textural qualities than boiling. • The losses of *vitamin A, folic acid and iron and retention was more in steamed noodles due to the* regular, smooth and close structure of steamed noodles. • The maximum loss of vitamin A, folic acid and iron was in the cooking stage in comparison to other unit operation.	Malahayati et al. (2017)
		For iron: Ferric sodium ethylene diamine tetra-acetic acid (NaFeEDTA: 4-5 mg/100 g) and ferrous sulphate (6,7,8 mg/100 g) For vitamin A: Retinyl acetate (700,800 and 900 μg/100 g)	• The vitamin A was retained to 89–90% while 56 to 62% of iron was retained after cooking. • Double fortification can meet the RDA for iron and vitamin A, when consumed in sufficient amount. • No significant difference was observed in the quality parameter of sensory analysis than control.	Sharma et al. (2020)
		Spinach and carrot (6–18%), millet (pearl millet and sorghum; 10–30%) to increase vitamin A and iron	• Cooking time decreased with increased level of incorporation, due to the physical disruption of gluten network that provides path for water absorption into pasta resulted in shorter cooking time. • Cooking loss increased but in acceptable limit. • The vitamin A was checked for losses, it was loss during kneading phase where water and lipoxygenase comes in contact with air and cause oxidation. Retention of more than 95% of vitamin A. • Iron was stable throughout the processing. • The incorporation of 12% spinach and carrot powder and 20% of pearl millet and sorghum were highly acceptable with excellent retention of vitamin A and iron.	Sharma et al. (2021)

5	Antioxidant rich pasta	Ground nut meal (10–20 g) and capsicum juice (14–30 mL)	• The pasta containing higher levels of capsicum juice showed higher antioxidant (18.12%) and phenolics (342.27 mg/100 g) value where in the pasta where ground nut meal was high, its protein (17.81%) content was high. • Pasta with high capsicum juice showed low cooking loss with high hydration ratio. • The optimized combination was 20 g groundnut meal, 30 mL capsicum juice and 89.99 g refined wheat flour.	Mridula et al. (2016)
		Pomegranate peel powder (3,5 and 7%)	• The polyphenolic and flavonoids content increased from 102.75 to 488.96 mg gallic acid equivalent/100 g and 139.72 to 190.82 mg quercetin equivalent/100 g with 7% incorporation. • Addition of pomegranate peel powder improves the dietary fiber content from 2.88 to 7.88%. • Cooking loss in the pasta was increased with peel powder percentage but within the acceptable limit. • The pomegranate powder can be added up to 7% for for improving the antioxidants in pasta products.	Essa and Mohamed (2018)
		Carrot pomace, finger millet and pearl millet	• The enriched pasta possesses 38.45% antioxidant activity and containing 0.67 mg gallic acid/g. • Minerals like Ca, Fe and Zn were improved. • The cooking time and loss was decreased in antioxidant rich pasta as it is added with gums that provides the mechanical strength. • Firmness increased. • The optimized blend was 20 g finger millet and 12 g pearl millet /100 g in 96 g of durum wheat and 4 g carrot pomace.	Gull et al. (2018)
		Spinach extract (0.4 to 40 mg/mL)	• The noodles were reported to have increased antioxidant activity with increased levels of extract fortification. • Fe, Mg and vitamin C was increased manifold in the spinach extract enriched noodle. • The optimum concentration was 4 mg/mL in terms of noodle sensory attributes.	Susanti et al. (2021)
		Wheatgrass juice (33,66 and 100%)	• The addition of wheatgrass juice increased the phenolic (31.13–85.28 mg/100 g), flavonoids (44.26–77.18 mg/100 g), chlorophyll (0.88 to 13.03 mg/100 g) and antioxidant activity (21.76 to 36.74%) of pasta. • Incorporation of juice significantly decreased the cooking time whereas water absorption and cooking loss increased but, well within the acceptable limit. • The color values were greatly impacted by the wheatgrass juice, especially a* value. • The firmness of the pasta increased. • The pasta containing 100% wheatgrass juice was acceptable.	Bawa et al. (2022)

mechanical strength to pasta. Therefore, during gluten free pasta preparation additives like hydrocolloids should be added in proper proportion to give good quality of product (Larrosa et al., 2013). Most commonly used raw materials for gluten free pasta are legume flour (faba bean and green lentil) (Laleg et al., 2016); amaranth, maize and sorghum (Paux & Rosentrater, 2018); rice, corn and flaxseed (Khosla et al., 2019); Rice and lupin flour (Albuja-Vaca et al., 2019); quinoa flour (Demir & Bilgicli, 2020) and brown rice and *Cladophora spp.* Powder (Choobthaisong & Oupathumpanont, 2021).

Cooking loss is an important criterion to check the quality of pasta, better the protein network forms, smaller is the cooking loss. Due to the lack of gluten in the gluten free pasta, the functional properties are majorly depending upon the starch, it gelatinizes while cooking and often functions as the only structural network of the final gluten free products (Lucisano et al., 2012). Marti et al. (2010) reported in absence of gluten addition of pre-gelatinized starch could provide structural properties to the gluten free pasta, amylose and amylopectin in rice plays a key role starch network, resulted in the textural integrity of pasta after cooking. Heating and cooling cycle of starch forms a rigid structure. Moreover, combination of starch gelatinisation and protein denaturation responsible for building a compact structure to increase the firmness of gluten free pasta (Foschia et al., 2017).

In the gluten free pasta, to acquire the good textural and cooking properties, additives like gums or pre-gelatinized starch are added to give the mechanical strength to the pasta. Hydrocolloids can mimic the function of gluten network in the gluten free pasta (Nilusha et al., 2019). The level of protein is an essential component in giving good cooking properties and nutritional value to pasta. In this regard, study conducted by Phongthai et al. (2017) to evaluate the effect of different protein such as rice bran protein, whey protein, soy protein and egg albumin on the rice flour based gluten free pasta in order to improve the cooking quality. They suggested in place of gluten egg albumen and soy protein can be used to improve the cooking quality of pasta.

Protein Enriched Pasta

Pasta are being enriched with protein sources to improve its nutritional quality as well as to meet the dietary requirements. Durum wheat semolina protein ranged between 9 and 14%, but it lacks essential amino acids, lysine is the limiting amino acid. Therefore, to improve the protein quality of pasta number of protein sources such as flours, protein concentrates, protein isolates and protein hydrolysates are added in the durum wheat partially for improvisation. Sometime the durum wheat is completely replaced by the protein rich flours like pulses and legumes flour added with gums. Additions of protein have a prominent effect on the nutritional, cooking, textural, sensory and morphological properties. These changes are majorly due to the dilution of starch and protein that will hamper the three-dimensional structure of gluten. In recent past, various studies have been conducted to improve the protein quality of pasta through enrichment with various soy protein isolate (Mishra &

Bhatt, 2017), oyster mushroom powder (Arora et al., 2018), proteins fish protein powder (Desai et al., 2018), soy protein concentrate (Khatkar & Kaur, 2018), pangas protein isolate (Reddy Surasani et al., 2019), oat and faba protein concentrate (Duta et al., 2019), quinoa protein isolate (Gupta et al., 2021) and pumpkin seed meal (Raina et al., 2022).

Protein affects the cooking properties of pasta. Cooking time of the pasta increased with addition of protein as it hinders the starch gelatinization by forming an inclusion complex with starch and thereby increasing cooking time (Chandrashekar & Kirleis, 1988). Oh et al. (1985) observed the direct correlation between protein content and cooking time. Several researchers reported increase in cooking time with protein inclusion (Kaur et al., 2013; Sharma et al., 2013; Gupta et al., 2021). They suggested protein interferes with the hydration and swelling capacity of starch by entrapping them and thus increasing the cooking time.

Moreover, Khatkar and Kaur (2018) reported if higher concentration of protein is present in the food matrix, water forms hydrogen bond with protein instead of starch due to its polarity thus hampering starch swelling and gelatinization. Cooking loss is the loss of solids in cooking water which should not be more than 9%. Cooking loss is increased by addition of protein in pasta at higher levels due to the dilution of gluten, weakening of starch and protein matrix and lesser interaction between starch and protein (Reddy Surasani et al. (2019). In addition, protein rich in albumins and globulins like fish protein, egg protein, quinoa protein, amaranth protein, unable to form continuous 3-D structure of gluten, resulted in increased cooking loss (Tazrart et al., 2016; López et al., 2018). Water absorption and swelling index of enriched pasta also increased due to the increased water holding and gelling ability of protein (Phongthai et al., 2017). Firmness reflects the mechanical strength and the bond integrity, so increased levels of protein increased the pasta firmness which might be due to the thermal denaturation of protein during cooking that enhances the firmness (Gupta et al., 2021). The raw material also impacts the color of the finished pasta. Thenutritional profile of the pasta ultimately improved in terms of protein quality, biological value, essential amino acid, and protein digestibility.

Fibre Enriched Pasta

The traditional raw-material durum wheat semolina contains less dietary fiber, therefore with the prior knowledge and understanding of the health benefits of fiber and increasing demand of health based and convenience food. Different sources can be incorporated to increase the fiber content of the bran like wheat, oat, barley and rice bran (Kaur et al., 2013); long chain-inlulin, short-chain inulin, glucagel, psyllium and oat (Foschia et al., 2015); oat β-glucan (Piwińska et al., 2016); turmeric flour (Wahanik et al., 2018) and insoluble fiber from wheat bran (Lei et al., 2021).

Quality parameters such as cooking time, swelling index, water absorption and cooking loss are affected by fiber supplementation. For instance, Foschia et al. (2015) reported increase in cooking time of pasta. On contrary some literature cites decrease in the cooking time of pasta with fiber incorporation (Kaur et al., 2013). This behaviour could be associates with the disruption of gluten network that allows water to penetrate inside and thus, shorten the cooking time. Bran enrichment increase the cooking loss of the pasta, as bran competes with starch for water, and disrupt the protein and starch matrix leads to uneven distribution of water within the pasta. The continuous three-dimensional network of gluten is also disrupted by bran thereby leading to loss of solids in cooking water. Sosulski and Wu (1988) reported enrichment of fiber in the wheat flour leads increase in water hydration values in proportional to the level of replacement.

The textural properties are also impacted by the fiber enrichment. Incorporation of low insoluble dietary fiber from wheat bran gradually decreased the springiness of the noodles while increased the hardness and chewiness. This can be associated with the noncovalent interaction between fiber and protein leads to formation of high molecular weight polymer, resulted in increased hardness. Additionally insoluble dietary fiber fills the gaps in gluten network that results in stiffness of the noodles (Lei et al., 2021). The color also gets affected depending upon the color of enriched material. Texture and color are the two important parameters along with cooking properties that decides the acceptability of pasta in the consumer forum.

Fiber incorporation and consumption in appropriate amount meeting the daily recommended daily allowance i.e., 25–30 g/day, will impart health benefits and possess the tendency to combat the lifestyle diseases like atherosclerosis, Type II diabetes, heart diseases, colon cancer, constipation and many more.

Micro-Nutrient Fortified Pasta

Despite of consuming micronutrients from different food sources, majorly the staple diet lacks micronutrients like consumption white rice from which bran is removed, which is a concentrated source of micronutrient, refined wheat flour or aata in which partial or complete removal of bran is observed. These can be the reason of micronutrient deficiency among people solely depending on rice and wheat. Children and pregnant women are mostly vulnerable to these deficiencies. These nutrients are needed in very small quantity but its absence impact the survival, physical and mental health and the overall well-being of the individual. Fortification is one such possibility to combat the processing loss of these micronutrients. This technology is economical and practical in long-term approach. Several products are also available in the market such as tata salt with iodine fortification, milk with vitamin D, flour with iron and zinc fortification etc. These days pasta is widely consumed and can be considered an excellent vehicle for food fortification, due to its convenience, ease of cooking and versatility. Several researchers have tried incorporating vitamin A, folic acid and iron to pasta in form of chemical powder or using a natural food source rich in micronutrients (Malahayati et al., 2017; Yulia et al., 2018; Sharma et al., 2020, 2021).

The cooking, textural and sensory properties are the parameter that are affected by the pasta enrichment withy any dietary ingredient, while as such no significant change in the abovesaid properties were observed during chemical supplementation (Malahayati et al., 2017; Sharma et al., 2020). But in case natural food source is added that will interferes with cooking, textural and sensory properties of the pasta. For instance, Sharma et al. (2021) enriched pasta with vegetable (spinach and carrot) powders and with millet (pearl millet and sorghum) flour to increase the iron and vitamin A content in the pasta. As apart from the micronutrient other constituents like protein, fiber, fat and starch are also present along with so they dictate the techno-functional properties of the pasta. This is good approach as with simple fortification we can make change in the diet that will ultimately improves the health on a long-term basis in preventing any micronutrient disease.

Antioxidant Rich Pasta

In the recent past the consumption of antioxidant rich food is increasing as they prevent the risk of chronic diseases. Thinking pasta as excellent vehicle for nutrients, several researchers have incorporated different antioxidant source to the durum wheat including ground nut meal and capsicum juice (Mridula et al., 2016); tomato peel (Padalino et al., 2017); pomegranate peel power (Essa & Mohamed, 2018); carrot pomace, finger millet and pearl millet (Gull et al., 2018); spirula (Hussein et al., 2021); spinach extract (Susanti et al., 2021) and wheatgrass juice (Bawa et al., 2022).

Fares et al. (2015) reported in the study that addition of β-glucan from barley and Bacillus coagulans BC30 into pasta products resulted in increased levels of phenolics in the pasta. Results have also revealed that careful milling to protect the aleurone layer cause a significantly improve the bioactive profile of the pasta. Agro industrial waste can also be utilized to increase the antioxidant potential of the pasta. Essa and Mohamed (2018) reported an increase in the polyphenols and flavonoids from 102.74–488.96 mg gallic acid /100 g and 139.72–190.82 mg quercetin/100 g. Moreover, increment in the dietary fiber was also seen from 2.88–7.86%. However, addition not only have the positive effect, but is affects the cooking properties of the pasta by increasing cooking loss, as pomegranate peel is rich in fiber it might have disruption the continuous structure of gluten protein resulting in more leaching loss. Susanti et al. (2021) reported an increase in the antioxidant properties of pasta with the addition of spinach extract, in addition to it, Fe, Mg, vitamin C were also observed to be improved in concentration with concomitant increase in the of spinach extract.

As cooking parameters are important to decide whether the pasta will have the consumer acceptability on not. Addition of antioxidant rich raw material will also add fiber, protein and fat to the pasta that will disturb the continuous 3-D structure of gluten and increase or decrease the cooking time, cooking loss, water absorption and swelling power that leads to change in the textural properties of the pasta. Gull

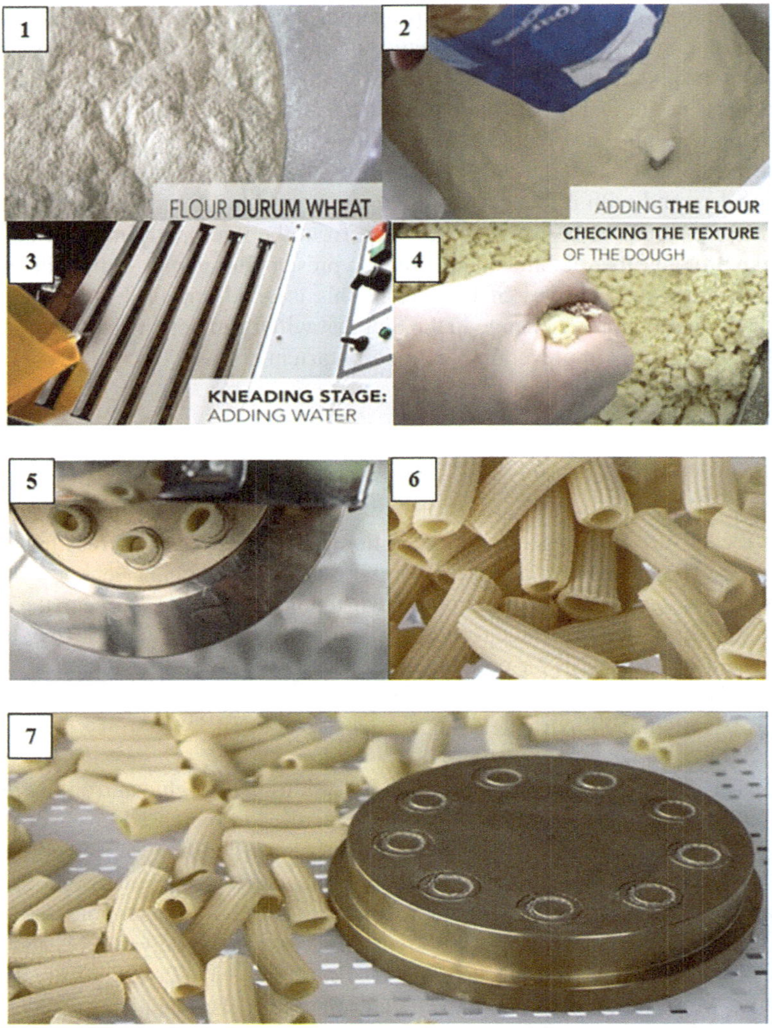

Fig. 9.5 Process for making pasta

et al. (2018) reported increased in phenolic content and antioxidant potential of the finished pasta with better cooking characteristics of pasta after the addition of carrot pomace, pearl millet and finger millet.

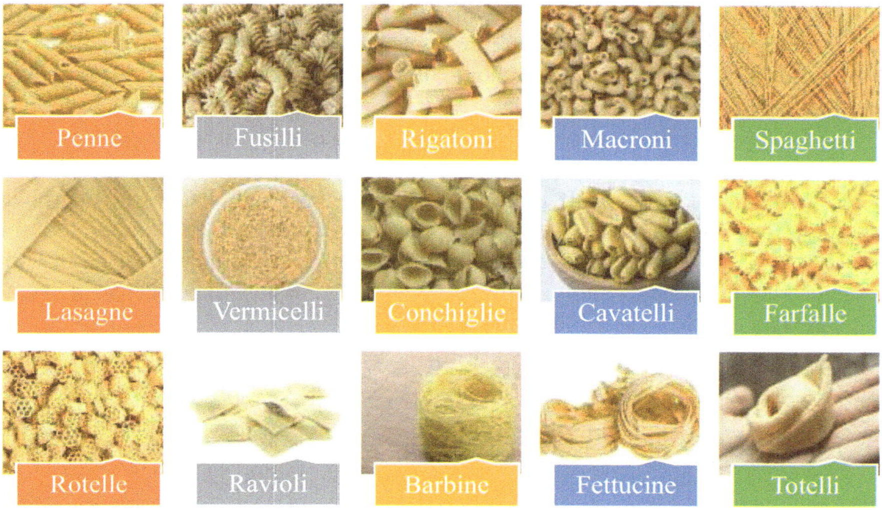

Fig. 9.6 Different shapes of pasta

5 Preparation Methods

Pasta products are manufactured by mixing durum wheat semolina, water, eggs (optional) and sometimes other ingredients like fruit juice, bran, antioxidant rich flour, protein rich flour etc. These ingredients are typically added into auger extruder which can be equipped with different die as per requirement of shape, and dried further to reduce the moisture content and to extend the storage stability and packaged for market. The general pasta manufacturing involves five major steps that are mixing, kneading, extrusion/shaping, drying and packaging. The process of making pasta is well picturized in Fig. 9.5. The raw material is transferred into the mixing chamber of extruder for mixing. In mixing operation, water is added @27–30% depending upon the flour and its particle size. Modern extruders are also equipped with vacuum chamber to remove air from the pasta before extrusion. The air removal possesses its significance, if air is present, it will diminish the mechanical strength and give final pasta a white and chalky appearance. During hydration of semolina, glutenin and gliadin forms a strong gluten network which is typical for wheat flour. This gluten network will form an excellent compact system with swelled starch granules during cooking.

Once the dough is mixed and kneaded it is forced through the die into desirable shapes like penne, fusilli, rigatoni, conchiglie, spaghetti, lasagne, vermicelli, macroni etc. (Fig. 9.6). Once the pasta is extruded it is subjected to drying which is the critical step. The prime objective of drying is to lower the moisture content from 30% to 7–8% in the finished product. Immediately after extrusion pasta is preliminary dried to prevent the pasta from sticking together and later on send for proper drying to required moisture content. During drying temperature and relative humidity in dryer make a prominent difference in the pasta quality. The outer layer of pasta dries more rapidly in comparison to the inside, this develops a moisture gradient in the pasta.

across the outer and inner surface. If pasta dried quickly, it will cause cracks in the pasta that will look very poor in appearance and due to which pasta posses' lower mechanical strength. If it dries too slowly, there are chances of microbial spoilage. Therefore, it is necessary to dry the pasta at standardized temperature for specific time (45–55 °C for 4–5 h). This will result in good quality of pasta with excellent cooking and technological properties. Then, pasta will be packaged in cellophane bag or boxes (Walsh & Gilles, 1977).

6 Quality Aspects of Pasta

The major parameters involved to evaluate the quality of pasta quality are minimum cooking time, cooking loss, firmness and cohesiveness. Other than this color, cooking tolerance, weight of cooked pasta, leaching loss of nutrients during cooking, sensory profile, morphological changes etc. are being characterised to evaluate the quality of pasta (Murray et al., 2017).

Cooking time is the minimum time required to cook the pasta to *Al Dente* stage. The standard procedure for analysing the cooking time is given in AACC (2000). The effect of non-conventional ingredients added will be studied. Like addition of protein decrease the cooking time (Petitot et al., 2010) while addition of dietary fiber may increase or decrease the time depending upon the source and type of fiber (Kaur et al., 2013; Biernacka et al., 2017). Cooking loss is also evaluated by AACC (2000) procedure. It determines the influence of ingredient and process parameters on the leaching of soluble starch and non-starch polysaccharides in the cooking water (Gupta et al., 2021). Generally, less than 9% of solids are desirable for high quality of pasta. Addition of protein as well as dietary fiber after certain percentage increase the cooking loss by weaking and disruption of the gluten network that weakened the structure of pasta (Desai et al., 2018). Moreover, due to the dilution of gluten may hamper the formation of continuous 3-D gluten and starch network (Reddy Surasani et al., 2019). Water absorption, swelling index, volume expansion pigment or nutrient loss are the other parameters taken into account to check the quality of pasta.

Consumer acceptability or sensory profile of the pasta is also an important check for the quality of the pasta in terms of overall acceptability (Larmond & Voisey, 1973). During sensory evaluation lighting, timing of evaluation, presentation and temperature of sample has a great impact on the result (D'Egidio & Nardi, 1996; Menger, 1979). The number of panellist required differ from study to study and country to country, like for Italy only three panellists are sufficient while other requires ten or more (Murray et al., 2017). The panellists are asked to check the quality parameters like appearance, color, taste, texture/mouthfeel, flavour and overall acceptability. The sensory analysis is coming under the category of subjective analysis, it includes humans and is unfortunately is expensive and time consuming. Moreover, result variation among the subjects is very high, leading many researchers to depend on chemical and instrumental analysis.

There is a lack of standardized procedure for many pasta quality parameters leading to inconsistency in the results of different labs. For instance, texture analysis by texture analyser which does not have any standardized procedure of AACC. It has been observed during the analysis that time of texture analysis is very important as the hot pasta having different firmness and stickiness value in comparison to the cool pasta. This is associated with the physical changes occurs during cooling of pasta and when is exposed to air. The firmness is defined as the impression of bond quality and the uprightness of the protein network in cooked pasta (Nilusha et al., 2019). The physical competition between the starch swelling and protein coagulation determines the textural properties of pasta (Cunin et al., 1995). During cooking when protein coagulation occurs over to starch swelling, resulted in firmness of the pasta. On contrary, when starch swelling occurs over protein coagulation, resulted in increasing stickiness of the pasta (Foschia et al., 2015).

Color is also an important parameter in selection of the pasta by the consumer on the market shelves. The yellowness in the pasta is most desirable and is greatly impacted by the raw material used, lipoxygenase activity, destruction of pigment during manufacturing pasta especially during extrusion and drying (Nilusha et al., 2019). The removal of air bubbles before extrusion is an important preliminary step as its presence of air oxidises the pigment while higher temperature during drying can result in destruction of pigment. The method by AACC (2000) is approved for characterising the color of pasta. The equipment used for analysing color is Hunter color lab, it gives L*, a* and b* value. The L* value varies between 0 to 100, 0 indicate the darkness (Black) and 100 represents lightness of the pasta sample. The a* value tells the redness (+) and greenness (−) while b* value indicates the yellowness (+) and blueness (−) of the pasta sample. Apart from these other parameters like morphological changes, thermal properties, rheological characterization, nutritional improvement are also being characterized to study the quality of pasta.

7 Conclusion and Future Trends

Pasta is regarded as a traditional Italian product and is consumed globally for its flavor, texture and nutritional benefits. Despite of established technology, the production process has to be improved taking into account of altered consumer lifestyle and demand of health-based diet. The current method for making pasta must therefore be re-examined, re-thought and re-adjusted in terms of contemporary interest in gluten free-, fiber rich and multigrain formulas. In consideration of the structure of pasta that can withstand extrusion, drying and cooking of pasta. More to this, there is an opportunity in exploring more alternative raw-material to meet the demand and promote economical and environmental sustainability, particularly in terms of upcoming climate change that might further restrict durum wheat availability.

References

Adegunwa, M. O., Bakare, H. A., & Akinola, O. F. (2012). Enrichment of noodles with soy flour and carrot powder. *Nigerian Food Journal, 30*, 74–81.

Albuja-Vaca, D., Yépez, C., Vernaza, M. G., & Navarrete, D. (2019). Gluten-free pasta: Development of a new formulation based on rice and lupine bean flour (Lupinus Mutabilis) using a mixture-process design. *Food Science and Technology, 40*, 408–414.

Amel, K., Amel, M., & Samir, F. (2015). Effect of the incorporation of wheat bran on pasta quality. *Journal of Klagenfurt, 22*, 225–234.

American Association of Cereal Chemists (AACC). (2000). *Approved methods of the AACC*. AACC.

Arora, B., Kamal, S., & Sharma, V. P. (2018). Nutritional and quality characteristics of instant noodles supplemented with oyster mushroom (*P. ostreatus*). *Journal of Food Processing and Preservation, 42*, 1–8.

Bawa, K., Singh, A., Brar, J. K., Reddy Surasani, V. K., & Pandiselvam, R. (2022). Influence of wheatgrass juice on techno-functional properties and bioactive characteristics of pasta. *Journal of Food Quality*, Article ID 3891983:1–12.

Biernacka, B. D. D., Gawlik-Dziki, U., Różyło, R., & Siastała, M. (2017). Physical, sensorial, and antioxidant properties of common wheat pasta enriched with carob fiber. *LWT-Food Science and Technology, 77*, 186–192.

Bresciani, A., Pagani, M. A., & Marti, A. (2022). Pasta-making process: A narrative review on the relation between process variables and pasta quality. *Food, 11*, 256.

Chandrashekar, A., & Kirleis, A. W. (1988). Influence of protein on starch gelatinization in sorghum. *Cereal Chemistry, 65*, 457–462.

Chillo, S., Laverse, J., Falcone, P. M., & Del Nobile, M. A. (2008). Quality of spaghetti in base amaranthus wholemeal flour added with quinoa, broad bean and chick pea. *Journal of Food Engineering, 84*, 101–107.

Choobthaisong, P., & Oupathumpanont, O. (2021). The nutritional quality improvement of dried gluten free pasta supplemented with Cladophora spp. *Emirates Journal of Food and Agriculture, 33*, 395–399.

Cunin, C., Handschin, S., Walther, P., & Escher, F. (1995). Structural changes of starch during cooking of durum wheat pasta. *LWT-Food Science and Technology, 28*, 323–328.

D'Egidio, M., & Nardi, S. (1996). Textural measurement of cooked spaghetti. In J. Kruger, R. Matsuo, & J. Dick (Eds.), *Pasta and noodle technology* (pp. 133–156). AACCI.

Demir, B., & Bilgicli, N. (2020). Utilization of quinoa flour (*Chenopodium quinoa* Willd.) in gluten-free pasta formulation: Effects on nutritional and sensory properties. *Food Science and Technology International, 27*(3), 1–9.

Desai, A., Brennan, M. A., & Brennan, C. S. (2018). The effect of semolina replacement with protein powder from fish (*Pseudophycisbachus*) on the physicochemical characteristics of pasta. *LWT- Food Science and Technology, 89*, 52–57.

Duta, D. E., Culetu, A., & Sozer, N. (2019). Effect of dry fractionated hybrid protein ingredients on the structural, textural, thermal and sensory properties of gluten-free oat and faba pasta. *International Journal of Food Science and Technology, 54*, 3205–3215.

Essa, R. Y., & Mohamed, E. E. (2018). Improvement of functional and technological characteristics of spaghetti by the integration of pomegranate peels powder. *American Journal of Food Technology, 13*, 1–7.

Fares, C., Menga, V., Martina, A., Pellegrini, N., Scazzina, F., & Torriani, S. (2015). Nutritional profile and cooking quality of a new functional pasta naturally enriched in phenolic acids, added with β-glucan and *Bacillus coagulans* GBI-30, 6086. *Journal of Cereal Science, 651*, 260–266.

Foschia, M., Peressini, D., Sensidoni, A., Brennan, M. A., & Brennan, C. S. (2015). How combinations of dietary fibres can affect physicochemical characteristics of pasta. *LWT-Food Science and Technology, 61*, 41–46.

Foschia, M., Beraldo, P., & Peressini, D. (2017). Evaluation of the physicochemical properties of gluten-free pasta enriched with resistant starch. *Journal of the Science of Food and Agriculture, 97*, 572–577.

Fuad, T., & Prabhasankar, P. (2010). Role of ingredients in pasta product quality: A review on recent developments. *Critical Reviews in Food Science and Nutrition, 50*, 787–798.

Gull, A., Prasad, K., & Kumar, P. (2018). Nutritional, antioxidant, microstructural and pasting properties of functional pasta. *Journal of the Saudi Society of Agricultural Sciences, 17*, 147–153.

Gupta, A., Sharma, S., Reddy Surasani, V. K., & Singh, A. (2021). Quinoa protein isolate supplemented pasta: Nutritional, physical, textural and morphological characterization. *LWT-Food Science and Technology, 135*(2021), 110045. https://doi.org/10.1016/j.lwt.2020.110045

Hussein, A., Ibrahim, G., Kamil, M., El-Shamarka, M., Mostafa, S., & Mohamed, D. (2021). Spirulina-enriched pasta as functional food rich in protein and antioxidant. *Biointerface Research in Applied Chemistry, 11*, 14736–14750.

Kaur, G., Sharma, S., Nagi, H. P. S., & Ranote, P. S. (2013). Enrichment of pasta with different plant proteins. *Journal of Food Science & Technology, 50*, 1000–1005.

Khatkar, A. B., & Kaur, A. (2018). Effect of protein incorporation on functional, thermal, textural and overall quality characteristics of instant noodles. *Journal of Food Measurement and Characterization, 12*, 2218–2229.

Khosla, R., Rathi, N., Gururani, P., Upadhyay, S., & Singh, M. (2019). Development and physicochemical analysis of gluten free pasta using Rice, corn and flaxseeds. *International Journal of Recent Scientific Research, 10*, 32300–32305.

Kovacs, M. I. P., Poste, L. M., & Butler, G. (1997). Durum wheat quality: Comparison of chemical and rheological screening tests with sensory analysis. *Journal of Cereal Science, 25*, 65–75.

Laleg, K., Cassan, D., Barron, C., Prabhasankar, P., & Micard, V. (2016). Structural, culinary, nutritional and anti-nutritional properties of high protein, gluten free, 100% legume pasta. *PLoS One, 11*, 0160721.

Larmond, E., & Voisey, P. (1973). Evaluation of spaghetti quality by laboratory panel. *Canadian Institute Food Science and Technology Journal, 6*, 209–211.

Larrosa, V., Lorenzo, G., Zaritzky, N., & Califano, A. (2013). Optimization of rheological properties of gluten-free pasta dough using mixture design. *Journal of Cereal Science, 57*, 520–526.

Lei, M., Huang, J., Tian, X., Zhou, P., Zhu, Q., Li, L., Li, L., Ma, S., & Wang, X. (2021). Effects of insoluble dietary fiber from wheat bran on noodle quality. *Grain & Oil Science and Technology, 4*, 1–9.

López, D. N., Galante, M., Robson, M., Boeris, V., & Spelzini, D. (2018). Amaranth, quinoa and chia protein isolates: Physicochemical and structural properties. *International Journal of Biological Macromolecules, 109*, 152–159.

Lucisano, M., Pagani, M., Mariotti, M., & Locatelli, D. (2008). Influence of die material on pasta characteristics. *Food Research International, 41*, 646–652.

Lucisano, M., Cappa, C., Fongaro, L., & Mariotti, M. (2012). Characterisation of gluten-free pasta through conventional and innovative methods: Evaluation of the cooking behaviour. *Journal of Cereal Science, 56*, 667–675.

Malahayati, N., Muhammad, K., Bakar, J., & Karim, R. (2017). The effect of processing method on fortified Rice noodle quality and Fortificant retention. *International Journal of Food and Nutritional Science, 4*, 30–37.

Manthey, F. A., & Twombly, W. (2006). Extruding and drying of pasta. In Y. H. Hui (Ed.), *Handbook of Food Science Technology and Engineering*. Taylor and Francis.

Marti, A., Seetharaman, K., & Pagani, M. A. (2010). Rice-based pasta: A comparison between conventional pasta-making and extrusion-cooking. *Journal of Cereal Science, 52*, 404–409.
Menger, A. (1979). Crucial points of view concerning the execution of pasta cooking tests and their evaluation. In: C. R. Symposium International Pates Alimentaires. (pp. 53–60)
Mercier, S., Des Marchais, L. P., Villeneuve, S., & Foisy, M. (2011). Effect of die material on engineering properties of dried pasta. *Procedia Food Science, 1*, 557–562.
Mishra, P., & Bhatt, D. K. (2017). Development of quality characteristics of dried pasta enriched with soya protein isolate powder. *Journal of Environmental Science, Toxicology and Food Technology, 11*, 1–6.
Mridula, D., Gupta, R. K., Bhadwal, S., & Khaira, H. (2016). Optimization of groundnut meal and capsicum juice for protein and antioxidant rich pasta. *Agricultural Research, 5*, 293–304.
Murray, J. C., Kiszonas, A. M., & Morris, C. F. (2017). Pasta production: Complexity in defining processing conditions for reference trials and quality assessment methods. *Cereal Chemistry, 94*, 791–797.
Nilusha, R. A. T., Jayasinghe, J. M. J. K., Perera, O. D. A. N., & Perera, P. I. P. (2019). Development of pasta products with nonconventional ingredients and their effect on selected quality characteristics: A brief overview. *International Journal of Food Science, 6750726*, 2019–2010.
Oh, N. H., Seib, P., Ward, B., & Deyoe, C. W. (1985). Noodles IV. Influence of flour protein, extraction rate, particle size, and starch damage on the quality characteristics of dry noodles. *Cereal Chemistry, 62*, 441–446.
Padalino, L., Conte, A., Lecce, L., Likyova, D., Sicari, V., Pellicanò, T. M., Poiana, M., & Nobile, M. A. D. (2017). Functional pasta with tomato by-product as a source of antioxidant compounds and dietary Fibre. *Czech Journal of Food Science, 35*, 48–56.
Paux, L., & Rosentrater, K. (2018). Development of gluten-free egg pasta based on Amaranth, maize and sorghum. *Journal of Food Research, 7*, 16–36.
Petitot, M., Boyer, L., Minier, C., & Micard, V. (2010). Fortification of pasta with split pea and fava bean flours: Pasta processing and quality evaluation. *Food Research International, 43*, 634–641.
Phongthai, S., D'Amico, S., Schoenlechner, R., Homthawornchoo, W., & Rawdkuen, S. (2017). Effects of protein enrichment on the properties of rice flour-based gluten-free pasta. *LWT – Food Science and Technology, 80*, 378–385.
Piwińska, M., Wyrwisz, J., Kurek, M., & Wierzbicka, A. (2016). Effect of oat β-glucanfiber powder and vacuum-drying on cooking quality and physical properties of pasta. *CyTA– Journal of Food, 14*, 101–108.
Pollini, C., Panto, F., Nespoli, A., Sissons, M., & Abecassis, J. (2012). Manufacture of pasta products. In M. Sissons, J. Abecassis, B. Marchylo, & M. Carcea (Eds.), *Durum wheat technology* (pp. 166–175). AACCI.
Raina, S., Gupta, A., Singh, A., Reddy Surasani, V. K., & Sharma, S. (2022). Functionality of pasta enriched with pumpkin seed meal: Cooking quality, techno-functional properties, textural and structural characterisation. *International Journal of Food Science and Technology, 58*, 2735–2743. https://doi.org/10.1111/ijfs.16077
Reddy Surasani, V. K., Singh, A., Gupta, A., & Sharma, S. (2019). Functionality and cooking characteristics of pasta supplemented with protein isolate from pangas processing waste. *LWT-Food Science and Technology, 111*, 443–448.
Resmini, P., & Pagani, M. (1983). Ultrastructure studies of pasta: A review. *Food Microstructure, 2*, 1–12.
Sharma, S., Kaur, A., Kaur, G., & Nanda, V. (2013). Influence of different protein sources on cooking and sensory quality of pasta. *International Journal of Engineering, 3*, 1757–1763.
Sharma, N., Sharma, S., Singh, B., & Kaur, G. (2020). Stability evaluation of iron and vitamin A during processing and storage of fortified pasta. *Quality Assurance and Safety of Crops & Foods, 12*, 50–60.

Sharma, S., Sharma, N., Singh, A., & Singh, B. (2021). Stability of iron and vitamin A in pasta enriched with variable plant sources during processing and storage. *Journal of Food Processing and Preservation, 00*, 1–15.

Sosulski, F. W., & Wu, K. K. (1988). High fiber breads containing field pea hulls, wheat, corn and Oat brans. *Cereal Chemistry, 65*, 186–191.

Susanti, S., Dwiloka, B., Bintoro, V. P., Hintono, A., Nurwantoro, N., & Setiani, B. E. (2021). Antioxidant status, nutrition facts, and sensory of spinach extract fortified wet noodles. *Food Research, 5*, 266–273.

Susanna, S., & Prabhasankar, P. (2012). A study on development of Gluten free pasta and its biochemical and immunological validation. *LWT-Food Science and Technology, 50*, 613–621.

Tazrart, K., Lamacchia, C., Zaidi, F., & Haros, M. (2016). Nutrient composition and in vitro digestibility of fresh pasta enriched with Vicia faba. *Journal of Food Composition and Analysis, 47*, 8–15.

UNAFPA. (2022). Estimate of world pasta production (tons). https://www.pasta-unafpa.org/newt/unafpa/default.aspx?IDCONTENT=102. Accessed 16 Nov 2022.

Wahanik, A. L., Neri-Numa, I. A., Pastore, G. M., Chang, Y. K., & Clerici, M. T. P. S. (2018). Turmeric (Curcuma longa L.): New application as source of fiber and antioxidants in pasta with whole wheat flour. *Revista Facultad Nacional de Agronomia Medellin, 71*, 8423–8435.

Walsh, D. E., & Gilles, K. A. (1977). Pasta technology. In N. W. Desrosier (Ed.), *Elements of food technology*. AVI Publishing Company.

Yulia, R., Yuslinaini, & Wildanawati (2018). Calcium fortification of dried noodle substituted by jackfruit seed (*Artocarpusheterophylus*) flour with broccoli (*Brassica oleracea* var.) pasta. *IOP Conference Series: Journal of Physics: Conference Series* 1232, (pp. 1–7).

Chapter 10
Noodles

Niaz Mahmud, Joinul Islam, and Reza Tahergorabi

1 Introduction

Noodles are considered one of the oldest types of food, and they are an essential-component of the diets of many Asian countries, including China, Malaysia, Japan, and Thailand. Noodles have been famous in many other regions in recent years. Their consumption increases with each passing day, for example, in developing countries such as Cambodia, Bangladesh, India, and Pakistan (Karim & Sultan, 2015). The extensive eating of noodles has resulted in substantial changes around the globe. Advanced and emerging economies such as the United States, Europe, and Brazil are becoming an increasingly appealing marketplace for noodles. More than 1 billion noodles-based meals were served globally in 2013, according to one estimate, demonstrating the market's fast expansion. Furthermore, as of 2020, instant noodles, a popular type of noodle in the modern world, had been served around 116,000 million times worldwide (Karim & Sultan, 2015; World Instant Noodles Association., n.d.). The debate over who invented the noodle is raging, with Chinese, Italians, and Arabs claiming credit. However, research found a crock of delicate noodles well-preserved for about 4000 years in a Yellow river slit, which may have tipped the scales in China's favor (Lu et al., 2005). It demonstrates that humans ate noodles at least a thousand years before they were documented in Europe and many centuries before they were distinguished in Asia. The thin, yellow noodles were about 0.5 m long and resembled *La-Mian*, an old-style Chinese noodle made by crushing wheat dough and then continue pulling and stretch it by hand. Dough sheet pieces in the soup were purportedly consumed 2000 years ago in China or even before that, according to early Chinese literature. Hot noodle soup in the

N. Mahmud · J. Islam · R. Tahergorabi (✉)
Food and Nutritional Sciences Program, North Carolina Agricultural and Technical State University, Greensboro, NC, USA
e-mail: rtahergo@ncat.edu

M. A. Shah et al. (eds.), *Cereal-Based Food Products*,
https://doi.org/10.1007/978-3-031-40308-8_10

winter and cold noodle soup in the summer were popular from the beginnings of the Tang dynasty (618–906 AD). The Yuan Dynasty (1279–1368 AD) saw the development of hand-made noodle technology. Noodle companies produced a variety of noodles in various forms, sizes, and tastes, including some slender hand-stretched noodles. Many of the noodles available now are descended from those produced during these periods. The hand-made Chinese noodle and their processing method were introduced to Japan some 1200 years ago (Hou et al., 2010). By tweaking Chinese hand-made noodle processing techniques, many salty noodles were developed in Japan. The Japanese revolutionized the magnitude of the manufacture of noodles with the advent of power-driven technology in 1884. During the start of the twentieth century, Chinese immigrants in Yokohama city gradually promoted alkaline salt noodles throughout Japan.

Different food manufacturing businesses then stepped forward to mass-produce and commercialize noodles, causing noodles to become a significant element of many people's diets (Hou et al., 2010). As stated earlier, noodles have had a long history of being processed by hand since their inception, even though the craft's tradition and heritage fluctuate by zone. With the advent of the industrial revolution, it accomplished the shift from a traditional handcraft industry to machine mass manufacture. Noodle machines initially appeared in the 1850s and are even nowthey can be seen in use today. China is now the world's largest consumer of noodles, with enormous yearly consumption and production levels. Noodle sales in China increased from 8.6 billion yuan in 2007 to 2026 Byuan in 2012. (Zhang & Ma, 2016). A few traditional noodles from China have been listed in Fig. 10.1. Machines now make most noodles. While the exact method for creating a particular variety of noodle may differ from region to land to meet local demands, the essential notions are nearly same. Many of these concepts are comparable to those used in the

Fig. 10.1 A couple of the traditional Chinese noodles — (**a**) Sichuandan noodles, (**b**) Qishan minced noodles, (**c**) Yunnan over-the-bridge rice noodles, (**d**) Beijing fried bean sauce noodles, and (**e**) Lanzhou hand-pulled noodles. The pictures are adopted from (Zhang & Ma, 2016) with permission

manufacture of traditional hand-made noodles. Hand-made noodles have survived and are still prevalent in China and Japan, despite their low productivity by today's standards. Hand-made noodles are widely regarded for their superior eating quality due to the technique of gluten formation. In the 1990s, innovations in technologies relating to noodle processing such as multi-layer sheeting, vacuum mixing, and waved rollers were built on the principles of gluten development in hand-made noodles (Fu, 2008).

Flour is the most fundamental raw element for producing noodles. As a result, it contains essentially the same elements as flour, such as protein, carbohydrates, minerals, and minimal fat. 100 g fine dry noodles, for example, comprise around 10 g protein, 75 g carbohydrates, just 0.7 g fat, 129 mg potassium, 19 mg sodium, 12 mg selenium, and so on (Zhang & Ma, 2016). Nutritional components such as eggs are included tothe practice ofnoodles manufacturing. As a result, noodles carry the nutritional content of eggs, with the amount variable depending on the number of eggs added.

Many initiatives have been launched in last few years to enhance the value of products containing functional substances. A recent study found that experimentally made mushrooms-fortified noodles had lower carbohydrate, fat, and sodium (Parvin et al., 2020). Salts, on the other hand, are added during the noodle-making process. Excessive salt consumption raises the risk of hypertension. Excessive salt consumption may contribute to the development of cardiovascular risk factors. A direct causal association between sodium intake and blood pressure (BP) is supported by factual data. A slight reduction in salt consumption is linked to a significant blood pressure reduction (Jaques et al., 2021). As a result, food manufacturers should attempt to use as little salt as feasible. We should consider the health impact of the art of cooking and other important factors, as well as the original nutritional content of noodles. Noodles are frequently paired with various meals in many cultures. People often eat noodles with sauce, which has been claimed to contain fat and salt, to increase the taste. To reduce the risk of chronic illnesses, we should use less gravy when eating noodles. To achieve the goal of diversity while also encouraging people's health, we may prepare noodles with a range of meals such as vegetables, eggs, and other components.

2 Ingredients

The most pressing issue in the noodle sector is assuring the production of a wide range of high-quality products with the rightingredients. Choosing the right ingredients is a keystageto produce noodles since it significantly impacts the noodles'preparation and quality. The formulation of ingredient lists differs depending on the type of noodles. Flour, water, salt, alkaline reagents, starch, oils, improvers, and preservatives, among other ingredients, are used to make traditional noodles. This section will review the common ingredients used in noodles production, their quality parameters, and their impact on the final product characteristics. A list of

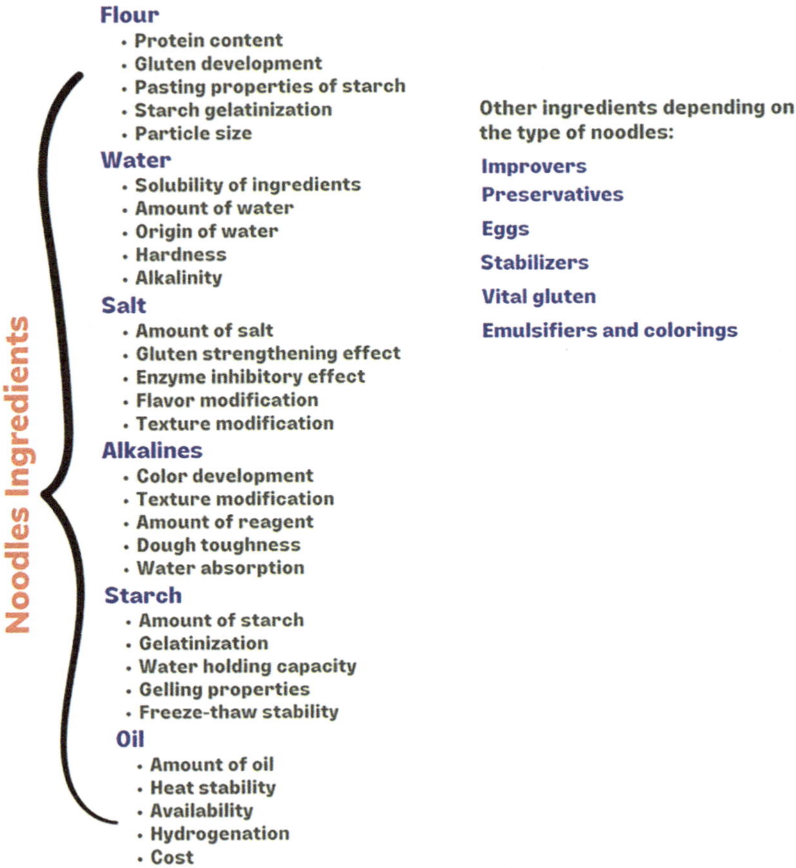

Fig. 10.2 Considerable variables relating to the fundamental ingredients of noodles posing an impact on the end product's quality

factors involving components is presented in Fig. 10.2. A detailed discussion on the functions of each ingredient will be discussed in Sect. 4.

Flour

The grade of flour used in noodle preparation varies based on the kind of noodle, geographical preference, preparation technique, and manufacturing environment. The starch pasting, kernel hardness, bran color, dough strength, protein content, etc. are crucial quality factors for wheat in noodles. The wheat quality milling technique is also critical to ensure the bright color, low ash level, and tiny particles of noodle flour. For noodle flour and dough to be strong enough to resist the sheeting process, they must have appropriate gluten strength and extensibility. Aside from gluten strength, maintaining an optimal protein concentration is vital for noodle texture. Dried noodles have a higher protein content than fresh noodles because they can withstand the drying process without disintegrating. Because it cuts down on the amount of fat that is absorbed during the frying process, the protein content is a crucial component in the manufacturing of instant noodles (Crosbie, 2005). Gentle, pliable, and dry salted noodles like Udons (Japanese and Korean kinds) require wheat flour with high starch paste viscosity and strong swelling properties (Asenstorfer et al., 2006). The quantity of flour that needs to be added differs according to the kind of noodles that are being prepared.

Water

Water is the subsequentessential element to the production of noodles, and it has a significant impact on the end-product quality. Water catalyzes physicochemical and biological processes that turn raw materials into finished goods. Gluten protein's viscoelastic characteristics cannot be observed without the presence of water. During the noodles manufacture, an appropriate amount of water is required to hydrate the flour and form a consistent dough sheet. Contingent on the flour weight, the water absorption rate for making noodles is between 30 and 40%.

Salt

Salt is a crucial element in noodles production, accounting for 1–8% of the total flour weight. Salt serves many purposes in the production of noodles. Salt improves the sheeting qualities of dough by strengthening the gluten (Dexter et al., 1979; Fu, 2008). Adding salt to the noodles improves their texture and taste. Salt improves tongue fullness and flavor balance and generates a purely salty taste. Finally, salt inhibits enzyme function and microorganism development.

Alkaline Reagents

During processing, alkaline salt, a qualitatively potential alkaline reagent, gives the alkaline noodles their distinct color, texture, and flavor. Alkaline salt can be used alone or in conjunction with other ingredients. The most prevalent alkaline salts are Na_2CO_3 (sodium carbonate) and K_2CO_3 (potassium carbonate). While 0.1–0.3% alkaline salts are only employed as a quality enhancer, 0.5–1.5% alkaline salts are often used for firm alkaline-flavored noodles.

Starch

The noodles'starches are employed to add texture. Starches are often put to noodles at a rate of 5–25% of the weight of the flour. The primary starches utilized in the production of noodles are tapioca and potato starches. These carbohydrates give the noodles a stretchy and chewy feel, which improves their consistency. The use of starch guarantees quick rehydration and a consistent texture for instant noodles. Tapioca starch works well for traditional, salty noodles, whereas potato starch is best for quick noodles. Additionally, to enhance its ability to hold water and its gelling qualities, chemically processed starch is utilized in the production of noodles.

Oils

After shaping and steaming, instant noodles are often fried in oil, making up 20% of the entire weight. Typically, frying mediums include canola oil, palm oil, and oil that has been partly hydrogenated. The cooking oil used in noodles should be chosen according to local preferences, as the flavor of the finished product is greatly influenced by the content of the frying oil. During frying, complicated chemical interactions cause oil quality to change. Food safety issues and sensory problems might result from extended heating. Therefore, while choosing frying oil, thermal stability is a crucial consideration.

Improvers

Improvers such as polyphosphate and hydrocolloids are generally used. Polyphosphate improves the starch gelatinization and water retention capability of noodles during processing. Additionally, polyphosphate functions as a chelating agent in dough systems, altering dough processing parameters and delaying the yellowing of new noodles. Generally, polyphosphate is used at 0.1% of flour weight by dissolving them in water before the dough mixing. Hydrocolloids like guar gum are frequently utilized in the production of instant noodles because of their hydrophilic and high-water binding capabilities. A modest bit of gum can enhance the texture of the finished product and help noodles rehydrate more easily during cooking. Usually, brine water and gums are combined before mixing the dough.

Preservatives

Several preservatives are utilized to suppress microbiological development and increase the shelf-life of noodles. Because they include a significant level of fat (15–22%), steamed and deep-fried instant noodles are very sensitive to oxidative rancidity. Antioxidants are employed as a preservative in frying oil to minimize oxidative rancidity. Antioxidants stop the chain process from continuing, resulting in mechanisms and chemicals that cause rancidity. The four most often used synthetic antioxidants are a) tertiary-butylhydroquinoneor TBHQ, b) propyl gallate, c) butylatedhydroxytolueneor BHT, and d) butylatedhydroxyanisoleor BHA. Among these, TBHQ is thought to be the most potent antioxidant.

Eggs

The functions of eggs in noodle processing are to improve the flavor and create elastic and firm textures. The addition of eggs in noodles darkens the product more rapidly than plain noodles. Fresh, frozen, and dried eggs can be added directly to the dough.

Stabilizers

While the noodles are boiling in water, stabilizers provide viscosity to increase hardness, body, and mouthfeel. Guar gum, alginate, locust bean gum, and carboxymethyl cellulose, which have extremely high molecular weight and are utilized in very small amounts, are the stabilizers most frequently used in the production of noodles (0.1–0.5%). Stabilizers can increase the pace of rehydration after cooking or soaking by decreasing the amount of oil that fried instant noodles absorb due to their capacity to bind water.

Vital Gluten

High protein flour is required to make Chinese-type noodles. Cooked noodles with low flour protein do not get the desired bite texture. Therefore, vital gluten powder is added to the dough to improve the noodle bite. However, using excessive gluten powder can cause noodle darkening.

Emulsifiers and Colorings

Emulsifiers are added to the noodles to reduce stickiness, cooking water cloudiness, and starch retrogradation. They can also help enhance starch gelatinization and rehydration. Mono and di-glycerides, sucrose fatty acids, and lecithin are the most prevalent emulsifiers in noodles. Natural and manufactured yellow colorants are employed in wet and fried instant noodles to improve their appearance. The most pervasivedyes in the noodles areβ-carotene, tartrazine, riboflavin, vitamin E, and sunset yellow.

3 Preparation Methods

Primary Preparation Methods

Mixing

Ingredients are combined usinghorizontal or vertical dough mixers and a variety of mixing techniques to make a crumb or dough, depending on the amount of water supplied (Fig. 10.3). Non-alkaline noodles often have saline or alkaline solutions combined at a 28–35% depending on the flour weight, while alkaline noodles have a saline or alkaline solution incorporated at 28–34%. Moisture additions may rise from 40 to 47% if a vacuum is utilized during the mixing process (Crosbie & Ross, 2016). The fundamental goal for making the first dough crumb is obtaining equal hydration of the dry components with low gluten development. The best method for determining how much water to add is to consider the crumb size, sheeting

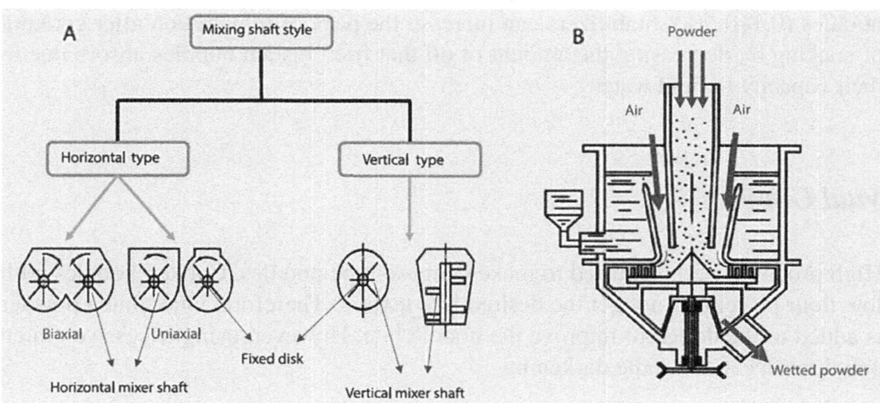

Fig. 10.3 (**a**)Dough mixers – horizontal and vertical. (**b**) Mechanical diagram of flow jet continuous dough mixer. They are reproduced from (Obadi et al., 2022) with permission

properties, and handling characteristics. The quantity of water injected has an impact on almost all manufacturing steps and final product quality (Crosbie & Ross, 2016; Fu, 2008; Obadi et al., 2022). It is advisable to use modest water addition and crumb formation when making noodles that will be dried before being distributed. Using more water during vacuum mixing provides benefits such as decreasing gluten damage following sheeting, combining, rolling operations, boosting raw noodle production, and shortening the cooking time for steaming or boilingpractices. During conducting the mixing steps of the noodling processing, maintaining tight control of dough temperature is critical, as too high or low temperature can negatively impact product quality (Crosbie & Ross, 2016; Fu, 2008; Obadi et al., 2022).

Initial Sheeting and Combining

Each half of the dough or crumb is split in two and sent through a sheeting roll. The two sheets are joined by a single pair of rollers to facilitate the production of gluten. While certain plants require longer rest times, the sheet is typically loosened for up to an hour. The rest period allows the gluten structure to unwind, minimizing gluten damage from subsequent drops in sheet thickness. It was investigated if greater noodle eating quality would come from a more developed protein matrix (Crosbie & Ross, 2016; Fu, 2008; Obadi et al., 2022).

Rolling of the Sheet

To gradually reduce the thickness of the sheet, it is passed through many rolls. The quality of eating the noodles is greatly influenced by this step of the process. Critical elements include the rate of sheet thickness decrease, sheet moisture content, sheeting direction, and roller temperature. The drop in sheet thickness should not be greater than 30% on each run through a pair of rolls; otherwise, the gluten structure may be disrupted and the noodle‘s eating quality may be jeopardized. The moisture content of the sheet is essential because greater moisture contents prevent the gluten structure from deteriorating. Due to the greater dough moisture level and subsequent kneading after mixing, homemade noodles have a better texture and more improved gluten development (Crosbie & Ross, 2016; Fu, 2008; Obadi et al., 2022). By contrast to normal sheeting, which produces unidirectional growth, kneading creates randomized gluten. The temperature of the sheeting rolls must be controlled, and water-jacketing the rolls helps. Temperature variations also have an impact on the elasticity and viscosity of sheets. These factors have an impact on the roll gap settings required to get the desired final sheet thickness. Low roll temperatures result in the noodle surface being rough, while high roll temperatures can lead dough sheets to hydrate out on the surface (Crosbie & Ross, 2016; Fu, 2008; Hou et al., 2010; Obadi et al., 2022).

Cutting

The noodles are sliced lengthwise via cutting rollers once the noodle sheet has been reduced to its ultimate thickness. Cutting rollers are offered to make noodles with widths ranging from 1.0 to 7.5 mm. The cutting rollers decide the final size of the noodle strands, while the sheeting process determines the thickness. These dimensions impact the time spent on secondary preparation processes, including drying, steaming, and boiling (Crosbie & Ross, 2016; Fu, 2008; Obadi et al., 2022).

Secondary Preparation Methods

Next, the noodle slices are subjected to secondary processing, comprising particular production stages about salt composition, processing techniques, and grain composition. Many phases and procedures are involved in the preparation of secondary processed noodles. This section will address distinct types of noodles, emphasizing how they are prepared, the methods and treatments necessary, etc. A short description of each kind of noodles is presented in Table 10.1. A short description of step-by-step manufacturing procedures are also presented in Figs. 10.4 and 10.5.

Noodles Preparation Based on Salt Composition

Regular-Salted

Regular salted noodles are made from a conventional flour and water dough that contains 2–8% salt by weight. The kind and processing of a noodle affect how much salt it contains. Although they only make up a small portion of the total noodles produced across Southeast Asia, regular salted noodles are well-known in Korea, Japan, and China. Fresh, dried, and cooked salty noodles are all options. They are also known by the terms extremely thin, thin, standard, and flat. The strand width of a typical ordinary salted noodle is 1.0–1.2, 1.3–1.7, 2.0–3.9, and 5.0–7.5 mm, respectively (Fu, 2008). The delicate noodles are frequently sold dry, as they are made by drying fresh noodle strands under regulated conditions. In Japan, ordinary salted noodles are still quite popular, despite most noodles being created by machines. Except in Hong Kong, Canton, and Hokkien regions, regular salted noodles are the most popular throughout China. Ordinary salted noodles should emerge from the boiling process with a smooth, glossy surface and a bright, clear hue spanning from white to creamy white. Most Japanese and Koreans like cooked, regular salted noodles that are soft, elastic, and have smooth surface. The majority of Chinese people, on the other hand, favor firm, elastic, and chewy noodles (Crosbie & Ross, 2016; Fu, 2008; Gulia et al., 2014).

Table 10.1 Short description of each type of noodles based on preparation methods

Noodles type	Short description
Salted (Regular)	Designed with clean wheat and water dough with 2 → 8% salt. The quantity of salt in a noodle is determined by its kind and handling.
Salted (Alkaline)	Alkaline salts, such as sodium carbonate or potassium carbonate, or a mixture of the two, are used to produce the noodles.
Fresh noodles	Without any extra processing, the noodles that emerge from the slitting rollers are cut into predetermined lengths for packaging.
Dried noodles	Fresh noodle strands are dried in a controlled room or the sun.
Steamed noodles	Alkaline noodles are made with fresh strands that are steam-cooked and then loosened by rinsing or steeping in water.
Boiled noodles	Fresh noodle strands are either parboiled (wholly cooked in 80–90 min) or fully cooked.
Instant noodles	The strands of fresh noodles are waved and cooked. Noodles are dehydrated by deep-frying in hot oil or hot blast air after cutting and shaping into a block.
Frozen noodles	In the manufacturing facility, fresh noodles are cooked, washed in cold water, drained, and then frozen in a hurry at −30 °C or below.
Rice noodles	Rice flour and water make a Chinese noodle with a mild, delicate flavor and a chewy texture. They are available in a range of sizes and shapes.
Starch noodles	Mung bean, yellow pea, and potato starches make starch noodles. This type of noodle is made by mixing starch with up to 50% water to produce dough, then extruding it directly into the boiling water using a cylinder-style extruder, chilling, then air-drying it.
Buckwheat noodles	These noodles are typically made from buckwheat flour and are boiled, washed, and served cold with various dipping sauces or in a hot, clear broth.

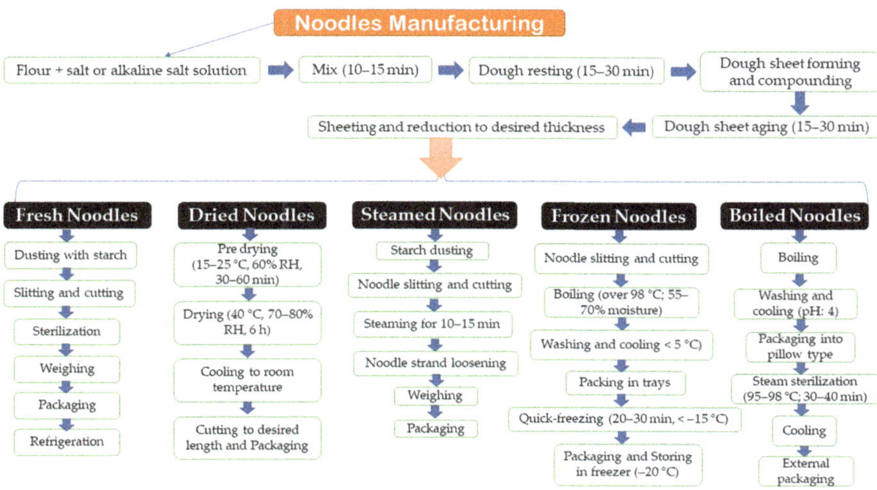

Fig. 10.4 Flowchart of manufacturing procedures involving different types of noodles based on processing methods

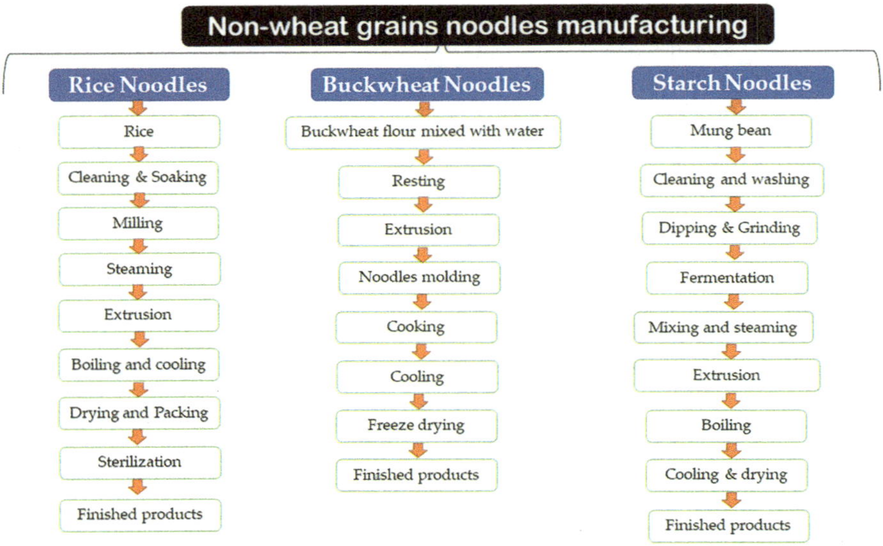

Fig. 10.5 Process flow diagram for making noodles from rice, buckwheat, and starch (mung bean)

Alkaline-Salted

Alkaline salt in noodle production dates back to China to extend noodle shelf-life by suppressing mold development. The most popular alkaline salts used today are potassium carbonate, sodium carbonate, or a mixture of the two. Other alkaline salts like sodium hydroxide or polyphosphates are widely used in the manufacturing of instant noodles and partially cooked noodles (Fu, 2008). Depending on the salts employed and their ionic strengths, the pH of alkaline salt-based noodles can range from 9 to 11. The flavor and aroma of alkaline noodles are distinctive, and they are yellow in color with a strong, elastic texture. In fresh alkaline noodles and steamed alkaline noodles, carbonates are frequently added at rates of 1.0–1.5% and 0.3–0.5%, respectively. Alkaline noodles that are partially cooked typically contain between 0.3 and 1.0% sodium hydroxide. Alkaline noodles were introduced to the area by immigrants from the Canton and Hokkien provinces, which were the initial homes of nearly all Chinese migrants to Southeast Asia. Alkaline noodles have become popular in Taiwan, Hong Kong, Thailand, Singapore, Indonesia, and Malaysia. Alkaline noodles are available in a variety of forms across Southeast Asia. The most popular types are Cantonese style fresh, Hokkien style slightly boiled, or steamed with egg as a component. Instant noodles are widely used throughout Southeast Asia. They may be steamed, deep-fried, boiled, and dried using hot air. Additionally, noodle producers distributed them across Japan. In Japan, alkaline noodles are now widely available and consumed. The most popular alkaline noodles in Japan are freshly made, highly moistened steamed noodles (Fu, 2008; Shiau & Yeh, 2001).

Noodles Preparation Based on Processing Methods

Fresh Noodles

Noodles that are made fresh are soft and raw. No further processing is carried out at the factory once the sheeted dough is cut into noodles of a specific length and breadth. Noodles are typically dusted with starch or fine flour just after being sliced in order to prevent the strands from sticking together while being handled and transported. To prepare for automated packaging or bulk placing in trays for retail outlets, a preset quantity of noodles is cut and sorted. The moisture content of fresh noodles ranges from 32 to 40%. The main drawback of fresh noodles is their short shelf-life, which can range from one to several days depending on storage and packing. Fresh alkaline noodles should have a firm, elastic, smooth texture that allows for clean, tooth-slicing cuts. Customers like noodles that seem tidy and vibrant. But depending on the region, different levels of yellowness are advised (Fu, 2008).

Dried Noodles

Uncooked, wet noodle strands that have been dried under controlled conditions are known as dried noodles. The ultimate moisture content of dried noodles is generally less than 14%. The noodle strands are produced, then cut to a length of 2–4 m and hung to dry. Additionally, noodles can be dried in a regulated setting that includes ventilation, relative humidity, and temperature. Because dried noodles contain little moisture, their shelf life is about 1–2 years. From flat, thick noodles to round, hair-thin noodles, dried noodles come in a variety of shapes and sizes. However, dried noodles require more time and care in the cooking than other varieties of noodles. The air cells in the noodles shrink throughout the drying process, making it more challenging for water to penetrate and absorb. Long-term cooking in rapidly boiling water results in a mushy and sticky surface on the noodles (Fu, 2008).

Steamed Noodles

Fresh noodles are partially cooked before being marketed by being exposed to saturated or unsaturated steam. Fresh noodles are heaped on a net conveyor that travels through a tunnel steamer in modern businesses, automating the steaming process. The moisture level of the finished product normally ranges from 28 to 65%. Noodles that have under 32% moisture content after steaming are partially dry. These low moisture steamed noodles are easy to handle and have a longer shelf life because of their low surface stickiness. About 60% of the moisture in steamed noodles with a high moisture content is water. The noodles are simultaneously steamed in a tunnel steamer and sprayed with hot water to hasten the starch gelatinization process. High moisture content steamed noodles have a sticky surface that has to be oil-coated in order to be packaged for distribution. Low moisture steamed noodles should be quickly boiled in boiling water before serving. The noodles are either added to a hot soup or stir-fried with meats or vegetables after being drained. Without using boiling water, highly moistened steamed noodles are stir-fried (Fu, 2008).

Boiled Noodles

Boiled noodles are precooked in boiling water. The two types of noodles offered are partially cooked noodles and totally boiled noodles. Automated factories run uncut noodles under a bath of boiling water. Noodles can also be divided into certain weights and fed into a basket on a conveyor belt that is moving through a boiling bath. Immediately after boiling, the noodles must be immersed in cold water. The extra water on the surface of the noodles is drained. The noodles are then covered in oil to stop them from adhering together. To prepare them for sale, noodles are either pre-packaged or supplied in wholesale to retail stores (Fu, 2008).

Frozen Noodles

When making frozen noodles, raw noodles are first boiled, then washed in cold water, drained, and immediately frozen at −30 °C or below. The majority of premium frozen noodles are made by mixing while under vacuum, which enables the addition of additional water. The additional water promotes the hydration and development of the gluten, reduces gluten breakage during sheeting, and hastens the boiling process by allowing heat to penetrate the noodle more rapidly. Additionally, quick freezing assures the formation of little water crystals, which are less likely to disrupt the gluten network than larger crystals produced at slower freezing rates. The quality of frozen noodles can be on par with that of fresh (raw) noodles. The process is essential, and if any of the key steps aren't adjusted, undesirable results may result (Crosbie & Ross, 2016).

Noodles Preparation Based on Non-wheat Grains

Rice Noodles

Rice noodles can be prepared and eaten as soup noodles in a broth or stir-fried with meats and veggies. Like wheat gluten, rice proteins are unable to produce continuous viscoelastic dough. A portion of rice flour is frequently pregelatinized to act as a binder for the remaining flour. The two main methods for making rice noodles are extrusion, which creates different types of strands, and batter sheeting, which creates sheets and flat noodles. Rice with a high amylose content is soaked in water for many hours before being wet-milled to produce rice vermicelli noodles. Balls are made from crushed, milled rice that has been filtered. To promote surface gelatinization, the balls are either steamed or precooked in boiling water for around 20 min.

To ensure that the gelatinized rice is equally distributed throughout the dough and serves as a binder, the partially cooked balls are next kneaded. The kneaded dough is extruded using a die. When the noodles are fully cooked, they are removed from the boiling water and allowed to float to the top. The freshly prepared noodles are then dropped into a tank of ice water to cool them down. The noodles are then put on drying racks. The strands of premium rice noodles are straight and regular, the color is white and translucent, and there are no broken strands. After soaking in hot water, premium noodles hydrate with a small turbidity and surface stickiness. The noodles may be made into fresh rice ribbon noodles by slicing them into strips that are 10 mm wide and selling them as sheets (approximately 1 mm thick). Flat rice noodles can also be dried before being sold. The key to getting the ideal noodle texture is pre-gelatinizing the rice flour. Some degree of gelatinization is necessary for extrusion to provide binding power, but too much gelatinization might be problematic. Traditionally, rice noodles have been made using long-grain rice that has an amylose level of between medium and high (>22% amylose). The ability of starch to swell and the amylose-amylopectin ratio affect the quality of rice noodles. For creating rice noodles, rice cultivars with a high amylose content and a firm gel consistency are best. High amylose rice creates vibrantly colored noodles with a low bulk density due to its weak swelling capability (Fu, 2008).

Starch Noodles

These noodles are frequently made with mung bean, yellow pea, and potatoes starches. The mung bean starch is separated by a wet-milling process. Mung bean seeds are first soaked in warm water for 6–9 h (30–40 °C), then in cold water for 8–16 h (5–10 °C). After draining, they are wet milled to get the starch granules out of the cells. Starches are separated from proteins and fibers by mechanical separation, and they are often further purified by enzyme treatment. The starches are then dried to a moisture content of 20–25%. A modest amount of starch (3–4% of the total) is totally gelatinized by boiling water. The raw starch, water, and pregelatinized starch are then mixed together to make a slurry that is subsequently extruded through a die with 0.7–0.8 mm diameter holes. A saucepan of boiling water is used to cook the formed noodles for a brief period of time. They are immediately cold and drained. To allow the starch to retrograde, the noodles are refrigerated (0–4 °C) or frozen (10 °C) for 12–24 h. They are defrosted and air- or sun-dried before packaging. Although potato starch differs from mung bean starch in terms of its physicochemical properties, it has been used to create appealing starch noodles. Potato starch is chosen over cereal starches for creating starch noodles because of its neutral taste, noticeably increased transparency, and flexibility of the noodles generated. Potato starch noodles are typically more significant than mung bean starch noodles due to their inferior cooking properties and texture (Fu, 2008).

Buckwheat Noodles

A three-sided, pointed buckwheat seed has a dark brown color. The four essential parts of the seed are the outer hull, embryo, seed coat, and endosperm tissue. The entire buckwheat grain has a gorgeous light-green kernel hidden inside the shell. When buckwheat blossoms are developing, chilly conditions are essential for their flavor and aroma to be at their best. Buckwheat is naturally covered in a hard, tightly adherent shell for a reason. After being kept for days, buckwheat's unique aroma and flavor disappear. The manner in which the seed is crushed and sorted after the outer layer is removed significantly determines the color, flavor, and texture of buckwheat flour. Buckwheat is susceptible to heat damage, thus it's crucial to avoid overheating the rollers when milling. The optimum procedure for milling buckwheat is frequently considered to be the stone-ground method. Korea, northeast China, and Japan are the main countries that produce buckwheat flour noodles. The manner in which the seed is crushed and sorted after the outer layer is removed significantly determines the color, flavor, and texture of buckwheat flour. Buckwheat is susceptible to heat damage, thus it's crucial to avoid overheating the rollers when milling. The optimum procedure for milling buckwheat is frequently considered to be the stone-ground method. Korea, northeast China, and Japan are the main countries that produce buckwheat flour noodles. In order to take use of the gluten found in buckwheat, which helps the flour bond, buckwheat flour and wheat flour are mixed. Similar to flour noodles, buckwheat noodles can be made by hand or machine. Due to the high paste viscosity of the buckwheat endosperm, the buckwheat noodles have a firm and elastic eating texture. Wheat flour ranges in viscosity from 450 to 950 BU, depending on how waxy it is, but buckwheat flour has a maximum viscosity of 1800 BU. In addition to having a superb protein composition (very high lysine), buckwheat noodles are also rich in dietary fiber, minerals, and vitamins (B1 B2) (Fu, 2008).

4 Quality Aspects

Many factors relating to ingredients and processing procedures impacts the quality of noodles. Due to consumer and governmental authority worries about quality and their industrial ambitions for progress, the extent of difficulties with noodle manufacture has increased. The quality factors involving ingredients and processes will be discussed in this section and how they affect the final quality. In addition, quality evaluation parameters for the end noodles will be addressed.

Quality Aspects Involving Ingredients

Flour

The typical flour used to produce noodles is prepared from healthy wheat grains that have not undergone sprouting, and it has an ash level of 0.5–0.6% and a medium protein concentration of 10–12.5%. The color of the flour, the level of polyphenol oxidase activity, and the extraction rate (milling) all have a direct impact on the end product's color. Noodles'brightness is adversely impacted by their ash concentration (Gulia et al., 2014; Ye et al., 2009). Noodle flour must have adequate dough strength and extensibility to sustain ripping, breakage, and shrinking. As discussed, the amount of flour extracted and the amount of ash in the flour significantly impact noodle appearance. Higher flour extraction yields duller noodles that are more prone to brown throughout dough processing. For the creation of noodles with a clean, dazzling appearance after cooking, low flour extraction and ash contents are recommended (Crosbie & Ross, 2016). The color of raw noodles fades over time. There is a connection between weather damage, Polyphenol oxidase (PPO) activity, flour extraction rate, and variety. The issue still exists despite the best efforts of wheat breeders in supplying nations to lower PPO levels and, therefore, raw noodle discoloration. PPO is located mostly in the bran and flour itself. More bran particles are present in the flour as a result of higher flour extraction, and there are more visible specks in both fresh and cooked noodles. If uncooked noodles are stored without being cooked for an excessive amount of time, PPO's actions lead the specks to discolor and become more obvious. Due to the presence of xanthophyll pigments, the flour makes the noodles look yellow, making them look different in color. In Japan, the most popular type of Udon is creamy-colored with a fair amount of yellow flour. When alkali is added to the flour, it changes the flavonoid compounds and yellow pigment in the flour, making the noodles yellow. Noodle brightness is also controlled by flour particle size, with smaller particles linked with brighter noodles at similar levels of starch degradation (Asenstorfer et al., 2006; Fu, 2008). Increased flour protein level reduces noodle brightness across the board but has minimal impact on noodle darkening over time (Crosbie & Ross, 2016; Fu, 2008). More specifically, cooked noodle stiffness is positively correlated with flour protein level, while noodle brightness is negatively correlated (Asenstorfer et al., 2006; C. Wang et al., 2004). Gluten, a wheat protein that has been hydrated, provides the fundamental framework for a strand of noodles. Enhanced gluten strength, which indicates the composition of the protein, and increased protein content in the flour have a significant impact on the processing and texture of noodles. In the literature, enhanced noodle hardness has been connected to greater protein content (Huang & Morrison, 1988; Yamauchi et al., 2007). It was also discovered that superior noodle texture (hardness and elasticity) is connected to high protein levels (Yamauchi et al., 2007). The surface of noodles is mainly determined by gelatinized starch; hence starch characteristics are crucial in evaluating quality (Békés et al., 2006). Variations influence noodle softness and viscoelastic attributes in starch properties (Békés et al., 2006; Crosbie, 1991; Ross, 2006). Noodle qualities are determined by flour swelling volume, peak viscosity, amylose amylopectin ratio, pasting components, and impaired starch. Noodle qualities are determined by flour swelling volume, peak viscosity, amylose amylopectin ratio, pasting components, and impaired starch (Baik & Lee, 2003). Reduced gelatinization temperatures in flours are preferred for cooking because they facilitate rapid hydrationReduced gelatinization temperatures in flours are preferred for cooking because they facilitate rapid hydrationReduced gelatinization temperatures in flours are preferred for cooking because they facilitate rapid hydrationReduced gelatinization temperatures in flours are preferred for cooking because they facilitate rapid hydration (Baik & Lee, 2003; Hatcher et al., 2002). Greater starch degradation is associated with major cooking loss, impaired noodle color, and severe exterior swelling. The decreased starch degradation caused by the fine particle flour makes for improved cooking qualities in the noodles (Hatcher et al., 2002).

Water

Another essential ingredient is water, which is required to develop gluten, provides the required viscoelastic qualities to the dough during noodle manufacturing. The volume of water necessary during noodle manufacturing is adjustedfor the hydration of the flour and generation of a consistentsheet (dough). Protein quantity, protein characteristic, impaired starch, and other bodilytraits of wheat all influence noodle water uptake (Park & Baik, 2002). Based on flour weight, the ideal water uptake level for making noodles is between 30 and 38%. The volume of water absorbed greatly affects both the color and the amount of processing required. Increased water absorption resulted in a considerable decrease in textural qualities (Hatcher et al., 1999; Park & Baik, 2002). Achieving the necessary processing parameters becomes challenging when water absorption differs more than 2–3% from the optimal level. Insufficient water leads in tight, non-cohesive dough and a less stretchable noodle sheet, whereas excessive water results in dough adhesion and processing difficulties (Hatcher et al., 2008b). Using a farinograph or mixograph, and knowledgeable individuals, the ideal water absorption may be determined based on the dough's handling qualities (Seib et al., 2000).

Salt and Alkaline Reagents

Typically, 1–3% weight of the total flour is used as salt in noodles. Due to salt's inhibitory effect on proteolytic enzymes or salt's direct interaction with grain proteins, gluten is strengthened and tightened. As a result, especially at elevated water uptake levels, it greatly improves dough sheeting properties. Salt reduces cooking time, enhances flavor, gives food a softer, more flex, and prevents the growth of bacteria and enzymes (Fu, 2008). Alkaline salt can be applied solely or in combination with other salts, relying on personal choice. Sodium and potassium carbonates are the most regularly utilized alkaline salts. Noodle quality is also affected by the alkaline salt used (Hatcher et al., 2008b). In the presence of alkali, endogenous flavonoids undertake a chromophoric shift, or change color to yellow, which is what gives ordinary wheat used to make noodles their unique yellow hue (Asenstorfer et al., 2006; Hatcher et al., 2008a). Dough that is harder, stiffer, and less extensible is produced as a result of changes that have a significant effect on the conduct of gluten proteins in dough features linked to an alkaline pH. The hardening of dough with the addition of alkali has a significant impact on the processing attributes and texture of the finished products. Alkali is applied to the noodle dough to increase its ability to absorb water, give the noodles a more sophisticated texture, increase their breaking and cutting forces, slow the gelatinization of the starch, and increase the viscosity of the starch paste (Gulia et al., 2014; Han et al., 2011).

Oil/Fat

Approximately 20% or so of the end product's weight is made up of oil. Due to its excellent frying performance, availability, heat stability, and low cost, palm oil is by far the most popular frying oil. Canola and soybean oils that have been partly hydrogenated can also be used for frying. The flavor of the noodles is greatly influenced by the frying oil; as a result, high-quality frying oil should be used (Gulia et al., 2014). Oil quality deteriorates during frying due to chemical reactions and extended and continuous heating, posing food safety risks and sensory failures. As a result, thermal stability is an important consideration when choosing frying oil. Approximately 20% or so of the end product's weight is made up of oil. Due to its excellent frying performance, availability, heat stability, and low cost, palm oil is by far the most popular frying oil. Canola and soybean oils that have been partly hydrogenated can also be used for frying. The flavor of the noodles is greatly influenced by the frying oil; as a result, high-quality frying oil should be used (Gulia et al., 2014; Marciniak-Lukasiak et al., 2019; Qi et al., 2019).

Improvers and Preservatives

Polyphosphates, hydrocolloids, emulsifiers, antioxidants, and enzymes are popular additives used in noodle processing to improve product quality. Polyphosphates operate as a chelating agent in the dough system, modifying the dough processing capabilities and delaying the discoloration of new noodles. Polyphosphates improve noodles'elasticity, flexibility, texture, and chewing qualities by accelerating gluten bonding. Additionally, they promote the gelatinization of starch during cooking, which increases water absorption and lessens cloudiness in the noodles. Before mixing the dough, they are mixed with water and typically employed at a concentration of 1.0% of the weight of the flour (Hourant, 2004). In the production of instant noodles, hydrocolloids such as guar gum, locust bean gum, alginates, and carboxymethyl cellulose (CMC) are commonly employed. Gums modify the texture and general "mouthfeel" of the finished product, increase the rehydration properties of noodles during frying, and lower the amount of fat absorbed during the frying of instant noodles since they have hydrophilic properties along with high capacity to bind water (Gulia et al., 2014). Biopolymers such as carrageenan, alginate, and chitosan have been used more frequently. These biopolymers have improved oxidative stability while also inhibiting microbial growth (Howard & Hung, 2010; Huang et al., 2007; Mahmud et al., 2021). However, further study into their impact on nutritional, sensory, and consumer acceptability aspects is still being conducted.

Quality Aspects Involving Processing

Dough Mixing

Along with managing the formulation of the individual noodles being mixed, managing the mixing temperature and duration time, dough temperature, and look is also necessary. When there is a divergence, such as a dough temperature declines, which can be referred to as too hot or cold, the excellent gluten formation is hampered, which leads to poor chewing capabilities for noodles. The dough's characteristics are observed, like if whether it is too dry or too moist. Changing the amount of water added to the dough is a practical approach to changing the dough's parameters. If the dough is overly moist, the amount of water in the next batch will be lowered. Wet dough makes the sheeting procedure difficult and leads to noodle sheet ripping. The normal amount of mixing water is 28–35 liters for 100 kilograms of flour. If starch is added to the formulation, dry dough may lead to inadequate starch gelatinization during steaming. Water sprayed on waved noodles before to steaming may help with starch gelatinization. Dough properties can be improved by gradually adding starch during the mixing process and increasing the amount of water in the formulation (Adejuwon et al., 2020; Gulia et al., 2014).

Dough Rolling

The two types of rollers are decreasing rollers and compounding rollers. To enhance the texture of the noodles, it is helpful to increase the number of reduction roller units. Lowering the dough's thickness should be done gradually and in stages. Noodle thickness shouldn't be decreased by more than 50% in any phase since doing so destroys the gluten structure and is unfavorable for the end product's texture. Since noodle sheet width impacts noodle weight and texture, it is frequently monitored and modified throughout production. Noodle thickness should be similar on both sides; however, achieving a consistent noodle thickness with oversized rollers is more challenging. In this situation, one side of the noodle cake may be heavier or lighter than the other. To regulate noodle sheet thickness, all rollers have to be in good functioning condition. Metal pieces in the dough that can harm the roller surface can be found using a metal detector before the reduction rollers (Adejuwon et al., 2020; Gulia et al., 2014).

Steaming

Noodles are steamed to pregelatinize starches, strengthen noodle blocks, fix noodle waves, and lessenthe cooking time before serving. If the fried noodle cakes/blocks are not adequately steamed, they will become fragile, and the supplied noodles will taste starchy after boiling or soaking in hot water. The fried noodle blocks lose their snackability if the noodles are over-steamed for an extended period. Steaming is done with saturated steam at around 100 °C. Steaming time can range from 1 → 5 min (Adejuwon et al., 2020; Gulia et al., 2014).

Frying

The purpose of frying is to pasteurize, or kill bacteria, while also swiftly removing moisture from steamed noodles. Critical factors include the length of time spent frying and the temperature at which it is finished. The frying temperature varies from 130→ 160 °C and the frying duration is less than 2 min, depending on the width and weight of the noodle. While a maximum moisture percentage of 8% is permitted in some conditions, finished products typically have a moisture level of less than 3%. To prepare snack noodles, steamed noodles are either tossed in the spice mixture or dipped in the mixture before being deep-fried. The finished products are tasty and crunchy. The process of frying improves the noodles'taste and appearance. The color of instant fried noodles should range from yellow to golden-brown. To get rid of the starchy flavor from the noodle cores, instant fried noodles that have not been properly steamed and fried need extra time to simmer in hot water before serving. Additionally, it has been noted that the noodles tend to get rancid more rapidly if they are not completely cooked (Adejuwon et al., 2020; Gulia et al., 2014).

Cooling

The cooling tunnel's purpose is to bring the temperature of the fried noodles down to room temperature. If the noodles are packaged while still hot, thermal and moisture condensation will result, which will degrade the seasoning sachet's quality and lead to mold growth (Adejuwon et al., 2020; Gulia et al., 2014).

Packaging

To prevent rapid oxidation and other changes, the product is either fried or dried before being chilled to room temperature. The soup base sachet is put in a bag with the cooled noodles. For the cup noodles, the powdered soup base is usually served and covered with shrinking film. The soup base contains seasonings and dried veggies. The higher the fat level, the more susceptible the noodles are to oxidative changes and the development of a rotten flavor. As a result, good packaging is critical in increasing the product's shelf life. The addition of antioxidants to the frying medium can postpone the onset of rancidity. Instant noodles'rancidity is exacerbated in the presence of ultraviolet light; as a result, noodles are primarily packaged in reddish-yellow or green packing material with no transparent sections. Bag noodles are packaged in polypropylene or polyethylene film, whereas cup noodles are packaged in polyester.

Quality Aspects Involving Final Products

Consumer preference is the most critical factor in determining the quality of noodles. These variables vary considerably across customers; nonetheless, features such as cooking and eating quality, color, flavor, texture, size, and shape are ubiquitous and influenced by basic material qualities, notably flour, formulation of ingredients, and handling variables. The most common parameters used to evaluate noodle quality are listed below (Moss et al., 1986; Wang et al., 2011).

Color

Color is still a vital featuresince it implies the product's initial visible attributes. Color choice is one of the most critical factors in determining noodle quality. Noodles should have a bright, creamy appearance or have a yellowish tone to them (Adejuwon et al., 2020). As previously stated, the color qualities of noodles are influenced by a variety of circumstances. Alkali-induced chromophoric shift of intrinsic flavonoids, the content oflutein pigment or xanthophyllsin wheat flour, protein content, food colorant addition by manufacturers, ash content of used flour, polyphenol oxidaseactivity, and flour extraction rate are just a few of them (Asenstorfer et al., 2006; Fu, 2008; Hatcher et al., 2008b; Moss et al., 1986; Zhao & Seib, 2005). The type of noodles processing utilized determines the color of the noodles. Color requirements vary depending on the type of noodle. Colorimetric analysis (Hunter Lab colorimeter) provides a better evaluation of noodle color by utilizing the L* (lightness →blackness), a* (redness → greenness), and b* (yellowness → blueness) values. Besides, sensory evaluation may be coupled with colorimetric analysis for better assessment.

Flavor

The flavor is a crucial factor in determining the quality of noodles, especially for instant noodles. It can comply with the taste during cooking, but the occurrence of putrid flavor after a prolongedstorage time possibly will cause the product to be rejected (O'Grady et al., 2001; St. Angelo et al., 1996). The flavor of instant fried noodles is also influenced by the deep-frying procedure (Choe & Min, 2007; Hau Fung Cheung et al., 2009). Seasoning has a complementing effect during cooking. In contrast, the existence of salt in the formula and the cookingprocedure contribute to flavor make-up in uncooked noodles (Gotoh et al., 2007). The most important factor in deciding whether the flavor of instantaneous fried noodles will remain pleasant over a lengthy period of storage is the quality of the oil. It may give shortening a pleasant aroma and tongue feels or give noodles an unpleasant rotten flavor (Adejuwon et al., 2020).

Texture

The texture of the noodles is one of its complex characteristics since factors like the quality of the wheat, the selection and quantity of salt or alkaline combinations, and processing variables such as dehydration techniques, sheeting, and steaming differ by region. According to a study, yellow alkaline noodles with appropriate textural characteristics may be produced using flours with a high protein content, good dough resilience, enough starch pasting strength, and low amounts of maltose (Gulia et al., 2014). The amount and grade of protein in the flour haveseveral effects on the quality of the noodles. In terms of texture, wheat flour with a lower protein concentration produced noodles with inferior textural attributes. Elasticity, cohesiveness, Stickiness, firmness, chewiness, softness, gumminess, and smoothnessare all words that can be used to characterize the texture of noodles (Guo et al., 2003; Konik et al., 1993; Kovacs et al., 2004; Ross, 2006). Both sensory and instrumental methods could assess noodle texture, although instrumental analysis has proven to be more reliable (Kovacs et al., 1997; Yue et al., 2017). Other studies have looked at various techniques for examining the textural properties of noodleand determined that a mix of systemassay and visual inspection is the best way to measure the texture of noodles (Ross., 2006; Ross & Crosbie, 2010). Texture Profile Analysis (TPA) is aneasy and effective way to evaluate texture. A single test provides a range of textural qualities such as firmness, gumminess, chewiness, and cohesion, with findings that are stated to be parallel to those obtained by mouth chewing (Gulia et al., 2014). Microscopy has recently been examined, and microstructure studies have been used extensively to investigate the impact of ingredients and formulations, and processing procedures on the texture of noodles. It was found that alkaline fresh noodles and fried instant noodles dough had similar microstructures, but different processing conditions, such as steaming, frying, and final cooking, caused further microstructural alterations in the noodles (Moss et al., 1986). Confocal laser scanning microscopy (CLSM) was utilized in a study to track microstructure changes caused by other processing conditions during instant noodle manufacture (Chewangkul et al., 2001). Researchers utilized CLSM to analyze the qualitative characteristics of white salted noodles that contained different cereal starches as well as the morphology of noodle dough sheets. Their research showed that the rheological properties of raw white salted noodles made from a combination of wheat starch and cereal starches are significantly influenced by the size of the starch granules. The rheological qualities of cooked white salted noodles, on the other hand, had a substantial impact on amylose content and smooth granules of amylopectin (Huang & Lai, 2010). In a related investigation, ultrasonography was utilized to identify variations in the microstructure of noodle dough made from various classes and varieties of wheat. Their research showed considerable variations in acoustic properties, proving that varietal differences influence the elasticity and hardness of noodles. The effectiveness of using ultrasound to measure dough strength was established, and the durability and dough development time of farinograph and ultrasonic doughs were assessed. Surprisingly, their findings contradicted popular belief that flours with more gluten or protein concentration produce more sturdy doughs (Diep et al., 2014). An ultrasonic approach was used in a similar study to uncover the effects of various additives on the textural factors of yellow alkaline noodles. They discovered that using ultrasonic technology, they could show changes in the textural features of yellow alkaline noodles made from different source materials. These microstructural alterations are responsible for many of the observed noodle quality parameters (Bellido & Hatcher, 2010).

Fat/Oil Uptake

During the frying of noodles, especially instant noodles, the cookingheattriggers moisture to escape from the noodles, which is eventually replaced by oil (Kim et al., 2013). Oil absorption in noodles may be justified by capillary forces (Pinthus & Saguy, 1994). Oil uptake is an essential fiscal factor in instant noodles because it may drastically raise the manufacturing cost. It also contributes to the formation of stale off-flavor in storing and is a component of the public wellbeing risk associated with elevated oil consumption. Quality and content of gluten are expected to have a part in noodle oil uptake during frying by generating a stiff structural network; hence, creating the dough more resilient implies that it will be less leaky to oil absorption (Gulia et al., 2014). Another study observed that noodles made from flour that are low in protein absorbed more oil than those made from high-protein wheat flour (Moss et al., 1986). Low-protein flour, the scientists reasoned, would enhance the formation of irregular globules during steaming, allowing oil to permeate the noodle easily. Furthermore, instant noodles made from high protein wheat flour with a large SDS sedimentation volume or less salt-soluble proteins have low free lipid levels (Park & Baik, 2004). This might be owing to the steaming process creating a uniform and thick noodle surface layout. Another factor that contributes to oil absorption during instant noodle frying is starch. Amylose may stimulate the formation of gaps around the surfaces of steamed noodle strands, resulting in pockets that facilitate oil uptake and accumulation, according to research that found a strong negative relationship between amylose and oil content in instant noodles (Park & Baik, 2004).

Cooking Quality

Several factors impact the cooking quality of noodles, including protein concentration, ash level, starch quality, starch denting, strand thickness, and cooking circumstances. Cooking quality and convenience of preparation are measured by rehydration value, cooking loss, and cooking time. Protein and amylose contents have a positive and negative relationship with the ultimate cooking period of noodles, respectively. (Adejuwon et al., 2020; Gulia et al., 2014). The frying period influences the cooking timetable and rehydration ratio when cooking instant noodles. Noodles preserve their starch formation as a multiplied3D network interconnected by crystallites made of amylose. Amylose networks grow due to amorphous areas' hydration after boiling in water. According to some studies, proteins function as a crucial underlyingelement in noodles in the course of cooking, triggering noodle strands to fit in and keep their structure (Adejuwon et al., 2020; Gulia et al., 2014).

5 Conclusion and Future Trends

The necessity for quick foods that are hygienic, safe, and rich in nutritional value has been emphasized as living standards have improved and the pace of modern life has accelerated. Noodles have emerged as a feasible solution to this issue. This chapter took a holistic approach to explain significant aspects of noodle manufacturing, such as history, production, quality factors, and preparation methods. Even though research on enhancing the quality of noodles has progressed significantly, research gaps still need to be filled to improve the production, processing, storage, and distribution of noodles. This part will discuss potential research gaps in the manufacturing of noodles and a list of future trends for more efficient and high-quality production of noodles. A few trends and outlooks are presented below:

- During the storage of noodles, quality variations in component quality are somewhat prevalent. We suggest evaluating potential issues that may impair the quality of noodles while in storage and looking into feasible solutions. Modern tools may be developed and used to regulate specific points that may affect the product's quality during processing. Furthermore, oxidation of lipids during the storage of noodles may produce volatiles and off-flavors. This topic should be thoroughly researched because there have been few improvements in this area. Establishing peroxide levels and thio-barbituric acid-reactive substances (TBARS) standards for each type of noodle could be the first step toward safer noodle production. Furthermore, usingactive substances with significant antioxidant effects may be necessaryfor delaying and suppressing noodle oxidation during storage. A variety of functional components should be added to the noodle formula to see if they survive and have a role in improving the final noodle product quality.

- Beyond the skill of making wheat flour noodles, understanding physiochemical reaction mechanisms and different processing phases will be helpful not only to the industry's bottom line and the consumer's health. The generation of hazardous compounds during the making of noodles could be a problem. The manufacture of noodles could be linked to the development of acrylamide, which is neurotoxic, genetically harmful, and carcinogenic (Friedman, 2003; Shrestha et al., 2016). The desire of consumers to limit the amount of fat, salt, and gluten in deep-fried foods, as well as consumer concern over the negative health consequences of products from Maillard Reaction or MRPs, compels food makers to look at a variety of physicochemical changes that take place during processing (glass transitions, gluten network, Maillard reactions, gelatinization). Finding the crust's advantageous organoleptic characteristics and color will result from this. Industries must thus come up with novel strategies for maximizing resource usage. To better understand the factors impacting and therefore offer application solutions, it may be possible to analyze the temperature of the deep-frying medium, mixing time, steaming time, heat supply, and other issues.
- Products made of fortified noodles have also become more popular. Micronutrient fortification of noodles is a distinct type of fortification. Micronutrients such as niacin, vitamin A, vitamin B1, vitamin B2, folic acid, iodine, and iron can be included after taking their processing steadiness and recommended daily quantities into account (Gulia et al., 2014). Noodle fortification was initiated in Indonesia in 1994. For instance, in countries like the Philippines, around 80% of the instant noodles produced are fortified willingly. Fortifying the seasoning insteadof the flour has the advantage of keeping the fortifiers out of the heat and moisture during the making of the noodles (Gulia et al., 2014). Additionally, the fortifiers are better protected because they are contained in a sachet. While it is very easy to add fortifiers, it might be difficult to maintain the mixture's nutritional stability while it is being manufactured (Bui & Small, 2007). Another study used chromatographical approaches (reversed-phase high-performance and capillary electrophoresis and) to test folic acid solidity in enriched instant noodles and found that it was durable during blending, sheeting, parboiling, and cooking of instant fried noodles (Hau Fung Cheung et al., 2009). Thiamin stability in noodles has been altered by processing, formulation, and storage, and it has been realized that the use of thiamin in noodles containing alkaline salts is hindered due to its volatility at higher pH. The shelf life is being increased, the formulation is being enhanced, and promotion of overall noodle fortification is being made. In order to increase nutritional qualities while companies fortify their own products, numerous studies are investigating whether noodle fortification may be utilized as a public health intervention.
- Despite the fact that numerous efforts have been made to improve the quality of the features, the vast majority of procedures are plagued by difficulties. In order

to improve the efficiency of the noodles manufacturing process, it is necessary to investigate grain type, particle size, mixing duration, amount of a component, dough temperature, and application of complex treatment with lower-cost alternatives. Because of the discoloration, fat oxidation, and spoilage that fresh noodles are subjected to, highly efficient preservation processes are required, particularly for fresh noodles. The use of acidification, high-pressure processing, and modified-atmosphere packing have all been shown to be beneficial for refined noodles; however, for other wheat-based noodles, certain modifications or entirely new procedures are still required in order to achieve the desired results.

- Noodles manufactured from non-wheat grains have also seen an increase in popularity. For instance, the demand for rice noodles and relevant products is rising because of the products' high nourishing content and delectableflavoring. Over the past several years, their market has progressively widened from East and South Asia to the remainder of the world (Li et al., 2021). On the other hand, the production of rice noodles has mostly remained low scalealong with dispersity. Manual manufacturing techniques include several drawbacks, such as inconsistent product quality, significant geographical variances, inadequate storage operation, and several possible safety concerns. The growth of the rice noodle market is significantly impacted by these worries. Currently, researchers are examining, among other things, the mechanisms of action of novel additives as well as the variables that affect the quality of rice noodles and the steps that may be taken to enhance that quality. Similar tendencies may be seen in non-wheat grain-based noodles. Three areas of the manufacture of non-wheat noodles may be the focus of future study. To better understand how different components interact with one another, a detailed analysis of the essential raw components is necessary. Investigating how external factors, such as ecological variables and handling methods, affect the excellence of non-wheat noodle products is crucial in this situation. For the purpose of choosing raw materials and creating production processes, the distinction between core and outer elements may be used as a theoretical guide. Second, it's important to look into the main alterations non-wheat noodles go through during storage and the countermeasures that may be used to lessen the consequences of these alterations. The development of control methods that will enhance the product's stability and storage performance includes identifying the bacteria and toxins that are likely to occur during the preparation and storage of noodles. A trustworthy assessment criterion for non-wheat noodles must also be developed. It will be necessary to develop a systematic and scientific evaluation system that combines objective technical methods with introspective sensory assessments of pertinent flavor markers. This will greatly enhance the manufacturing of non-wheat noodles by allowing for uniformity, mechanization, and automation.
- The molecular mass distribution, protein content, secondary structures, and degree of chemical cross-linking of various varieties of wheat flour varies significantly from one another. When substances are combined, differences in chemical structure will produce several different consequences. More investigation into

the mechanism-building, application, and relevance of various flours and appropriate processing methods is required.

- Because of the significant population pressure in many countries, the enhancement of grain supply has long been a major objective for many nations' grain breeding projects. Farmers are less likely to accept varieties with low processing quality than those with strong processing quality in recent years due to the importance of processing quality growing. In order to attain maximum production and quality, high grain production and outstanding industrial accuracy must be integrated into innovative combinations as well as broad adaptability and tolerance to a wide variety of biotic and abiotic challenges. The development of noodle grain types has advanced significantly, but there is still more to be done to raise the standard of noodles as a whole (Z. He et al., 2010). The future quality of noodles might be improved using one of the following schemes. In order to distribute varieties with outstanding noodle quality, noodle quality testing should be included into the different development and release procedures. In the majority of breeding operations nowadays, especially in commercial breeding projects, only progressed lines with improved varieties are examined for end-use caliber. We advise that varieties of outstanding quality only be issued if their yield performance is on par with the control variables in order to promote the development of high-quality noodle variations. Second, while improving dough suppleness and starch consistency is crucial for noodle grade breeding, dough strength and color are also important. The objective of enhancing dough strength should have been the top priority from the start of the quality improvement initiative. This will lead to better grain quality and more evenly distributed dough properties. The extensibility of dough will thus need to be improved by considerably more effort in the future. Desirable genes must be incorporated into existing types, and both biochemical and molecular indicators can be crucial to gene integration. The potential for creating delicate grain, especially for wheat types with middling to firm gluten quality, should be looked at as a third area to enhance the quality of noodles. Even though this kind of germplasm is prevalent, many nations with a long tradition of enhancing wheat quality may not have access to it. China affirmed that Asian noodle and mill professionals gave soft wheat with ore essential gluten and remarkable brightness favorable reviews. In Asia, this has long been the case. Therefore, one of the most important objectives in the effort to raise the caliber of noodles products might be the creation of a breed with softer husk wheat, mild to robust gluten strength, and vivid color (He et al., 2004, 2010).

References

Adejuwon, O. H., Jideani, A. I. O., & Falade, K. O. (2020). Quality and public health concerns of instant noodles as influenced by raw materials and processing technology. *Food Reviews International, 36*(3), 276–317. https://doi.org/10.1080/87559129.2019.1642348

Asenstorfer, R. E., Wang, Y., & Mares, D. J. (2006). Chemical structure of flavonoid compounds in wheat (Triticum aestivum L.) flour that contribute to the yellow colour of Asian alkaline noodles. *Journal of Cereal Science, 43*(1), 108–119. https://doi.org/10.1016/j.jcs.2005.09.001

Baik, B.-K., & Lee, M.-R. (2003). Effects of starch amylose content of wheat on textural properties of white salted noodles. *Cereal Chemistry Journal, 80*(3), 304–309. https://doi.org/10.1094/CCHEM.2003.80.3.304

Békés, F., Keměny, S., & Morell, M. (2006). An integrated approach to predicting end-product quality of wheat. *European Journal of Agronomy, 25*(2), 155–162. https://doi.org/10.1016/j.eja.2006.04.008

Bellido, G. G., & Hatcher, D. W. (2010). Ultrasonic characterization of fresh yellow alkaline noodles. *Food Research International, 43*(3), 701–708. https://doi.org/10.1016/j.foodres.2009.11.010

Bui, L. T. T., & Small, D. M. (2007). The contribution of Asian noodles to dietary thiamine intakes: A study of commercial dried products. *Journal of Food Composition and Analysis, 20*(7), 575–583. https://doi.org/10.1016/j.jfca.2007.03.010

Chewangkul, L., Naivikul, O., Garnjanagoonchorn, W., Oates, C. G., Siroth, K., & Suwonsichon, T. (2001). Microstructural changes in instant noodles during production via triple staining and confocal laser scanning microscopy and scanning electron microscopy. *Kasetsart Journal, Natural Sciences, 35*(3), 311–317.

Choe, E., & Min, D. B. (2007). Chemistry of deep-fat frying oils. *Journal of Food Science, 72*(5), R77–R86. https://doi.org/10.1111/j.1750-3841.2007.00352.x

Crosbie, G. B. (1991). The relationship between starch swelling properties, paste viscosity and boiled noodle quality in wheat flours. *Journal of Cereal Science, 13*(2), 145–150. https://doi.org/10.1016/S0733-5210(09)80031-3

Crosbie, G. B. (2005). Defining and developing tests to meet the key wheat quality requirements of Asian foods. In C. L. Blanchard, H. Truong, H. M. Allen, A. B. Blakeney, & L. O'Brien (Eds.), *Cereals 2005: Proceedings of the 55th Australian cereal chemistry conference and AACCI pacific rim symposium* (pp. 32–36). Royal Australian Chemical Institute.

Crosbie, G. B., & Ross, A. S. (2016). Noodles: Asian wheat flour noodles. In *Encyclopedia of food grains* (pp. 72–78). Elsevier. https://doi.org/10.1016/B978-0-12-394437-5.00122-4

Dexter, J. E., Matsuo, R. R., & Dronzek, B. L. (1979). A scanning electron microscopy study of Japanese noodles. *Cereal Chemistry, 56*(3), Article 3.

Diep, S., Daugelaite, D., Strybulevych, A., Scanlon, M., Page, J., & Hatcher, D. (2014). Use of ultrasound to discern differences in Asian noodles prepared across wheat classes and between varieties. *Canadian Journal of Plant Science, 94*(3), 525–534. https://doi.org/10.4141/cjps2013-043

Friedman, M. (2003). Chemistry, biochemistry, and safety of acrylamide. A review. *Journal of Agricultural and Food Chemistry, 51*(16), 4504–4526. https://doi.org/10.1021/jf030204+

Fu, B. X. (2008). Asian noodles: History, classification, raw materials, and processing. *Food Research International, 41*(9), 888–902. https://doi.org/10.1016/j.foodres.2007.11.007

Gotoh, N., Iwasawa, A., Watanabe, H., Osato, R., & Wada, S. (2007). Oxidation of fats and oils in instant noodles stored under various conditions. *Journal of Food Lipids, 14*(4), 350–365. https://doi.org/10.1111/j.1745-4522.2007.00091.x

Gulia, N., Dhaka, V., & Khatkar, B. S. (2014). Instant noodles: Processing, quality, and nutritional aspects. *Critical Reviews in Food Science and Nutrition, 54*(10), 1386–1399. https://doi.org/10.1080/10408398.2011.638227

Guo, G., Jackson, D. S., Graybosch, R. A., & Parkhurst, A. M. (2003). Asian salted noodle quality: Impact of amylose content adjustments using waxy wheat flour. *Cereal Chemistry Journal, 80*(4), 437–445. https://doi.org/10.1094/CCHEM.2003.80.4.437

Han, H. M., Cho, J. H., & Koh, B. K. (2011). Processing properties of Korean rice varieties in relation to rice noodle quality. *Food Science and Biotechnology, 20*(5), 1277–1282. https://doi.org/10.1007/s10068-011-0176-5

Hatcher, D. W., Kruger, J. E., & Anderson, M. J. (1999). Influence of water absorption on the processing and quality of oriental noodles. *Cereal Chemistry Journal, 76*(4), 566–572. https://doi.org/10.1094/CCHEM.1999.76.4.566

Hatcher, D. W., Anderson, M. J., Desjardins, R. G., Edwards, N. M., & Dexter, J. E. (2002). Effects of flour particle size and starch damage on processing and quality of white salted noodles. *Cereal Chemistry Journal, 79*(1), 64–71. https://doi.org/10.1094/CCHEM.2002.79.1.64

Hatcher, D. W., Dexter, J. E., & Fu, B. X. (2008a). Investigation of amber durum wheat for production of yellow alkaline noodles. *Journal of Cereal Science, 48*(3), 848–856. https://doi.org/10.1016/j.jcs.2008.06.009

Hatcher, D. W., Edwards, N. M., & Dexter, J. E. (2008b). Effects of particle size and starch damage of flour and alkaline reagent on yellow alkaline noodle characteristics. *Cereal Chemistry Journal, 85*(3), 425–432. https://doi.org/10.1094/CCHEM-85-3-0425

Hau Fung Cheung, R., Hughes, J. G., Marriott, P. J., & Small, D. M. (2009). Investigation of folic acid stability in fortified instant Asian noodles by use of capillary electrophoresis. *Food Chemistry, 112*(2), 507–514. https://doi.org/10.1016/j.foodchem.2008.05.110

He, Z. H., Yang, J., Zhang, Y., Quail, K. J., & Peña, R. J. (2004). Pan bread and dry white Chinese noodle quality in Chinese winter wheats. *Euphytica, 139*(3), 257–267. https://doi.org/10.1007/s10681-004-3283-z

He, Z., Xia, X., & Zhang, Y. (2010). Breeding noodle wheat in China. In G. G. Hou (Ed.), *Asian noodles* (pp. 1–23). Wiley. https://doi.org/10.1002/9780470634370.ch1

Hou, G. G., Otsubo, S., Okusu, H., & Shen, L. (2010). Noodle processing technology. In G. G. Hou (Ed.), *Asian noodles* (pp. 99–140). Wiley. https://doi.org/10.1002/9780470634370.ch5

Hourant, P. (2004). General properties of the alkaline phosphates: – Major food and technical applications. *Phosphorus Research Bulletin, 15*(0), 85–94. https://doi.org/10.3363/prb1992.15.0_85

Howard, B. M., & Hung, Y.-C. (2010). Formulation of pasta noodles made with Peanut flour. *Peanut Science, 37*(2), 95–99. https://doi.org/10.3146/PS09-009.1

Huang, Y.-C., & Lai, H.-M. (2010). Noodle quality affected by different cereal starches. *Journal of Food Engineering, 97*(2), 135–143. https://doi.org/10.1016/j.jfoodeng.2009.10.002

Huang, S., & Morrison, W. R. (1988). Aspects of proteins in chinese and British common (hexaploid) wheats related to quality of white and yellow chinese noodles. *Journal of Cereal Science, 8*(2), 177–187. https://doi.org/10.1016/S0733-5210(88)80028-6

Huang, J., Huang, C., Huang, Y., & Chen, R. (2007). Shelf-life of fresh noodles as affected by chitosan and its Maillard reaction products. *LWT – Food Science and Technology, 40*(7), 1287–1291. https://doi.org/10.1016/j.lwt.2006.08.004

Jaques, D. A., Wuerzner, G., & Ponte, B. (2021). Sodium intake as a cardiovascular risk factor: A narrative review. *Nutrients, 13*(9), 3177. https://doi.org/10.3390/nu13093177

Karim, R., & Sultan, M. T. (2015). Background and introduction. In I. R. Karim & M. T. Sultan (Eds.), *Yellow alkaline noodles* (pp. 1–3). Springer International Publishing. https://doi.org/10.1007/978-3-319-12865-8_1

Kim, Y., Kim, Y., Bae, I. Y., Lee, H. G., Hou, G. G., & Lee, S. (2013). Utilization of preharvest-dropped apple powder as an oil barrier for instant fried noodles. *LWT – Food Science and Technology, 53*(1), 88–93. https://doi.org/10.1016/j.lwt.2013.02.022

Konik, C. M., Miskelly, D. M., & Gras, P. W. (1993). Starch swelling power, grain hardness and protein: Relationship to sensory properties of Japanese noodles. *Starch – Stärke, 45*(4), 139–144. https://doi.org/10.1002/star.19930450406

Kovacs, M. I. P., Poste, L. M., Butler, G., Woods, S. M., Leisle, D., Noll, J. S., & Dahlke, G. (1997). Durum wheat quality: Comparison of chemical and rheological screening tests with sensory analysis. *Journal of Cereal Science, 25*(1), 65–75. https://doi.org/10.1006/jcrs.1996.0069

Kovacs, M. I. P., Fu, B. X., Woods, S. M., & Khan, K. (2004). Thermal stability of wheat gluten protein: Its effect on dough properties and noodle texture. *Journal of Cereal Science, 39*(1), 9–19. https://doi.org/10.1016/S0733-5210(03)00058-4

Li, C., You, Y., Chen, D., Gu, Z., Zhang, Y., Holler, T. P., Ban, X., Hong, Y., Cheng, L., & Li, Z. (2021). A systematic review of rice noodles: Raw material, processing method and quality improvement. *Trends in Food Science & Technology, 107*, 389–400. https://doi.org/10.1016/j.tifs.2020.11.009

Lu, H., Yang, X., Ye, M., Liu, K.-B., Xia, Z., Ren, X., Cai, L., Wu, N., & Liu, T.-S. (2005). Millet noodles in late Neolithic China. *Nature, 437*(7061), 967–968. https://doi.org/10.1038/437967a

Mahmud, N., Islam, J., & Tahergorabi, R. (2021). Marine biopolymers: Applications in food packaging. *PRO, 9*(12), 2245. https://doi.org/10.3390/pr9122245

Marciniak-Lukasiak, K., Zbikowska, A., Marzec, A., & Kozlowska, M. (2019). The effect of selected additives on the oil uptake and quality parameters of fried instant noodles. *Applied Sciences, 9*(5), 936. https://doi.org/10.3390/app9050936

Moss, H. J., Miskelly, D. M., & Moss, R. (1986). The effect of alkaline conditions on the properties of wheat flour dough and cantonese-style noodles. *Journal of Cereal Science, 4*(3), 261–268. https://doi.org/10.1016/S0733-5210(86)80028-5

O'Grady, M. N., Monahan, F. J., & Brunton, N. P. (2001). Oxymyoglobin oxidation and lipid oxidation in bovine muscle—Mechanistic studies. *Journal of Food Science, 66*(3), 386–392. https://doi.org/10.1111/j.1365-2621.2001.tb16115.x

Obadi, M., Zhang, J., He, Z., Zhu, S., Wu, Q., Qi, Y., & Xu, B. (2022). A review of recent advances and techniques in the noodle mixing process. *LWT, 154*, 112680. https://doi.org/10.1016/j.lwt.2021.112680

Park, C. S., & Baik, B.-K. (2002). Flour characteristics related to optimum water absorption of noodle dough for making white salted noodles. *Cereal Chemistry Journal, 79*(6), 867–873. https://doi.org/10.1094/CCHEM.2002.79.6.867

Park, C. S., & Baik, B.-K. (2004). Relationship between protein characteristics and instant noodle making quality of wheat flour. *Cereal Chemistry Journal, 81*(2), 159–164. https://doi.org/10.1094/CCHEM.2004.81.2.159

Parvin, R., Farzana, T., Mohajan, S., Rahman, H., & Rahman, S. S. (2020). Quality improvement of noodles with mushroom fortified and its comparison with local branded noodles. *NFS Journal, 20*, 37–42. https://doi.org/10.1016/j.nfs.2020.07.002

Pinthus, E. J., & Saguy, I. (1994). Initial interfacial tension and oil uptake by deep-fat fried foods. *Journal of Food Science, 59*(4), 804–807. https://doi.org/10.1111/j.1365-2621.1994.tb08132.x

Qi, J., Wang, X., Wang, X., Akoh, C. C., & Jin, Q. (2019). Effect of oil type and emulsifier on oil absorption of steam-and-fried instant noodles. *Journal of Oleo Science, 68*(6), 559–566. https://doi.org/10.5650/jos.ess18217

Ross, A. S. (2006). Instrumental measurement of physical properties of cooked Asian wheat flour noodles. *Cereal Chemistry Journal, 83*(1), 42–51. https://doi.org/10.1094/CC-83-0042

Ross, A. S., & Crosbie, G. B. (2010). Effects of flour characteristics on noodle texture. In G. G. Hou (Ed.), *Asian noodles* (pp. 313–329). Wiley. https://doi.org/10.1002/9780470634370.ch13

Ross. (2006). Sheeting characteristics of salted and alkaline Asian noodle doughs: Comparison with lubricated squeezing flow attributes. *Cereal Foods World*. https://doi.org/10.1094/CFW-51-0191

Seib, P. A., Liang, X., Guan, F., Liang, Y. T., & Yang, H. C. (2000). Comparison of Asian noodles from some hard white and hard red wheat flours. *Cereal Chemistry Journal, 77*(6), 816–822. https://doi.org/10.1094/CCHEM.2000.77.6.816

Shiau, S.-Y., & Yeh, A.-I. (2001). Effects of alkali and acid on dough rheological properties and characteristics of extruded noodles. *Journal of Cereal Science, 33*(1), 27–37. https://doi.org/10.1006/jcrs.2000.0344

Shrestha, P., Rai, K. P., & Anal, A. K. (2016). Interactions of bio-macromolecules during processing of instant Asian noodles: A review. *Journal of Food Science and Technology Nepal, 9*, 1–10. https://doi.org/10.3126/jfstn.v9i0.16197

St. Angelo, A. J., Vercellotti, J., Jacks, T., & Legendre, M. (1996). Lipid oxidation in foods. *Critical Reviews in Food Science and Nutrition, 36*(3), 175–224. https://doi.org/10.1080/10408399609527723

Wang, C., Kovacs, M. I. P., Fowler, D. B., & Holley, R. (2004). Effects of protein content and composition on white noodle making quality: Color. *Cereal Chemistry Journal, 81*(6), 777–784. https://doi.org/10.1094/CCHEM.2004.81.6.777

Wang, L., Hou, G. G., Hsu, Y.-H., & Zhou, L. (2011). Effect of phosphate salts on the Korean non-fried instant noodle quality. *Journal of Cereal Science, 54*(3), 506–512. https://doi.org/10.1016/j.jcs.2011.09.008

World Instant Noodles Association. (n.d.). Retrieved February 3, 2022, from https://instantnoodles.org/en/

Yamauchi, H., Ito, M., Nishio, Z., Tabiki, T., Kim, S.-J., Hashimoto, N., Noda, T., Takigawa, S., Matsuura-Endo, C., Takata, K., Ohta, K., Fukushima, M., Miura, H., & Ism, Z. (2007). Effects of high-molecular-weight Glutenin subunits on the texture of yellow alkaline noodles using near-isogenic lines. *Food Science and Technology Research, 13*(3), 227–234. https://doi.org/10.3136/fstr.13.227

Ye, Y., Zhang, Y., Yan, J., Zhang, Y., He, Z., Huang, S., & Quail, K. J. (2009). Effects of flour extraction rate, added water, and salt on color and texture of Chinese white noodles. *Cereal Chemistry Journal, 86*(4), 477–485. https://doi.org/10.1094/CCHEM-86-4-0477

Yue, F.-L., Guo, X.-N., & Zhu, K.-X. (2017). Impact of characteristics of different wheat flours on the quality of frozen cooked noodles. *Cereal Chemistry Journal, 94*(5), 881–886. https://doi.org/10.1094/CCHEM-04-17-0082-R

Zhang, N., & Ma, G. (2016). Noodles, traditionally and today. *Journal of Ethnic Foods, 3*(3), 209–212. https://doi.org/10.1016/j.jef.2016.08.003

Zhao, L. F., & Seib, P. A. (2005). Alkaline-carbonate noodles from hard winter wheat flours varying in protein, swelling power, and polyphenol oxidase activity. *Cereal Chemistry Journal, 82*(5), 504–516. https://doi.org/10.1094/CC-82-0504

Chapter 11
Cereal Based Fermented Products

Manzoor Ahmad Shah, Rifat Jabeen, Irfan ul Majid, and Shabir Ahmad Mir

1 Introduction

Cereals are an important part of our diet as they represent the staple foods in almost every continent. Cereals provide us with main energy source, fibres in addition to other nutrients. Global food security revolves round the production of cereals as they are the main energy source in almost every diet due to their high carbohydrate content (~75%) (Laskowski et al., 2019). Moreover cereals are rich in other bioactive components e.g., phenolic compounds, dietary fibre, fatty acids, phytoestrogens, macro & micronutrients (Fernandes et al., 2018; Ignat et al., 2020). The micro and macronutrients are essential for daily life activities (Fernandes et al., 2018), whereas other bioactive compounds help in prevention and fighting against various diseases like cancers, CVDs, type-2 diabetes, obesity, improve gut health and other inflammatory diseases (Fardet, 2010). Cereals can be converted into different products by different processing techniques and fermentation is an important processing technique used to prepare a large number of diverse cereal-based products throughout the world (Table 11.1). The interest in production and consumption of cereal-based fermented products has been increased due to low quantity of fat and high concentration of minerals, dietary fibers, and phytochemicals (Ray et al., 2016).

Fermentation is an age old as well as economical processing technique. Fermentation helps preserve foods for a longer tenure. In addition to increasing shelf life and providing probiotic functions, fermented foods especially cereals have better palatability, nutritional value (increased phenolic and vitamin content),

M. A. Shah (✉) · R. Jabeen · I. ul Majid
Department of Food Science and Technology, Government Degree College for Women, Anantnag, Jammu & Kashmir, India

S. A. Mir
Department of Food Science and Technology Government College for Women, Srinagar, Jammu & Kashmir, India

M. A. Shah et al. (eds.), *Cereal-Based Food Products*,
https://doi.org/10.1007/978-3-031-40308-8_11

Table 11.1 Fermented cereal products

Name	Cereal used	Microorganism
Idli	Rice Black gram	*Leuconostoc mesenteroides, Streptococcus faecalis, Lactobacillus delbrueckii, Lb. fermenti, Lb. lactis, Pediococcus cerevisiae*
Dosa	Rice Black gram	*Leuconostoc mesenteroides, S. faecalis, Torulopsis candida, T. pullulans*
Dhokla	Rice, Bengal gram	*L. mesenteroides, S. faecalis, T. candida, T. pullulans*
Uttapam	Rice, urad dahl	*Lactobacillus pentosus, L. plantarum*, yeasts
Kishk	Wheat	*Lb. plantarum, Lb. casei, Bacillus subtilis, Lb. brevis*, yeasts
Tarhana	Wheat	*L. plantarum, L. casei, B. subtilis, L. brevis*, yeasts
Ogi	Maize, sorghum or millet	*L. plantrium, Corynebacterium, Saccharomyces, Candida* species
Kenkey	Maize	*Lb. fermentum, Lb. reuteri, Candida, Aspergillus, Penicillium, Fusarium* species
Pozol	Maize	*Lactococcus lactis, S. suis, Lb. plantarum, Lb. casei, Lb. delbruekii, Lb. alimentarium, Clostridium sp.*
Injera	Sorghum, tef, corn, finger millet or barley	*Pullaria sp., Aspergillus sp., Rhodotorula sp., Penicillium sp., Hormodendrum sp., Candida sp.*
Kisra	Sorghum or pearl millet	*Lactobacillus sp., Acetobacter sp., S. cerevisiae.*
Sake	Rice	*Aspergillus oryzae*, LAB
Bouza	Wheat	LAB, yeasts
Chicha	Maize	*S. cerevisiae, Lactobacillus sp., Leuconostoc sp., Aspergillus sp., Acetobacter sp.*
Mahewu	Maize, millet malt, sorghum or wheat	*Lb. lactis sub sp. Lactis*
Boza	Wheat, maize, rye, millet, andother cereals	*Lb. plantarum, Lb. acidophilus, Lb. coprophilus, Lb. fermentum, Leuconostoc raffinolactis, Ln. mesenteroides and Ln. brevis, Candida tropicalis, C. glabrata, S. cerevisiae, Geotrichum penicillatum, G. candidum*
Yosa	Oat bran	*Bifidobacterium spp., Lactobacillus spp.*, other species of LAB
Amazake	Rice	*Lb. sakei, Aspergillus oryzae*
Obushera	Sorghum or millet, maize	*Lc. lactis, Lb. plantarum, Lb. fermentum, Lb. delbrueckii, Weissella confuse*
Gowe	Sorghum and maize	*Lb. fermentum, W. confusa, W. kimchii, Lb. mucosae, Pediococcus acidilactici, P. pentosaceus*
Champus	Maize, rye, wheat	*S. cerevisiae, Issatchenkia orientalis, Pichia fermentans, Pichia kluyveri, Zygosaccharomyces fermentati, Torulospora delbruekii, Galactomyces geotrichum and Hanseniaspora spp.*
Kvass	Rye, barley	*L. casei, L. mesenteroides, S. cerevisiae*
Togwa	Maize, sorghum, finger millet	*Lb. plantarum, Ln. brevis, Lb. fermentum, L. cellobiosus, W. confusa, P. pentosaceus, Issatchenkia orientalis, S. cerevisiae, Candida pelliculosa, C. tropicalis*
Uji	Maize, finger millet, sorghum	*Lb. plantarum, Lb. fermentum, Lb. cellobiosus, Lb. buchneri, P. acidilactici, P. pentosaceus, L. paracasei*

digestibility, modified dietary fibres, antibiotic content, reduced anti-nutritional compounds and allergens, improved bioavailability and food security (Kohajdová, 2017; Tsafrakidou et al., 2020; Khatun & Ray, 2022; Chelule et al., 2010). Most of the food fermentations are carried out using lactic acid bacteria (LAB). These fermentation processes are being used from ancient times and economically more feasible for the production and preservation of food for humans (Blandino et al., 2003; Holzapfel, 2002; Adegoke & Babalola, 1988; Caplice & Fitzgerald, 1999). The most common strains used include species of *Lactobacillus, Leuconostoc, Pediococcus, Micrococcus, Bacillus, Fusarium, Saccharomyces, Pencillium, Streptococcus, Bifidobacterium, Pullaria, Rhodotorula, Hormodendrum, Zygosaccharomyces, Acetobacter, Aspergillus, Candida, Geotrichum*, etc. (Blandino et al., 2003; Kohajdová, 2017; Altay et al., 2013; Phiri et al., 2019; Chibuzor-Onyema et al., 2021). It has been reported that fermented foods and beverages form a significant part of human diets globally (Nyanzi & Jooste, 2012).

2 Fermented Foods

Fermented foods have been prepared using various raw materials which mostly include fruits and vegetables. However, different cereals have been employed in fermentation processes to yield different fermented food products. The kind and percentage of cereals along with the fermentation methods and conditions also vary widely (Blandino et al., 2003; Ghosh et al., 2015a, b). The general process of preparing fermented foods commence with the preparation of cereals, making batter to a semi-solid consistency with the addition of other ingredients. The fermentation process initiates either naturally or is induced using fresh starters or by back slopping method. Fermentation process induces food specific changes thereby changing the properties of the mix and improving the overall nutrient content (Yousif & Tinay, 2003; Chavan & Kadam, 1989). Once the fermentation process is completed, some of the products are consumed as such i.e., they don't undergo any further processing while as other products need further processing in terms of frying, steaming, baking or cooking. The fermented foods undergo these processing techniques in order to obtain a final product with unique/desired characteristics like aroma, taste, texture etc. (Savitri & Bhalla, 2007).

Idli

Idli is popular savoury cake type product most widely consumed product in India especially in South India (Blandino et al., 2003). Idli is a cereal based fermented product prepared from rice (*Oryza sativum*) and black gram dhal (*Phaseolus mungo*). Idli is nutritious and low caloric and thus mostly consumed as a breakfast food. Both the rice and black gram are soaked in water separately and then converted into slurry by grinding. A thick batter is prepared by mixing both the slurries. The batter is then fermented overnight at room temperature. After fermentation the batter is poured on idli pan and steamed. The resulting product idli is spongy in texture (Smid & Lacroix, 2013) (Fig. 11.1).

Dosa

Dosa is a flat thin and crispy product prepared by fermentation from rice and black gram and is popular throughout India. The ingredients are soaked in water separately followed by grinding. A smooth batter is prepared by adding water and then fermented overnight at room temperature. After fermentation the batter is poured on a flat heated surface in thin layer. After few minutes a round crispy dosa is removed from the heating surface (Blandino et al., 2003).

Dhokla

Dhokla is a soft and spongy product prepared by fermentation of rice and Bengal gram and is popular in India. Sometime vegetables are also to increase its palatability. The ingredients are soaked and then turned into thick batter. Fermentation of the batter carried out overnight at room temperature. The batter is then poured into trays greased with oil and steamed in open conditions (Ray et al., 2016).

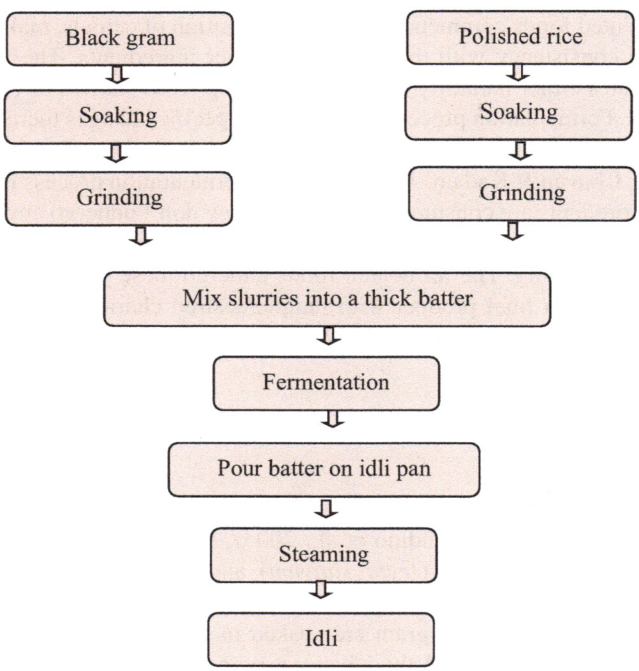

Fig. 11.1 Idli preparation process

Uttapam

Uttapam is a fermented cereal based product made from rice and urad dahl (*Vigna mungo*) and is thick and slightly crispy pan cake popular in India. The ingredients rice and urad dahl are taken in 3:1 ratio and soaked in water. The soaked ingredients are converted into batter by grinding. The batter is fermented for 5–6 hours at room temperature. The uttapam is prepared by spreading the batter on a heated pan and cooked over a low flame (Blandino et al., 2003).

Injera

Injera (Enjera) is the fermented cereal based product popular in Ethiopia (Chavan & Kadam, 1989). It is prepared from diverse cereals, such as sorghum, tef, corn, finger millet and barley, but tef (*Eragrostis tef*) is the main component in Ethiopian injera. For preparation of injera the dehulled grains are converted into flour by milling. The flour and water is mixed to prepare a dough. The starter (ersho), obtained from previous fermented dough, is added to the fresh dough and fermented for 2–3 days. The dough is then changed into thick batter and spread on a greased pan. A lid is kept over the pan for steam and it ready within 2–3 min. Normally, injera is round, soft, spongy with slightly sour flavor (Blandino et al., 2003) (Fig. 11.2).

Kisra

Kisra is cereal based fermented food prepared in Arabian Gulf, Sudan and Iraq. It is prepared from sorghum or pearl millet flour and is similar to injera. The flour is converted into dough and fermented (Thakur et al., 1995). Thin sheets are prepared from the fermented dough and then baked (Blandino et al., 2003).

Kishk

Kishk is a cereal based fermented food made from wheat from milk. It is prepared by mixing strained yoghurt to bran free parboiled wheat (bulgur). This mixture is fermented at ambient temperature for different time periods. The wheat grains are milled and the bran is removed. Milk is soured separately, concentrated and then the added with the moistened wheat flour. Lactic acid fermentation occurs in the mixture resulting in a paste like consistency which is then dried and ground into a powder. The product is prepared as dried balls with brown color and rough surface (Blandino et al., 2003).

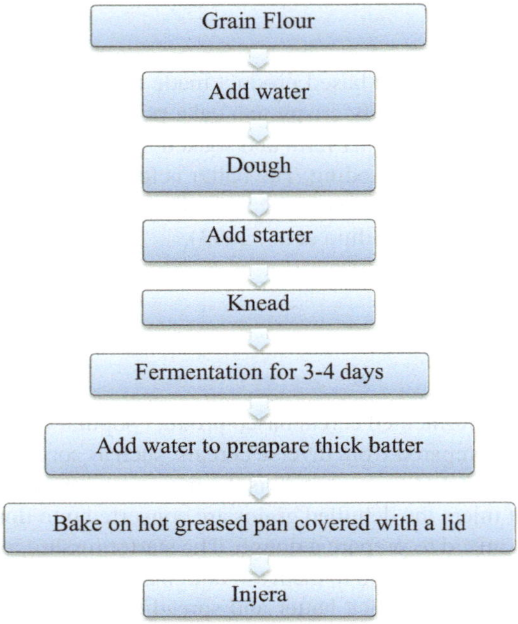

Fig. 11.2 Injera preparation process

3 Ogi

Ogi is a fermented cereal product popular in West Africa. It is prepared by fermentation of maize, sorghum or millet. Ogi is a staple food and is used as weaning food for infants. Traditionally, ogi is made from corn kernels which are soaked in water for 1–3 days. These soaked kernels are then grinded by wet milling process and sieved to remove bran, hulls and germ (Odunfa, 1985). The filtrate obtained is fermented for 2–3 days and the resulting product is known as ogi.

4 Fermented Beverages

Fermented beverages especially cereal beverages have been manufactured and consumed all over the world. Fermentation processes are broadly classified as per the end product into acetic acid, lactic acid, alcoholic, and alkali fermentation. The type/strain of microorganisms to be used depends upon the substrate and the end product as well as the fermentation conditions (Blandino et al., 2003). Numerous beverages have been prepared from a cereal crop singly (whole grain or a part of it) or in combination with other cereals or foods which include: Bouza from wheat;

Busaa, Kachasa, Seketeh, Tesguino, Tobwa, Chicha, Koko, Liha and Pozo from maize; Boza from wheat and other cereals; Khaomak, Sake, Chongju, Tapaipulut and Tapuy from rice; soymilk from soybeans; Darassum, Mangisi and Thumba from millet; Kwunu-zaki, Kunu, Togwa, Uji and Mahewu from sorghum, millet and maize; Chikokivana from maize and millet, Merissa & Bushera from sorghum and millet; Otika from sorghum; Kvass from rye and many more (Blandino et al., 2003; Tsafrakidou et al., 2020).

Fermented cereal beverages can be broadly divided into two groups: alcoholic beverages and non-alcoholic beverages. Some of the fermented cereals are detailed below:

5 Alcoholic Cereal Beverages

Sake

Sake is one of the oldest and most widely consumed alcoholic beverages esp. in China and Japan. Sake also known as rice wine is made from rice. The beverage is transparent with an alcoholic content of 13% to 17% by volume (Zhang et al., 2020). It has a good nutritional quality in terms of ample amounts of amino acids and vitamins. The initial microorganism involved in Sake production is *Aspergillus oryzae*, which is used in the preparation of koji wherein action of amylases is required. Later on brewing is carried out by the yeast Saccharomyces cerevisiae. The fermentation carried out in sake production is a slow process done below 15 °C for 2 weeks to 1 month.

Boza

Boza is an alcoholic traditional Turkish drink; however it is consumed in many Asian and African countries as beer. Boza is made using millet, maize, wheat, rice, barley, oats but millet is the preferred cereal (Fig. 11.3). The cereals after cooking and sugar addition are fermented with previous Boza, yogurt or sourdough and kept 30 °C for 24 h (Arici & Daglioglu, 2002). However, the shelf life of the drink is relatively short lesser. The microbiota involved on the production of Boza include LAB and some yeasts including: *Leuconostoc paramesenteroides, L. Mesenteroides, L. oenos, Lactobacillus sanfrancisco, Lb. confusus, Lb fermentum, L. coryniformis, Saccharomyces cerevisiae, S. uvarum, S. carlsbergensis, Micrococcus* spp., *Streptococcus* spp., and other *Lactobacillus* spp. The chemical composition of the drink majorly varies with the substrate used (Arici & Daglioglu, 2002). Boza is a high alcoholic drink constituting of about 7% alcohol content by volume (Kohajdova, 2014).

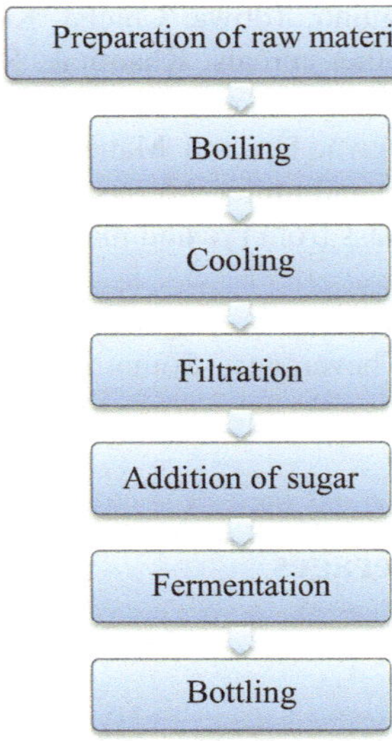

Fig. 11.3 Generalized method of Boza preparation

Champus

Champus is a low alcoholic (2.5–4.2% alcohol content) drink made from maize, wheat and rye. It is widely consumed in Colombia and several other South American countries. The microorganisms involved include *Saccharomyces cerevisiae*, *Issatchenkia orientalis*, *Pichia fermentans*, *Pichiakluyveri* var. *kluyveri*, *Torulospora delbruekii*, *Galactomyces geotrichum*, *Zygosaccharomyces fermentati* and *Hanseniaspora* spp. (Osorio-Cadavid et al., 2008). Champus is a rich drink both nutritionally as well as in terms of sensory properties. Different fruits and spices have also been added to improve taste.

6 Non-alcoholic Cereal Beverages

Amahewu or Mahewu

Amahewu or Mahewu is a non-alcoholic cereal beverage. It is a refreshing and nutrient dense drink consumed in South Africa and some Gulf countries by adults as well as children (Chelule et al., 2010). It is majorly prepared using white or yellow maizes, however cassava and sorghum have also been used to make amahewu. In some cases millet and sorghum are added while as for enrichment and fortification different food materials have also been added e.g. aloe vera (Oyeyinka et al., 2021).

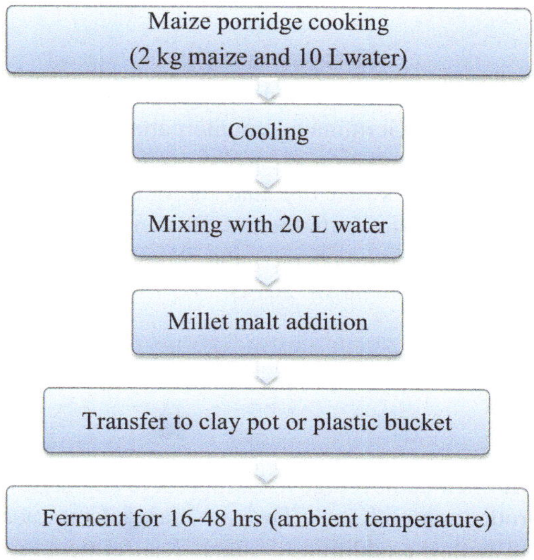

Fig. 11.4 Traditional Amahewu production process (Oyeyinka et al., 2021)

Different starters have been employed to prepare Amahewu, however, traditionally the back-slopping method is used. The microbiota associated with Amahewu production include predominantly lactic acid bacteria LAB (*Lactobacillus brevis, Lb. mesenteroides, Lb. fermentum, Lb. acidophilus, Lb. lactis* and *Wessella confusa*) however, *Lactobacillus casei, Saccharomyces cerevisiae, S. pombe and Candida glabrata* are also used (Simatende et al., 2019) (Fig. 11.4).

Shalgam Juice

Shalgam juice is consumed in turkey esp. in the Mediterranean region. It is a red colored sour juice. The cereal crop used is wheat in the form of black carrot bulgur (broken wheat) flour, along with turnip, salt and water. Altay et al., 2013 gave a general composition of Shalgam juice as 2.0–4.0% total dry matter, 0.09–0.018% protein, 1.1–2.2% salt, 0.578–8.05 g/L lactic acid, 1.06–9.1 g/L titrable acidity, 1.46–2.06% ash, 0.79–6.41 ethanol, a pH between 3.15 and 4.25 at a temperature of 10 °C and 35 °C for 3–5 days. Shalgam juice is reported be a nutrient rich drink with a good amount of vitamin, mineral and polyphenol content. Shalgam juice is made traditionally firstly using sourdough method (flour, salt, water & sourdough) followed by fermentation of carrot. However, commercial practices involve only second fermentation only. Species of *Lactobacillus*, *Leuconostoc* and *Pediococcus* are the predominant microorganisms involved in the production of Shalgam juice.

Bushera

Bushera is a fermented drink consumed in Uganda. It is prepared by using the cereals sorghum and millet. Germinated counterparts are also used in the preparation of Bushera and the fermentation is carried out for 12–24 hrs. Two versions of Bushera are common one is sweet one and the other being sour. Back slopping method is commonly used to manufacture Bushera. Species of *Lactobacillus, Lactococcus, Leuconostoc, Enterococcus*, and *Streptococcus* are the predominant microorganisms involved (Kohajdova, 2014).

Kvass

Kvass is a popular fermented beverage in Europe. Rye and barley are the main cereals used for the production of Kvass. The raw materials are added in the form of bread, flour or malt with the addition of sugar. The fermentation is carried out at 12–30 °C for 12–24 h and latter stopped by cooking. The predominant microorganisms include LAB and yeasts viz.: *L. casei, L. mesenteroides*, and *S. cerevisiae* (Kohajdova, 2014).

Togwa

Togwa is traditional African non-alcoholic drink. It is manufactured using maize, finger millet and sorghum. Back slopping method is employed and the fermentation is carried out at 35 °C for 12–24 h. The microbiota involved in the fermentation include *Lb. plantarum, Lb. brevis, Lb. fermentum, Lb. cellobiosus, P. pentosaceus, Issatchenkia orientalis, S. cerevisiae, Candida pelliculosa, W. confusa* and *C. tropicalis* (Kohajdova, 2014).

Uji

Uji is another fermented beverage made using maize, finger millet and sorghum. It is also consumed in Africa. Back slopping method is used and the fermentation is carried out at ambient temperature for 1–3 days. *Lactobacillus plantarum, Lb. fermentum, Lb. cellobiosus, Lb. buchneri, P. pentosaceus, L. paracasei, P. acidilactici*, are the principal microorganisms involved in the fermentation (Kohajdova, 2014).

7 Health Benefits

Cereals contain several components termed as anti-nutritional factors e.g. phytic acid, lectins, enzyme inhibitors, tannins, saponins. These antinutritional compounds/factors can lead to reduced absorption and bioavailability of nutrients (esp. minerals), disfunctioning of some major enzymes, intestinal adhesions and gastrointestinal discomfort (Fardet, 2010). Phytic acid and tannins are majorly responsible for mineral chelation leading to diminished absorption and bioavailability of minerals present in the diet. Fermentation has successfully reported to decrease the antinutritional components among the cereals thereby increasing the absorption and making them more bioavailable. Moreover, inactivation of enzyme inhibitors leads to increased digestibility of food components (Tsafrakidou et al., 2020). Furthermore many enzymes (e.g. amylases, lipases, proteases) are activated which hydrolyse food components making them more digestible (Kohajdova, 2014). This paves a way to incorporate these foods in geriatric and infant formulations. Fermentation processes affect the starch digestibility thus having an impact on glycemic index which is a major concern for diabetic patients and consumers that are health conscious. Several fermented foods especially sourdough has reported to have a low glycemic index owing to changes in starch due to fermentation. During the fermentation processes in cereals lactic acid gets produced which leads to association of starch and gluten thus lowering its digestion and thereby glycemic index (Tsafrakidou et al., 2020). Fermented foods and beverages have been said to improve memory and mental health and help in stress management due to the production of neurotransmitters and immune-modulatory compounds (Marco et al., 2017). Thus the fermented foods can be of great help in patients with neurodegenerative diseases esp. Alzheimer's disease (Mota de Carvalho et al., 2018). Furthermore various strains of LAB are involved in production of vitamins, thus they can be used for bio-fortification in various foods (Tsafrakidou et al., 2020). In addition, LAB possess antimicrobial and anti-tumour properties and also they ameliorate immunity and lipid profiles (Kohajdova, 2014). Fermentation processes have reported to increase the phenolic content in foods owing to their metabolic pathways. These compounds possess anti-oxidant, anti-inflammatory, anti-carcinogenic properties and thus help reduce cardiovascular diseases, neurodegenerative diseases, type-2 diabetes, inflammatory bowel disorders, (Kohajdova, 2014; Marco et al., 2017).

Idli has been reported to be useful in reducing the risk of various diseases. It has been used in weight loss management and as a dietary supplement for children. The carbohydrate and the dietary fiber present in idli improve digestion and prevent constipation (Blandino et al., 2003). Fermentation results in increase in several vitamins and bioactive components. There is a reduction in antinutrients and increase in the bioaccessibility of Zn and Fe (Gupta & Tiwari, 2014). Dosa is suitable for individuals with wheat allergies as it is gluten free. It has a low glycemic index and thus suitable for diabetic people. It has been used in treatment of rheumatism and neural disorders (Blandino et al., 2003; Gupta & Tiwari, 2014). Dhokla is also a low glycemic product and helps to lower down the blood cholesterol level and helps to prevent cardiovascular diseases (Roy et al., 2009; Moktan et al., 2011). Uttapam is recommended for high sugar and cholesterol patients as it is cholesterol-free product. It is easy to digest and can be used in weight management (Saraniya & Jeevaratnam, 2014). Injera is a rich source of Ca and Fe (Zegeye, 1997). Fermentation increases riboflavin content in kisra, (Mahgoub et al., 1999). Kishk has high nutrional value and has higher B vitamins than the either ingredients (Abd-el-Malek & Demerdash, 1993; Mahmoud, 1993).

8 Conclusion

Cereals are the main source of energy, fibres and other nutrients in our diet. They are also rich in various bioactive components which play an important role in prevention of many diseases. Cereals are being processes using different processing techniques and fermentation is one of the processing techniques used to prepare a large number of cereal-based products throughout the world. The interest in production and consumption of cereal-based fermented products has been increased due to low quantity of fat and high concentration of minerals, dietary fibers, and phytochemicals.

References

Abd-el-Malek, Y., & Demerdash, M. (1993). Egyptian kishk. In K. H. Steinkraus (Ed.), *Handbook of indigenous fermented foods* (pp. 295–299). Marcel Dekker.

Adegoke, G. O., & Babalola, A. K. (1988). Characterisation of microorganisms of importance in the fermentation of fufu and ogi, two Nigeria foods. *Journal of Applied Bacteriology, 65*, 449–453.

Altay, F., Karbancioglu-Güler, F., Daskaya-Dikmen, C., & Heperkan, D. (2013). A review on traditional Turkish fermented non-alcoholic beverages: Microbiota, fermentation process and quality characteristics. *International Journal of Food Microbiology, 167*, 44–56.

Arici, M., & Daglioglu, O. (2002). Boza: A lactic acid fermented cereal beverage as a traditional Turkish food. *Food Reviews International, 18*(1), 39–48.

Blandino, A., Al-Aseeri, M. E., Pandiella, S. S., Cantero, D., & Webb, C. (2003). Cereal-based fermented foods and beverages. *Food Research International, 36*(6), 527–543.

Caplice, E., & Fitzgerald, G. F. (1999). Food fermentations: Role of microorganisms in food production and preservation. *International Journal of Food Microbiology, 50*, 131–149.

Chavan, J. K., & Kadam, S. S. (1989). Nutritional improvement of cereals by fermentation. *Critical Reviews in Food Science and Nutrition, 28*, 349–400.

Chelule, P., Mbongwa, H., Carries, S., & Gqaleni, N. (2010). Lactic acid fermentation improves the quality of amahewu, a traditional south African maize-based porridge. *Food Chemistry, 122*(3), 656–661.

Chibuzor-Onyema, I. E., Ezeokoli, O. T., Sulyok, M., Notununu, I., Petchkongkaew, A., Elliott, C. T., Adeleke, R. A., Krska, R., & Ezekiel, C. N. (2021). Metataxonomic analysis of bacterial communities and mycotoxin reduction during processing of three millet varieties into Ogi, a fermented cereal beverage. *Food Research International, 143*, 110241.

Fardet, A. (2010). New hypotheses for the health-protective mechanisms of whole-grain cereals: What is beyond fibre? *Nutrition Research Reviews, 23*, 65–134.

Fernandes, C. G., Sonawane, S. K., & Arya, S. S. (2018). Cereal based functional beverages: A review. *Journal of Microbiology, Biotechnology and Food Sciences, 8*, 914–919.

Ghosh, K., Ray, M., Adak, A., Dey, P., Halder, S. K., Das, A., Jana, A., Parua (Mondal), S., Das Mohapatra, P. K., Pati, B. R., & Mondal, K. C. (2015a). Microbial, saccharifying and antioxidant properties of an Indian rice based fermented beverage. *Food Chemistry, 168*, 196–202.

Ghosh, K., Ray, M., Adak, A., Halder, S. K., Das, A., Jana, A., Parua (Mondal), S., Vagvolgyic, C., Das Mohapatra, P. K., Pati, B. R., & Mondal, K. C. (2015b). Role of probiotic lactobacillus fermentum KKL1 in the preparation of a rice based fermented beverage. *Bioresource Technology, 188*, 161–168.

Gupta, A., & Tiwari, S. K. (2014). Probiotic potential of lactobacillus plantarum LD1 isolated from batter of dosa, a south Indian fermented food. *Probiotics Antimicrob Proteins, 6*, 73–81.

Holzapfel, W. H. (2002). Appropriate starter culture technologies for small-scale fermentation in developing countries. *International Journal of Food Microbiology, 75*, 197–212.

Ignat, M. V., Salant, L. C., Pop, O. L., Pop, C. R., Tofana, M., Mudura, C. T. E., Bors, A., & Pasqualone, A. (2020). Current functionality and potential improvements of non-alcoholic fermented cereal beverages. *Food, 9*, 1031.

Khatun, L., & Ray, S. (2022). A comprehensive review on rice based fermented food and beverages in north East India. *The Pharma Innovation Journal, SP-11*(2), 370–375.

Kohajdova, Z. (2014). Fermented cereal products. In D. Montet & R. C. Ray (Eds.), *Microorganisms and fermentation of traditional foods*. CRC Press.

Kohajdová, Z. (2017). Fermented cereal products. In Current Developments in Biotechnology and Bioengineering (pp. 91–117). Elsevier.

Laskowski, W., Górska-Warsewicz, H., Rejman, K., Czeczotko, M., & Zwolinska, J. (2019). How important are cereals and cereal products in the average polish diet? *Nutrients, 11*, 679.

Mahgoub, S. E. O., Ahmed, B. M., Ahmed, M. M. O., & El Agib El Nazeer, A. A. (1999). Effect of traditional Sudanese processing of kisra bread and hulu-mur drink on their thiamine, riboflavin and mineral contents. *Food Chemistry, 67*, 129–133.

Mahmoud, S. A. Z. (1993). Egyptian kishk. In K. H. Steinkraus (Ed.), *Handbook of indigenous fermented foods* (pp. 295–299). Marcel Dekker.

Marco, M. L., Heeney, D., Binda, S., Cifelli, C. J., Cotter, P. D., Foligné, B., Gänzle, M., Kort, R., Pasin, G., Pihlanto, A., Smid, E. J., & Hutkins, R. (2017). Health benefits of fermented foods: Microbiota and beyond. *Current Opinion in Biotechnology, 44*, 94–102.

Moktan, B., Roy, A., & Sarkar, P. K. (2011). Antioxidant activities of cereal-legume mixed batters as influenced by process parameters during preparation of dhokla and idli, traditional steamed pancakes. *International Journal of Food Sciences and Nutrition, 62*, 360–369.

Mota de Carvalho, N., Costa, E. M., Silva, S., Pimentel, L., Fernandes, T. H., & Pintado, M. E. (2018). Fermented foods and beverages in human diet and their influence on gut microbiota and health. *Fermentation, 4*, 90.

Nyanzi, R., & Jooste, P. J. (2012). Cereal-based functional foods. In E. Rigobelo (Ed.), *Probiotics* (pp. 161–196). InTech. https://doi.org/10.5772/50120

Odunfa, S. A. (1985). African fermented foods. In: Microbiology of Fermented Foods, Vol 2, Wood, B.J. (ed) Elsevier Applied Science Publishers. London & New York.

Osorio-Cadavid, E., Chaves-Lopez, C., Tofalo, R., Paparella, A., & Suzzi, G. (2008). Detection and identification of wild yeasts in Champús, a fermented Colombian maize beverage. *Food Microbiology, 25*(6), 771–777.

Oyeyinka, A. T., Siwela, M., & Pillay, K. (2021). A mini review of the physicochemical properties of amahewu, a southern African traditional fermented cereal grain beverage. *LWT – Food Science and Technology, 151*, 112159.

Phiri, S., Schoustra, S. E., van den Heuvel, J., Smid, E. J., Shindano, J., & Linnemann, A. (2019). Fermented cereal-based Munkoyo beverage: Processing practices, microbial diversity and aroma compounds. *PLoS One, 14*(10), e0223501.

Roy, A., Bijoy, M., & Prabir, S. K. (2009). Survival and growth of foodborne bacterial pathogens in fermentating batter of dhokla. *Journal of Food Science and Technology, 46*, 132–135.

Ray, M., Ghosh, K., Singh, S., & Mondal, K. C. (2016). Folk to functional: an explorative overview of rice-based fermented foods and beverages in India. *Journal of Ethnic Foods, 3*(1), 5–18.

Saraniya, A., & Jeevaratnam, K. (2014). Purification and mode of action of antilisterial bacteriocins produced by Lactobacillus pentosus SJ65 isolated from uttapam batter. *Journal of Food Biochemistry, 38*, 612–619.

Savitri, & Bhalla, T. C. (2007). Traditional foods and beverages of Himachal Pradesh. *Indian Journal of Traditional Knowledge, 6*, 17–24.

Simatende, P., Siwela, M., & Gadaga, T. H. (2019). Identification of lactic acid bacteria and determination of selected biochemical properties in emasi and emahewu. *South African Journal of Science, 115*(11–12), 1–7.

Smid, E. J., & Lacroix, C. (2013). Microbe-microbe interactions in mixed culture food fermentations. *Current Opinion in Biotechnology, 24*, 148–154.

Thakur, S., Prasad, M. S., & Rastogi, N. K. (1995). Effect of xanthan on textural properties of idli (traditional south Indian food). *Food Hydrocolloids, 9*, 141–145.

Tsafrakidou, P., Michaelidou, A.-M., & Biliaderis, C. G. (2020). Fermented cereal-based products: Nutritional aspects, possible impact on gut microbiota and health implications. *Food, 9*(6), 734.

Yousif, N. E., & Tinay, A. H. E. (2003). Effect of natural fermentation on protein fractions and in vitro protein digestibility of rice. *Plant Food for Human Nutrition, 58*, 1–8.

Zegeye, A. (1997). Acceptability if injera with stewed chicken. *Food Quality and Preference, 8*, 293–295.

Zhang, K., Wu, W., & Yan, Q. (2020). Research advances on sake rice, koji, and sake yeast: A review. *Food Science & Nutrition, 00*, 1–9.

Chapter 12
Pastry

Saqib Farooq, Manzoor Ahmad Shah, and Shabir Ahmad Mir

1 Introduction

Pastries are a type of baked good that is enjoyed all over the world. They are made from a mixture of flour, fat, sugar, and come in a wide range of shapes, sizes, and flavors (Cauvain, 2017). From sweet fruit tarts to savory meat pies, pastries are versatile and can be enjoyed as a dessert, snack, or meal. It involves combining a different ingredients such as flour, butter, sugar, and eggs to create a dough that is then transformed into an array of exquisite treats. The origins of pastries can be traced back to ancient civilizations, such as the Greeks and Romans, who were known for their sweet and savory baked goods. Over time, pastry-making techniques spread throughout Europe, and different regions developed their own unique styles and recipes (Fig. 12.1). Today, pastries are enjoyed in many parts of the world, and each culture has its own take on these delicious treats (Renzetti et al., 2016).

One of the most important elements of pastry-making is the dough. There are many types of pastry dough, each with its own unique characteristics and uses. Some of the most common types of pastry dough include puff pastry, shortcrust pastry, choux pastry, and filo pastry (Deligny & Lucas, 2015). The most appealing things about pastries is their versatility. They can be filled with a wide range of ingredients, from sweet jams and fruit to savory meats and vegetables (Oliveira

S. Farooq
Department of Food Technology, Islamic University of Science & Technology, Awantipora, Jammu and Kashmir, India

M. A. Shah
Department of Food Science and Technology, Government Degree College for Women, Anantnag, Jammu & Kashmir, India

S. A. Mir (✉)
Department of Food Science and Technology, Government College for Women, M. A. Road, Srinagar, Jammu & Kashmir, India

M. A. Shah et al. (eds.), *Cereal-Based Food Products*,
https://doi.org/10.1007/978-3-031-40308-8_12

Fig. 12.1 Different types of pastries

et al., 2019). Some of the most popular pastry dishes include apple turnovers, fruit tarts, quiches, meat pies, and cream puffs.

2 Types of Pastry

Pastry is a broad category of baked goods that can be sweet or savory and come in a variety of forms.

1. **Puff pastry**

   Puff pastry is a light, flaky pastry made by repeatedly folding butter and dough together. It's typically made from flour, sugar, salt, fat and acids (lime juice, tartaric acid and vineager). Puff pastry is often used in sweet and savory recipes, such as croissants, turnovers, and quiches. When baked, the layers of butter and dough create a light and airy texture.

2. **Shortcrust pastry**

   Shortcrust pastry is a crisp, crumbly pastry made by rubbing together flour, butter, and water. It's often used as the base for pies and tarts, and can be sweet or savory. Shortcrust pastry is typically made from flour, butter, water, and salt, and is easy to make. When baked, shortcrust pastry has a crumbly texture that provides a nice contrast to the filling.

3. **Choux pastry**

   Choux pastry is a light, airy pastry made by cooking flour, fat and eggs. It's often used to make pastries such as eclairs, cream puffs, and profiteroles. Choux pastry is unique in preparation that it starts out as a paste that's cooked on the stove, then piped onto a baking sheet and baked until it puffs up. The resulting pastry is hollow on the inside and has a delicate, crisp exterior.

4. **Filo pastry**

   Filo pastry is a thin, flaky pastry made from layers of phyllo dough. It's often used in Mediterranean and Middle Eastern cuisine. Filo pastry is typically made from flour, fat and requires a delicate touch to work with as it can tear easily. When baked, filo pastry is crispy and flaky.

5. **Sweet pastry**

   Sweet pastry is a type of pastry that has been sweetened with sugar and is often used for desserts. It can be made from a variety of different doughs, such as shortcrust, puff, or choux pastry, and can be flavored with ingredients such as vanilla or cinnamon. Sweet pastry is often used as the base for pies, tarts, and cakes.

6. **Savory pastry**

   Savory pastry is a pastry that is not sweetened and is often used for appetizers or main dishes. It can be made from a variety of different doughs, such as shortcrust or puff pastry, and is often filled with ingredients such as cheese, vegetables, or meat. Savory pastry is often served as a finger food or as a main course.

3 Ingredients Used in Pastry

There are many ingredients used in pastry making but some of the most common ones include:

1. Flour: Flour is the backbone of most pastries, as it provides structure and stability to the dough or batter. Pastry flour, which has a lower protein content than all-purpose flour, is commonly used in pastry making as it produces a tender and delicate crumb. Other types of flour used in pastry making include all-purpose flour, cake flour, and bread flour.
2. Butter: Butter is a key ingredient in many types of pastry, as it adds richness, flavor, and moisture to the dough. Butter is often used in laminated pastries such as croissants, puff pastry, and Danish pastries, as it creates a flaky and tender texture.
3. Sugar: Sugar is used in pastry making for sweetness and to help create a caramelized crust. Granulated sugar is commonly used in doughs and batters, while powdered sugar is used for dusting or making icings and glazes.

4. Eggs: Eggs are an important ingredient in many types of pastry, as they provide structure, richness, and flavor to the dough or batter. Whole eggs, egg yolks, and egg whites are all used in different pastry recipes, and they can be whisked, beaten, or separated as needed.
5. Milk or cream: Milk or cream is often used in pastry making to add moisture, richness, and flavor to the dough or batter. Whole milk, skim milk, or heavy cream can be used, depending on the recipe.
6. Yeast: Yeast is used in some types of pastry, such as croissants and Danish pastries, to create a light and flaky texture. Yeast ferments the dough, producing carbon dioxide gas that causes it to rise.
7. Baking powder and baking soda: Baking powder and baking soda are leavening agents commonly used in pastry making to create a light and airy texture. Baking powder is a combination of baking soda and an acidic ingredient, while baking soda needs an acidic ingredient to react with.
8. Salt: Salt is used in pastry making to enhance the flavor and to balance the sweetness. It also helps to control the fermentation of the yeast in some types of pastry.
9. Chocolate: Chocolate is a versatile ingredient used in many types of pastry, such as cakes, brownies, and tarts. It adds flavor, richness, and texture to the pastry, and can be used in different forms like chocolate chips, cocoa powder, or melted chocolate.
10. Fruits: Fruits are commonly used in pastry making to add flavor, texture, and color to the dough or filling. Some popular fruits used in pastry making include apples, strawberries, blueberries, raspberries, and peaches.
11. Nuts: Nuts are often used as a filling or topping in pastries, such as pecan pie or almond croissants. They add crunch, richness, and flavor to the pastry, and can be used in different forms like chopped, sliced, or ground.
12. Shortening: Shortening is a type of fat commonly used in pastry making, especially for pie crusts, as it creates a tender and flaky texture. Vegetable shortening, which is solid at room temperature, is often used instead of butter, as it has a higher melting point and can withstand higher temperatures.
13. Vanilla extract: Vanilla extract is a popular flavoring agent in pastry making, as it adds a sweet and floral aroma and taste to the dough or batter. Other flavoring agents used in pastry making include almond extract, lemon zest, orange zest, and spices like cinnamon and nutmeg.

4 Preparation Methods

Pastry can be prepared in many different ways, depending on the type of pastry. Here are some common preparation methods used in pastry making:

1. **Rubbing in method**

This method is used for making shortcrust pastry, which is a common pastry used for pies and tarts. The method involves rubbing cold butter or shortening into flour using your fingertips, until the mixture resembles coarse breadcrumbs. Water is then added to bring the mixture together into a dough.

2. **Creaming method**

   This method is used for making cakes and cupcakes, which require a light and fluffy texture. The method involves creaming together softened butter and sugar until light and fluffy, then adding eggs and flour in alternating batches, and mixing until smooth.

3. **Folding method**

   This method is used for making laminated pastries, such as croissants and puff pastry, which require multiple layers of butter and dough to create a flaky texture. The method involves rolling out the dough into a rectangle, placing a layer of butter on top, then folding the dough over the butter to create a sandwich. The dough is then rolled out and folded multiple times, creating layers of butter and dough.

4. **Whisking method**

   This method is used for making meringues, which are used as toppings for pies and other desserts. The method involves whisking egg whites until stiff peaks form, then gradually adding sugar and continuing to whisk until the mixture is glossy and stiff.

5. **Kneading method**

   This method is used for making yeast-based doughs, such as bread dough and pizza dough, which require gluten development to create a chewy texture. The method involves kneading the dough by hand or with a stand mixer, until it becomes smooth and elastic.

6. **Press-in method**

   This method is used for making crumb crusts, which are commonly used for cheesecakes and other desserts. The method involves mixing together crumbs (usually graham crackers or cookies), sugar, and melted butter, then pressing the mixture into a pan to create a crust.

7. **Blind-baking method**

   This method is used for pre-baking a crust before adding a filling, to prevent the crust from becoming soggy. The method involves lining a pastry crust with parchment paper or aluminum foil, then filling it with weights (such as dried beans or rice) to keep the crust from puffing up during baking. The crust is then baked for a short time until it is set, before adding the filling and baking the pastry again.

5 Conclusion

Pastry is a versatile and delicious component of many desserts and baked goods. The key ingredients used in pastry making include flour, fat, sugar, which can be combined in a variety of ways to create different types of pastry with unique textures and flavors. The preparation methods used in pastry making can vary widely depending on the type of pastry being made, but some common methods include rubbing in, creaming, folding, whisking, kneading, pressing in, and blind-baking.

References

Cauvain, S. P. (2017). *Baking problems solved*. Woodhead Publishing.

Deligny, C., & Lucas, T. (2015). Effect of the number of fat layers on expansion of Danish pastry during proving and baking. *Journal of Food Engineering, 158*, 113–120.

Oliveira, S., Fradinho, P., Mata, P., Moreira-Leite, B., & Raymundo, A. (2019). Exploring innovation in a traditional sweet pastry: Pastel de Nata. *International Journal of Gastronomy and Food Science, 17*, 100160.

Renzetti, S., de Harder, R., & Jurgens, A. (2016). Puff pastry with low saturated fat contents: The role of fat and dough physical interactions in the development of a layered structure. *Journal of Food Engineering, 170*, 24–32.

Chapter 13
Cereal Based Functional Products

Luis Martín Sánchez Magaña, Liliana León López, Edith Oliva Cuevas Rodríguez, and Dulce María Domínguez Arispuro

1 Introduction

The term "functional food" first appeared in Japan in 1984. The Japanese government defined a new product category, Food for Specific Health Uses (FOSHU), as "food containing an ingredient with functions for health and officially approved to claim their physiological effects on the human body" and produced a dedicated legislative framework. "Functional food" generally refers to products that provide specific health benefits beyond basic nutrition. Nowadays, the terms nutraceutical and functional foods are often confused and interchanged. Different attempts have been made to propose a commonly agreed upon definition. The last proposal was that presented by the Functional Food Center (FFC) in 2018, describing functional foods as "natural or processed foods that contain biologically active compounds, which, in defined, effective, and non-toxic amounts, provide clinically proven and documented health benefits utilizing specific biomarkers for the prevention, management, or treatment of a chronic disease or its symptoms" (Alongi & Anese, 2021).

Whole grain (WG) cereals are considered one type of functional food with an array of nutritionally important and healthy constituents for human consumption, such as carotenoids, inulin, β-glucan, lignans, vitamin E-related compounds, tocols, phytosterols, bioactive carbohydrate fractions, peptides, and phenolic compounds, which are beneficial. The major sources of these phytochemicals are phenolic compounds (PCs), due to the high concentrations of bioactive constituents in the bran and germ layer and the fact that they are largely one of the most important dietary sources of energy intake worldwide (Adebo et al., 2019).

L. M. S. Magaña (✉) · L. L. López · E. O. C. Rodríguez · D. M. D. Arispuro
Facultad de Ciencias Químico Biológicas, Universidad Autónoma de Sinaloa. Blvd. de las Américas y Josefa Ortiz de Dominguez, Ciudad Universitaria, Culiacán, Sinaloa, Mexico
e-mail: luis_magana@uas.edu.mx

M. A. Shah et al. (eds.), *Cereal-Based Food Products*,
https://doi.org/10.1007/978-3-031-40308-8_13

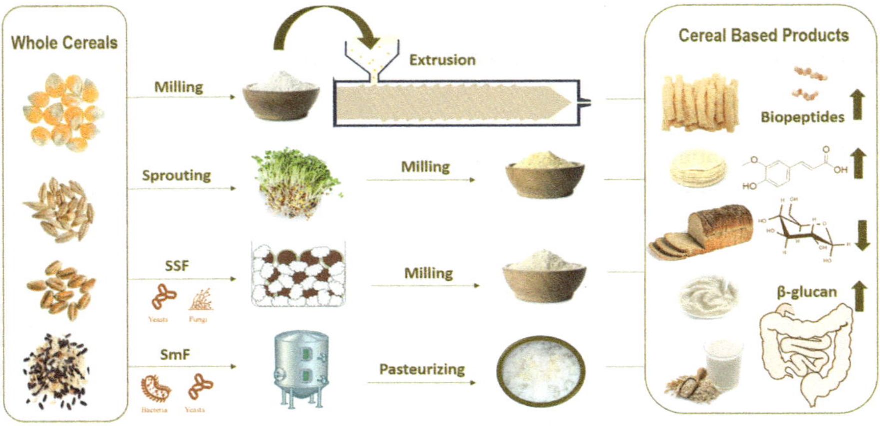

Fig. 13.1 Food processing technologies to develop functional cereal based products. SSF (Solid-state fermentation); SmF (Submerged fermentation)

Although many plant foods may be consumed raw, cereals require transformation to become edible. WG may be processed efficiently and innovatively to create products that would be considered as healthy. Traditional and nontraditional methods (steeping, malting, roasting, fermentation, boiling, baking, frying, nixtamalization, micronization, extrusion, etc.) (Fig. 13.1) have been used to process whole grains into food. The preparation of grains into finished food products usually requires a combination of several strategies (Basinskiene & Cizeikiene, 2020). Ultimately, research urgently needs to focus the application of conventional and unconventional processes, and the exploitation of processing wastes as bioactive ingredients to increase the environmental and economic sustainability of food production, while enhancing its functionality, to obtain what can be called a "functional" food (Alongi & Anese, 2021). Alarge amount of research has shown the importance of processing (germination, fermentation, extrusion) and supplementation of cereals to generate novel food products with improved nutritional and nutraceutical properties (Rochín-Medina et al., 2015a, b; Espinoza-Moreno et al., 2016, 2021; Milán-Carrillo et al., 2017; Félix-Medina et al., 2020, 2021; Gómez-Favela et al., 2021). Recent technologies utilizing ultrasound, micro- or nano-capsule polymers, and infrared processes are also evaluated for their effectiveness in improving the phenolics content and bioaccessibility (Kasote et al., 2021). In addition, food science and technology have progressed from designing nutritional foods with properties to improve gastrointestinal health, such as probiotics, prebiotics, and synbiotics. These foods have been the most important segment within functional foods (Salmeron, 2017).

2 Importance of Functional Cereal-Based Products

Cereal stands out as the main food grown around the world with a global production of 2.98 billion tons in 2017, according to FAO, emphasizing the consumption of corn, wheat, and rice. Advances in the industrial processes have led to the consumption of refined grain products, which present greater storage stability (Menis-Henrique et al., 2020). During the last century, dietary habits have considerably changed, leading to an excessive consumption of highly refined sugars and high-calorie-density foods. Such dietary changes, together with sedentary lifestyle, have been correlated with the increasing incidence of non-communicable diseases (NCDs). Among these, one of the most alarming is type 2 diabetes (Alongi et al., 2019). In 2021, it was estimated that 537 million adults globally (20–79 years) are living with diabetes; this number is predicted to rise to 643 million by 2030 and 783 million by 2045 (IDF, 2021). Consequently, the main directions of research aimed at nutritional improvement have to face either excess or deficiency in the diet. In the latter case, nutrient-rich foods with long shelf-life are needed to prevent malnutrition, whereas in developed countries it is mostly required to decrease the energy value, sucrose, salt, and increase dietary fiber content of foods to prevent obesity and nutrition-related chronic diseases, such as cardiovascular disease, hypertension, and diabetes mellitus (Pasqualone & Summo, 2021). In this context, the research interest towards functional food experienced a steep increase only in the twenty-first century. This globally growing attention has tremendously influenced their market, which was estimated at USD 162 billion in 2018 and was projected to reach USD 280 billion by 2025 with an annual growth rate of around 8% (Alongi & Anese, 2021). Products containing dietary fiber, vitamins, and minerals as the major bioactives occupy the largest market space. Maximum number of functional foods manufactured today are dairy based followed by the bakery and cereal based. The latter are promising due to the intake of whole grain or bran-added cereals, which are promoted as a source of dietary fiber in the human diet and which can be consumed in main meals, such as pasta and breakfast cereals, and as expanded snacks between meals. The expected growth of the functional food market between 2014–2024 shows that there is a steady increase in the global demand for functional food to which the food industry needs to cater (Menis-Henrique et al., 2020).

3 Ingredients

Given the importance of cereal products as a primary source of food in human nutrition, the development of cereal-based foods that may provide health-promoting benefits to consumers is becoming a new trend in the food industry. The health-promoting properties of the cereal-based products as functional foods are based on their bioactive components content, which may be naturally occurring or added externally into a regular cereal-based product formulation in order to enhance specific health-promoting characteristics (Rawat & Indrani, 2015). Various ingredients have been used in order to modify nutritional attributes and enhance the health-improving properties of cereal-based products; some of the most used are described below.

Dietary Fiber

Defined as the food fraction that is not decomposed by endogenous enzymes of the human gastrointestinal tract (Augustin et al., 2020) and constituted by several carbohydrate polymers, such as homopolysaccharides, heteropolysaccharides, lignins, oligosaccharides, resistant starches, gums, and mucilages (Ahmad & Khalid, 2018; Schmidt, 2022), dietary fiber (DF) is an important part of our diet due to its influence on gastrointestinal tract physiology (Williams et al., 2019). DF is classified as insoluble fiber and soluble fiber based on their solubility in water. Is suggested that soluble and insoluble fiber bring differing outcomes due to their differences in physicochemical properties (Lee et al., 2022). Soluble fiber is known to improve cholesterol and blood glucose control by interfering with their absorption, followed by a subsequent prebiotic effect contributing to the microbiome modulation (Aoe et al., 2018). Fiber-fortification into refined cereal-based products lacking in fiber (e.g., wheat white bread) results in the improvement of their potential capacity to reduce the risk of health diseases. The impact of fiber-fortified food consumption on body composition, blood pressure, blood lipid-lipoprotein, and glycemic-related markers has been recently reviewed and analyzed by Lee et al. (2022). The authors meta-analysis found that soluble fiber fortification was more beneficial for improving cardiometabolic outcomes (triglycerides and insulin), while insoluble fiber fortification led to greater improvements in anthropometric outcomes (body weight, body mass index).

Protein and Bioactive Peptides

Cereals, such as barley, maize, millet, oat, rice, and wheat, are staple foods in many regions around the world and contribute to more than half of human energy requirements. Food proteins have been investigated as sources of food peptides. They can be obtained by the hydrolysis of cereal-based food protein. Food peptides can exert biological functions, promote health, and prevent oxidative stress at the cellular level (Esfandi et al., 2019). Peptides (DNIQGITKPAIR, IAFKTNPNSMVSHIAGK, and IGVAMDYSASSKR) from broken rice protein can be potentially used as a natural anti-inflammatory alternative, acting as inhibitors of nitric oxide, IL-1β, IL-16 and TNF-α in RAW 264.7 cells (Qu et al., 2022). The oral administration of foxtail millet (*Setaria italica*) protein hydrolysatesgenerates peptides (LPF, ANP, PY, YW and IPP) and reduce the NF-κB phosphorylation, TNF-α, and IL-6, suggesting that these cereals can be used to prevent inflammatory bowel diseases (Zhang et al., 2022).

Polyunsaturated Fatty Acids (PUFAs)

PUFAs can be divided into omega-6 (ω-6) and omega-3 (ω-3) fatty acids based on the position of the first double bond from the terminal methyl carbon from the fatty acid molecule (Yadav et al., 2020). In ω-6 fatty acids, this double bond occurs between the sixth and seventh carbon atoms, whereas in ω-3 fatty acids, it is located between the third and fourth carbon atoms (Patel et al., 2022). Linoleic acid (LA; 18:2 n-6) and α- linolenic acid (ALA; 18:3 ω3) are the precursors of other omega-6 and omega-3 PUFAs, respectively, and are considered essential fatty acids (Yadav et al., 2020).

The three types of ω-3 fatty acids involved in human physiology are ALA; found in vegetable oils from flaxseed, canola, soybean, walnut, and wheat germ, sunflower, and eicosapentaenoic acid (EPA; 20:5ω3) and docosahexaenoic acid (DHA; 22:6 ω3); both commonly found in fish oil and other marine food products such as algae (Yadav et al., 2020). Consuming an optimal ω-6/ω-3 ratio (1:1 to 5:1) is crucial for a healthy lifestyle; however, western diets are associated with higher consumption of ω-6 and lower consumption of ω-3 fatty acids. The observed changes in fatty acids consumption ratio coincide with a significantly increased prevalence of coronary heart disease, hypertension, cancer, diabetes, obesity, rheumatoid arthritis, and autoimmune or neurodegenerative disorders. To balance the effect of ω-6 fatty acids, a sufficient intake of ω-3 fatty acids is required. A sustainable and cost-effective way of reaching a large population with essential ω-3 fatty acids is the fortification of staple foods (Patel et al., 2022).

During bakery products making, fats addition is required in regular formulations, moreover, some bakery products have relatively low water activity, making this type of food a suitable choice for ω-3 fatty acid incorporation (Feizollahi et al., 2018).

Containing a high content of ALA (50–60%) (Yadav et al., 2020), flaxseed is the most used vegetable source of ω-3 fatty acids for the enrichment of bread and cereals. Both flaxseed oil and flaxseed flour have been used (Murugkar et al., 2021). The use of chia flour to improve the fatty acid profile in wheat bread has been also studied (Giaretta et al., 2018). Fish oil has been used successfully too. However, a major difficulty in fortifying this type of food products with fish oil is the high susceptibility to oxidation of some of its components (EPA and DHA) and its unpleasant flavors. To overcome this problem and keep fish oil oxidative stable for the shelf-life of the food product, the fish oil can be added as fish-oil-encapsulated powders (Santhanam et al., 2015; Feizollahi et al., 2018) ω-3 enriched bakery products as breads, cookies, and cakes can be also obtained by using bio fortified products, such as ω-3 enriched eggs (obtained by feeding hens with ω-3 oils) and dairy ingredients biofortified or fortified with ω-3 PUFAs (Patel et al., 2022).

Most of the reports about bakery products enriched with ω-3 fatty acids are focused on the quality of the products and their consumer acceptability (Giaretta et al., 2018; Murugkar et al., 2021). Only some of them have been focused on demonstrating the beneficial health properties of the products developed. Petrovna et al. (2022) developed a ω-3-enriched bread by adding liposomal polyunsaturated fatty acids (PUFAs) concentrate into a conventional bread recipe and evaluated its hypolipidemic and antioxidant effects in an experimental rat model of diet-induced dyslipidemia. The authors demonstrated improvements in the lipid metabolism by decreasing total cholesterol and the pro-atherogenic fractions of lipoproteins (LDL and VLDL).

Antioxidants

Although they were initially used just as food preservatives, nowadays it is known that food antioxidants inhibit oxidation processes, not only in foods but also in human metabolism, preventing oxidative stress and related diseases (Granato et al., 2020). In recent years, there has been a global trend toward the use of naturally occurring antioxidants from vegetable sources. Phenolic compounds, particularly phenolic acids, are the main antioxidants found in cereals (Horvat et al., 2020). These compounds are concentrated in the bran and aleurone layer of the grains, explaining the lower antioxidant activity observed in cereal-based products made with refined flour compared with those derived from whole grain flours. Cereal-based products (e.g., white wheat bread) can be enriched using natural sources rich in antioxidants, such as pseudo-cereal flours, spices, herbs, parts of green plants, non-cereal seeds, fruit or vegetable products, and agro-industrial by-products (Dziki et al., 2014). Some health-promoting polyphenols are found in other plant sources with limited application possibilities, such as in non-edible plant parts or food waste processing products; therefore, the extraction of the antioxidant compounds and their subsequent addition into a cereal-based product formulation is an alternative (Czubaszek et al., 2021).

Many scientific studies had been published outlining the development of antioxidant-enriched cereal-based products, such as breads, biscuits, cookies, tortillas, and pasta, using diverse sources of vegetable antioxidants (Chávez-Santoscoy et al., 2016; León-López et al., 2019; Czubaszek et al., 2021). Most of them are focused on increasing the antioxidant content and antioxidant capacity as well as on maintaining the textural and sensory properties expected from the products. Although in vitro chemical and cell-based methods may provide a quantitative result of the antioxidant content and antioxidant capacity of the antioxidant-enriched products developed, clinical trials and research studies are needed to confirm their potentially health-promoting effects.

Probiotics

Conventionally, dairy products have been used as probiotic vehicles; however, lactose intolerance and high cholesterol content associated with them, coupled with the current vegan/vegetarian consumer trend, motivated the search for non-dairy products as probiotic carriers. Beverages formulated with fruit and vegetable juices can also serve as carriers of probiotic bacteria. Cereals are likewise potentially viable substrates, as they hold nutrients easily assimilated by probiotics (Salmeron, 2017).

Fermentation of cereals and their derivatives using lactic acid bacteria (LAB) is a very promising and cheap alternative to enhance the nutritional value of cereal-based products either to improve or change their taste. Cereal-based fermented food products should even be taken into account as a promising source of new probiotics (Tomasik & Tomasik, 2020). Bacterial strains most commonly used as probiotics belong to the genera *Lactobacillus* and *Bifidobacterium,* but other organisms are also applied, such *as Lactococcus* and *Enterococcus* (Longoria-García et al., 2018; Tomasik & Tomasik, 2020). In the case of bakery products, due to the harsh thermal stress during baking, sourdough technology (a mixture of flour and water that is fermented by LAB, with or without yeasts), along with microencapsulation of probiotics, has been studied. Sourdough technology provides the benefits of probiotics in human health by supplying bioactive compounds produced during dough fermentation without the actual delivery of living probiotic cells (Longoria-García et al., 2018).

On the other hand, the microencapsulation technique offers protection to probiotic microorganisms against thermal stress during baking, increasing cell viability after food processing. Different microencapsulation techniques and coating materials have been used. Encapsulating *lactobacilli* in calcium alginate has been found to improve their heat tolerance and increase cell survival. The combination of calcium alginate with prebiotics, such as resistant starch, improves both the viability of probiotics and the capsule structure (Seyedain-Ardabili et al., 2016). Most of the research conducted in the development of potential functional bakery products is done by including probiotics and prebiotics (synbiotic food products) to show an enhanced health benefit to its consumers (Longoria-García et al., 2018).

Food Industrial By-products

By-products from fruits, vegetables, cereals, legumes, nuts, seeds, brewery, winery, and distillery are rich sources of dietary fiber, proteins, vitamins, minerals, and bioactive compounds (Martins et al., 2017); therefore, one reason to incorporate by-products into cereal-base products formulations is to increase their nutritional value and health-promoting properties by improving the bioactive compounds content (Subiria-Cueto et al., 2021). Fruit processing by-products are recognized for their high total dietary fiber content and well-balanced proportion of insoluble dietary fiber and soluble dietary fiber, as well as for the presence of polyphenols, flavonoids, and carotenes (Quiles et al., 2016). The incorporation of fruit pomace (a mixture of pulp, skin, seeds, and stems) has been proposed as an additive in the formulation of bakery products to enhance their contents in natural antioxidants and dietary fibers (Majerska et al., 2019; Socas-Rodríguez et al., 2021).

The percentage of substitution with by-product flour in bakery product formulations depends on the by-product characteristics (Subiria-Cueto et al., 2021). A balance in the incorporation of functional ingredients from by-products has to be found in order to obtain desirable bakery products that can compete with the traditional ones (Martins et al., 2017).

4 Methods to Produce Functional Ingredients and Cereal-Based Products

Several strategies and processing technologies have been proposed in literature and applied in the preparation of processed cereal foods to deal with deficiencies of nutrients, dietary fiber, and bioactive compounds contents of refined flour and to enrich the end products of cereal and cereal-based foods. Processing technologies, such as germination, fermentation, and enzymatic treatments, have been recommended for improving the bio accessibility and bioavailability of nutrients and bioactive compounds of cereal foods (Saleh et al., 2019). In addition, innovative processing technologies as a tool for development of novel functional foods. Over the past two decades, innovative processing technologies (e.g., high hydrostatic pressure, pulsed electric fields, ultrasounds) have emerged as suitable food-processing alternatives. These sustainable technologies provide better preservation of native nutrients in fruits and vegetables, avoid microbial growth, and utilize less energy, and they can be applied to the exploitation of by-products while being eco-friendly (Granato et al., 2020). Despite the importance of selecting the appropriate food-processing conditions to improve the nutritional characteristics and increase the number of bio components of the final product; research in this area is still limited. Thus, novel studies focusing on the optimization of thermal and no-thermal operations have a vast potential for the food industry (Reyes-Moreno et al., 2018).

Fermented Grains

Fermented food products, especially those derived from cereals and legumes, are important contributors to diet diversity globally (Adebo et al., 2022). Grain fermentation improves flavor, aroma, texture, nutritional value, and functional properties besides detoxifying and destroying undesirable substances or pathogen microbes present in raw grains (Adebo et al., 2019; Funck et al., 2019; Ahnan-Winarno et al., 2021; Robledo-Márquez et al., 2021). Different types of fermentations (alcoholic, acetic, lactic, butyric, butanedioic, propionic, etc.) are carried out by different microorganisms, such as bacteria, yeast, and mycelial fungi, which are naturally present or intentionally added, as with enzymes. It is used as a technique for food preservation, increasing the shelf-life of a broad range of foods such as milk, meat, fish, cereals, fruit, vegetables, legumes, roots/tubers, alcoholic beverages, and non-alcoholic beverages (Robledo-Márquez et al., 2021). Conventional techniques of fermentation include (i) natural (also called spontaneous) occurrences through the actions of endogenous microorganisms, (ii) back slopping involves using plenty of successful previous fermentation batches) and (iii) controlled fermentation, which entails the inoculation of starter cultures/specific strains. Fermentation could also be classified as either a solid-state fermentation (SSF) or submerged/liquid fermentation (SmF). The SSF process involves the growth of microorganisms on moist substrates in the absence of free-flowing water, while SmF occurs in the presence of free-flowing liquid medium/water (i.e., SmF has more fluids compared to SSF) (Adebo et al., 2022). Recently, SSF has gained considerable attention for the production and extraction of bioactive compounds from plant materials, mainly pulses and cereals. In this process, different carbohydrases, like cellulases, β-glucosidase, xylanase, pectinases, β-xylosidase, β-galactosidase, α-amylases, esterase, others, are synthetized by the microorganisms and liberate free amino acids, short chain fatty acids, and free phenolics (Dey & Kuhad, 2014; Sandhu & Punia, 2017; Chen et al., 2020). SSF with fungi (*Aspergillus* spp. and *Rhizopus* spp.), bacteria (*Bacillus subtilis*) and lactic acid bacteria (LAB) spp.), and yeast (*Saccharomyces cerevisiae*) significantly increased the bioactive compounds and antioxidant activities (AOXA) in cereal and legume grains and by-products (Table 13.1) (Chen et al., 2020; De Villa et al., 2021). The increase in bioactive compounds and AOXA depends on the substrate, microorganism, and the fermentation conditions used during the process. In particular, microorganisms, such as *Aspergillus* spp. and *Rhizopus* spp., have been used in the fermentation of soybean and other legumes (Table 13.1). In cereal grains, *Aspergillus awamori*, *A. oryzae*, and *A. sojae* are the most traditional cultures used to start the SSF. With *A. oryzae*, the Asian culture obtains fermented foods, namely douche, doenjang, kanjang, kecap, ketjap, meju, miso, and shoyu. Cultures of *Rhizopus oligosporus* and *Rhizopus oryzae* have been primarily used in the production of Indonesian fermented soybean products, known as tauco and tempeh (De Villa et al., 2021). Yeasts are predominantly responsible for the production of ethanol (e.g., beers and wines, breads), while lactic acid bacteria produce mainly lactic acid (e.g., cereals and fermented milk products) (Von Mollendorff et al., 2016). Compared with lactic acid bacteria, filamentous fungi have some advantages in the production of functional fermented foods. During fermentation, these fungi can produce many enzymes with high activities, such as β-glucosidase, esterase, phytase, and xylanase. These enzymes can hydrolyze insoluble-bound nutrients and release free nutrient substances, e.g., polyphenols (Cai et al., 2014; Chen et al., 2020). Enrichment of phenolic compounds through SSF by using filamentous fungi has been reported in wheat bran (Tu et al., 2020), wheat (Sandhu et al., 2016; Starzyńska-Janiszewska et al., 2019), oat (Cai et al., 2014; Chen et al., 2020), rye bran (Koistinen et al., 2016), barley (Sandhu & Punia, 2017), pearl millet (Salar et al., 2016; Salar et al., 2017), rice and rice bran (Janarny & Gunathilake, 2020), and maize (Milán-Carrillo et al., 2017; Sánchez-Magaña et al., 2019). The increase of these bioactive compounds has been implicated for improved antioxidant, anti-inflammatory, anti-carcinogenic, anti-diabetic, and angiotensin-converting-enzyme

(ACE) inhibitory effects in fermented products (Table 13.1) (Rochín-Medina et al., 2015b; Milán-Carrillo et al., 2017; Sánchez-Magaña et al., 2019; De Villa et al., 2021). Hence, fermented soy (tempeh) can be used to fortify nutraceutical cereal bars (De Melo et al., 2020) and cookies (Leite et al., 2013). Recently, *R. oligosporus* has been used to ferment maize and common beans through SSF to produce mixtures of tempeh flours, which were used to formulate nutraceutical beverages (Rochín-Medina et al., 2015b; Milán-Carrillo et al., 2017; Reyes-Moreno et al., 2014). Also, during oat tempeh production, γ-amino butyric acid (GABA), total phenolics content (TPC), and flavonoids increased rapidly. The *Aspergillus oryzae*-fermented oats had the highest GABA content, whereas *Rhizopus oryzae*-fermented oats had the highest TPC (Cai et al., 2014). In addition to fungus, *Lactobacilli* with the potential of producing large amounts of GABA were screened from traditional fermented food (Jin et al., 2013). Coda et al. (2010) reported how *L. plantarum* C48 and *Lactococcus lactis* subsp. *lactis* PU1, used for sourdough fermentation of cereal, pseudo-cereal and leguminous flour, were able to synthesize high levels of GABA. Clinical studies have related increased intake of GABA or analogues to several health benefits, including lowering of blood pressure of mildly hypertensive animals and humans (Jin et al., 2013). Several authors studied the impact of bioprocessing with baker's yeast, LAB, and fungi on phenolic acids release in wheat bran (Manini et al., 2014), rye bran (Koistinen et al., 2016), and rice bran (Abd Razak et al., 2015). In all cases, substantial increases, especially of ferulic acid, were observed. Ferulic acid has antioxidant properties and have been ascribed to its ability to inhibit lipid peroxidation and LDL oxidation greater than other hydroxycinnamic acids (Verni et al., 2019).

A recent study demonstrates the ability of sourdough lactic acid bacteria to release antioxidant peptides through the proteolysis of native cereal proteins (Galli et al., 2018). In addition, yeast strains singly inoculated in barley and wheat (Palla et al., 2020), wheat bran (Călinoiu et al., 2019), rice flour (Ilowefah et al., 2017), significantly increased the AOXA, the soluble, free, and conjugated forms of phenolic acids, and anthocyanins of fermented substrates. According to Palla et al. (2020) such increases could be ascribed also to other compounds, i.e., small bioactive peptides produced by the action of proteolytic yeasts during dough fermentation. The specific selection of yeast strains is important in order to enhance the pro-technological, nutritional, and nutraceutical traits of fermented doughs (Palla et al., 2020). Co-fermentation of cereals with fungus and lactic acid bacteria can increase the nutritional values and improve health-promoting properties of the final products, such as beverages, bread, biscuits, porridges and breakfast cereals (Table 13.1). (Freire et al., 2017; Wang et al., 2018a; Menezes et al., 2018; Wu et al., 2018; Ripari et al., 2019). Sourdough is a mixture of cereal flour and water, fermented by a complex biological ecosystem, comprising lactic acid bacteria (LAB) and yeasts, which interact with each other, often establishing stable associations and contributing to the beneficial properties of sourdough breads (Palla et al., 2020). The sourdough fermentation may not only influence the features of leavened baked goods, but can also have positive effects on human health (Rizzello et al., 2016; Koistinen et al., 2018). LAB and yeast can produce Se-enriched breads with improving antioxidant activity. Selenium is an antioxidant trace mineral that can be found in both organic and inorganic forms. Selenite (SeO_3^{-2}) and selenate (SeO_4^{-2}) are the main inorganic forms, and selenomethionine (SeM) and selenocysteine (SeC) are considered the main organic moieties associated with foods. However, the uptake of SeM by colon microbiota is more efficient than the uptake of inorganic form (Guardado-Félix et al., 2019). The most popular way to produce Se-enriched leavened breads is by the use of yeast (*S. cerevisiae*). Several species of lactic acid bacteria (LAB) efficiently convert inorganic Se to organic forms. The addition of Na_2SeO_3-$5H_2O$ to the flour did not inhibit the LAB, used as starter cultures in bread dough formulation, and favored the production of breads with high biological value and improved quality (Zlateva et al., 2017; Guardado-Félix et al., 2019).

Formulation of probiotics, prebiotics, and symbiotic products is another strategy for producing functional cereal-based foods through SmF. In this regard, LAB, including the genera *Lactobacillus* and *Bifidobacterium*, are commonly used in probiotic preparations (Freire et al., 2017). Among LABs, *Lactobacillus plantarum* is a prominent species which has been proven to enhance the immune system and promote nutrient absorption by populating the human colon. Prebiotics are a group of carbohydrates which cannot be decomposed or absorbed in the upper digestive tract but can selectively enhance the growth of one or more probiotics (Wang et al., 2018a). Cereals (e.g., wheat, maize, oat, barley), abundant sources of dietary fiber, can have several beneficial physiological effects on the gut and can be used as prebiotics, providing specific non-digestible carbohydrates, and as encapsulation materials, improving the viability and stability of probiotics (Min et al., 2019; Jia et al., 2019).

As shown in Table 13.1, cereals can be applied as nutrient sources that, besides promoting several beneficial physiological effects, can stimulate the growth of lactobacilli and bifidobacteria (Basinskiene & Cizeikiene, 2020). In recent years, β-glucan derived from oat or other cereals has been intensively studied for its prebiotic effects. Barley β-glucan can improve the growth rate of *L. acidophilus*, *L. plantarum*, and *L. fermentum*. It was reported that oat-based products stimulate the bifidobacteria in humans. Symbiotic foods, which combine probiotics with prebiotics, might be more efficient than the individual components in the colon (Wang et al., 2018a). The co-culture of LAB and yeasts to ferment plant substrates in substitution to milk have been proposed in order to obtain non-dairy symbiotic beverages as an alternative to the traditional dairy products with improved health and sensory characteristics (Silva et al., 2016; Freire et al., 2017; Menezes et al., 2018; Wang et al., 2018a; Funck et al., 2019). Cereal-based products (kvass, boza, bushera, pozol, tejuino, Tanzanian togwa, mahewu), fermented with LAB, could be a perfect alternative for people who are allergic to milk proteins (Basinskiene & Cizeikiene, 2020).

The information presented in the previous section showed how SSF and SmF can be used to enhance bioactive content of whole cereals and cereal bran under appropriate conditions and selected microorganisms, conferring the highest functional properties to fermented foods.

Table 13.1 Functional cereal-based products develop by fermentation processes

Cereal based product	Microorganisms and fermentation conditions	Beneficial effect	Reference
Red sorghum sourdoughs	Two strain combinations, *L. plantarum* and *L. casei*or *L. fermentum* and *L. reuteri*, Sourdoughs were fermented for 24 h at 34 °C.	Increasing: caffeoylglycerol, dicaffeoylglycerol, coumaroyl-caffeoylglycerol, coumaroyl-feruloylglycerol, phenolic acids, phenolic acid esters, flavonoid glucosides	Svensson et al. (2010)
Sourdough fermentation of various cereal flours	A pool of selected lactic acid bacteria sourdoughs was fermented for 24 h at 37 °C.	Water/salt-soluble extracts from sourdoughs containing antioxidant peptides that increase AOXA	Coda et al. (2012)
Cookies added with tempeh flour	*R. oligosporus* Soya beans cotyledons were incubated at 32 °C for 24 h.	The concentrations of linoleic and linolenic unchanged. Significant increase aglycone levels in the cookies made with mixed flour.	Leite et al. (2013)
Oat	*A. oryzae var. effuses* 3.2825 *A. oryzae* 3.5232 *R. oryzae* 3.275 SSF at 25 °C for 24, 36, 48 and 60 h.	The oats fermented by *A. oryzae* had the highest amount of GABA. *R. oryzae*-fermented oats had the highest TPC and AOXA.	Cai et al. (2014)
Brown rice, maize, oat, wheat	*Aspergillus oryzae, Aspergillus awamori, Rhizopus oligosporus and Rhizopus oryzae* SSF for 72 h at 30 °C	Increased: AoxA, TPC, 4-hydroxy benzoic acid, 4-hydroxy-3-methoxybenzoic, caffeic acid, ferulic acid, gallic acid, protocatechuic acid, trans-cinnamic acid	Dey and Kuhad (2014)
Wheat	*Rhizopus oryzae* RCK2012 SSF for 3 days at 30 °C.	Improved AoxA (ABTS, DPPH HPSC, FRAP) and *in vivo* antioxidant capacity using *Saccharomyces cerevisiae*.	Dey and Kuhad (2014)
Yogurt-like beverage containing 25% of oat flakes	*L. plantarum* LP09 The enzyme (xylanase and α-amylase) addition favored the growth of the starter, shortened the time needed to reach pH 4.2 to ca. 8 h at 30 °C.	Fermentation increased the polyphenols availability and the AoxA (25% and 70% higher, respectively) and decreased the hydrolysis index *in vitro*.	Luana et al. (2014)
Rice bran	*Lactococcuslactis, Pediococcusacidilactici Pediococcuspentosaceus*	Fermentation with *P. acidilactici* showed a two-fold increment in ɣ-oryzanol and α-tocopherol and a higher concentration of ferulic acid.	Abd Rashid et al. (2015)

(continued)

Table 13.1 (continued)

Cereal based product	Microorganisms and fermentation conditions	Beneficial effect	Reference
Beverage prepared from 60% fermented QPM maize +40% fermented common beans	*R. oligosporus*. SSF for fragmented maize and bean cotyledons was performed at 35 °C/84 h and 38 °C/100 h, respectively.	SSF increased TPC, AoxA (ORAC, ABTS) and angiotensin converting enzyme–inhibitory (ACE-I) activity of quality protein maize and common beans.	Rochín-Medina et al. (2015b)
Bread made using 20% (w/w) of quinoa sourdough	*Lactobacillus plantarum* T6B10 and *Lactobacillus rossiae* T0A16. Bread was fermented at 30 °C for 16 h.	Fermentation increased organic acids, free amino acids, soluble fibers, total phenols, phytase and AoxA (DPPH), and *in vitro* protein digestibility markedly.	Rizzello et al. (2016)
Nondairy fermented beverage from a blend of cassava and rice based on Brazilian indigenous beverage cauim	*L. plantarum* CCMA 0743 (from cauim), yeast *Torulaspora delbrueckii* (from tarubá),and the commercial probiotic, *L. acidophilus*. SmF at 37 °C for 48 h.	Co-culturewith the yeast strain *T. delbrueckii* can improve the product's digestibility by reducing starch, increasing the AoxA (DPPH, ABTS), and further stimulating LAB growth during fermentation.	Freire et al. (2017)
Brown rice flour	Commercial baker's yeast *Saccharomyces cerevisiae* Fermented for 12 h at 25 °C	Effect on TPC differed between commercial baker's yeast. Increase in ferric reducing ability power.	Ilowefah et al. (2017)
Beverage prepared from 60% fermented blue maize +40% fermented black common beans	*Rhizopus oligosporus* SSF for fragmented maize and bean cotyledons was performed at 37 °C/90 h and 38 °C/100 h, respectively.	SSF increased AoxA, TPC and ACE inhibition of blue maize and black common beans.	Milán-Carrillo et al. (2017)
Pearl millet	*Aspergillus sojae* SSF for 10 days at 25 °C	Extract from 6th days exhibited the highest TPC, AoxA potential and DNA damage protection activity.	Salar et al. (2017)
Tejuino (A typical Mexican beverage of fermented nixtamalized maize)	*Leuconostoccitreum, Weisellacibaria, Chryseobacteriumbovis, and Bacillus safensis) isolated from tejuino.* Nixtamalized maize mass in water was stored 12 h, cooked and fermented for 72 h at room temperature.	*Leu. citreum* could be a potential probiotic strain due to its tolerance to gastrointestinal conditions, antagonistic activity towards foodborne pathogens, short-chain fatty acids production and adhesion to HT-29 cells.	Silva et al. (2017)

(continued)

Table 13.1 (continued)

Cereal based product	Microorganisms and fermentation conditions	Beneficial effect	Reference
Wheat bread (17.4 g of $ZnSO_4 \cdot 7H_2O$ and 88.028 mg of $Na_2SeO_3 \cdot 5H_2O$ were added to 100 kg of flour.)	*Lactococcus lactis ssp., Leuconostoc mesenteroides ssp., L. delbrueckii, Streptococcus thermophilus.* Dough fermentation at 37 °C for 6 h.	Zn and Se-enriched breads with increased biological value and improved quality.	Zlateva et al., 2017
Wheat and barley by-products	*Pediococcus acidilactici* LAB suspension was used for cereal by-products solid state (SSF) and SmF for 72 h at 27 °C.	The SmF could be more preferred in order to reduce the losses of alkylresorcinols and reduce aflatoxin. Under SSF higher contents of matairesinol and secoisolariciresinol were observed.	Bartkiene et al. (2018)
Bread (sourdough)	131 isolates from different Italian sourdoughs and belonging to different species of *Lactobacillus,* both hetero- and homofermentative	Increased AoxA and anti-inflammatory activities on cultured cells (RAW 264.7 murine macrophage, murine H-end endothelium cells and human intestinal Caco-2 cells) by assaying ROS content, NFkB/IkB expression level and Interleukin-1β production.	Galli et al. (2018)
Oat	*R. oryzae*-fermented oats(RFO), *L. plantarum*-fermentedoats (LFO) and coinoculated fermented oats (CFO). SSF for 72 h at 30 °C	CFO and RFO presented higher ACE inhibitory activities, the protein extracts from CFO exerted the best IC_{50} value (0.42 mg protein per mL).	Wu et al. (2018)
Oat-based beverage	*L. plantarum* and prebiotic (inulin) were fermented by SmF for 12 h at 30 °C	Symbiotic oat-based beverage is a low fat and high dietary fiber symbiotic food.	Wang, Liang, et al. (2018a)
Low tannin (LT)-sorghum and high tannin (HT)-sorghum to produce porridges	*L. fermentum* FUA 3321 was used to spontaneously fermenting HT and LT sorghums for 72 h at 28 °C and 24 h at 34 °C, respectively.	Fermentation reduces all the mycotoxins (fumonisin zearalenone, alpha-zearalenol and beta-zearalenol) by 98%	Adebo et al. (2019)
Wheat bran, oatbran	*Saccharomy cescerevisiae* (SSF for 6 days at 30 °C)	Increasing: AOXA, TPC, ferulic acid, vanillic acid, 2-hydroxybenzoic acid, apigenin-glucoside, avenanthramide	Călinoiu et al. (2019)

(continued)

Table 13.1 (continued)

Cereal based product	Microorganisms and fermentation conditions	Beneficial effect	Reference
Defatted rice bran (DRB).	*Rhizopus oryzae* DRB was fermented at 30 °C for 5 days	Fermentation enhanced ORAC and cellular AoxA. Bioaccessibile phenolics increased in fermented DRB after *in vitro* digestion compared with raw DRB.	Chen et al. (2019)
Cereal bars added with tempeh flour	*R. oligosporus* Soya beans cotyledons were incubated at 30 °C for 40 h.	Cereal bars added with tempeh flour resulted in substantial is of lavone aglycones and PUFA contents	De Melo et al. (2020)
Oat dairy beverage (40% milk powder, and 30% soluble extract of oats (16.6% oat bran in water).	*Streptococcus thermophilus Lactobacillus delbrueckii* ssp. *Bulgaricus* and probiotic culture *L. curvatus* P99. SmF process was monitored until reaching pH 4.6.	*L. curvatus* remains viable for 35 days in the oat dairy beverage and was able to survive in simulated gastrointestinal tract conditions. *E. coli* O157:H7 showed sensitivity to *L. curvatus* P99.	Funck et al. (2019)
Fiber	*Aspergillusoryzae* FIBRA is a fermenting mixture of brown rice and rice bran	FIBRA prevented carcinogen-induced carcinogenesis in almost all representative rodent models except mammary carcinogenesis in rats.	Mori (2019)
Wheat sourdough	Co-fermentation of *Lactobacillus plantarum and Lactobacillus hammesii* SSF at 30 °C for 24 h	Co-fermentation achieved release of bound ferulic acid and conversion of the resultant free ferulic acid to dihydroferulic acid and volatile metabolites.	Ripari et al. (2019)
Whole maize	*Rhizopus oligosporus* SSF for 108 h at 35 °C	Increased: AoxA (ORAC, ABTS), inhibition of alfa amylase and alfa glucosidase, bound and free phenolics, ferulic acid, hydroxybenzoic acid, caffeic acid, syringicacid,	Sánchez-Magaña et al. (2019)
Spelt wheat	*Rhizopus oligosporus* SSF for 30 h at 31 °C	Increased: AoxA, ferulic acid, hydroxybenzoic acid, caffeic acid, syringicacid, amino acids, peptides	Starzyńska-Janiszewska et al. (2019)

(continued)

Table 13.1 (continued)

Cereal based product	Microorganisms and fermentation conditions	Beneficial effect	Reference
White rice varieties and red rice varieties	*Rhizopus oryzae* SSF for 96 h at 30 °C.	SSF increased TPC, TF, TAC, TC, AoxA (FRAP, TAC) and anti-inflammatory properties whereas anti-diabetic properties of all samples decreased with SSF.	Janarny and Gunathilake (2020)
Extruded brown rice (EBR)	Co-culture *L. plantarum, L. fermentum, Saccharomyces cerevisiae*. EBR was incubated at 37 °C for 24 h.	Enhanced free, conjugated and bound phenolics and flavonoids. More than tenfold increase in vanillic acid and quercetin contents. Ferulic, p-coumaric and chlorogenic acids were increased and improved AoxA (ORAC)	Khan et al. (2020)
Wheat varieties and hull-less spring barley variety	139 yeast strains isolated from cereal-based fermented foods and drinks. Substrates were used to produce doughs fermented at 30 °C for 24 h	Hull-less barley showed the highest AoaX (FRAP, ABTS). SPA and CWPA contents were highest in blue-grained wheat. T two selected yeast strains possessing the highest phytase and AoaX.	Palla et al. (2020)
Wheat bran	*Fomitopsispinicola* SSF at 26 °C for 2, 4 and 6 days	TAC increased 5.73 times by the 4th day. TPC and alkylresorcinols increased 5.91 and 1.55 times, respectively by the 6th day.	Tu et al. (2020)
Oat	*Cordyceps militaris* (SSF for 8 days at 25 °C) *and Monascusanka, Bacillus subtilis* (SSF for 12 days at 30 °C).	Increased: TPC, AoxA (DPPH, ABTS), gallic acid, p-hydroxybenzoic acid, caffeic acid,vanillin, p-coumaric acid, ferulic acid, avenanthramides, luteolin, apigenin and protection against oxidative DNA damage	Xiao et al. (2015); Chen et al. (2020)

AOXA antioxidant activity, *ABTS+* 2,2′-azinobis (3-ethylbenzothiazoline-6-sulfonic acid), *DPPH* 2,2-diphenyl-1-picrylhydrazyl, *FRAP* ferric reducing antioxidant power, *ORAC* oxygen radical absorbance capacity, *HFRSA* hydroxyl free radical scavenging activity, *CUPRAC* cupric reducing antioxidant capacity, *HPSC* hydrogen peroxide scavenging capacities, *ACE* angiotensinconvertingenzyme, *SPA* soluble phenolic acids, *CWPA* cell wall-bound phenolic acids, *ROS* reactive oxygen species, *PUFA* polyunsaturated fatty acid

Sprouted Grains

"Sprouts" (Regulation (EC) No 208/2013) are "the product obtained from the germination of seeds and their development in water or another medium, harvested before the development of true leaves and which is intended to be eaten whole, including the seed". "Sprouted grains" are defined by the American Association of Cereal Chemists (AACC) with the endorsement of the United States Department of Agriculture (USDA) as follows: "malted or sprouted grains containing all of the original bran, germ, and endosperm shell be considered whole grains as long as sprout growth does not exceed kernel length and nutrient values have not diminished. These grains should be labeled as malted or sprouted whole grain" (Benincasa et al., 2019). The complex physical and metabolic events during germination can be grouped into 3 phases mainly linked to water uptake by the seed. In phase I, the seed is steeped in water, and cell material and matrices become fully hydrated (imbibition). Further water uptake is only limited in phase II, which consists of the activation of the endogenous metabolism necessary for mobilizing reserve material and radicle growth. The radicle emerges at the end of phase II, and the germination "*sensu stricto*" is terminated. In phase III, the seed takes up further water, major mobilization of reserve material occurs, and the seedling starts to grow (Lemmens et al., 2019).

Germination is a natural economical process and an efficient method that increases nutritional bioavailability and content of bioactive compounds of grains (Kasote et al., 2021; Aparicio-García et al., 2021a, b). It has been found that germinated wheat and brown rice have more phenolic compounds, both free and bound, than un-germinated grains (Bei et al., 2018; Xu et al., 2021). Among the phenolics, the amount of bound ferulic acid and p-coumaric acid rose significantly during wheat and rice sprouting (Bei et al., 2018). In wheat, it is reported that the germination temperature and length have a beneficial effect on the accumulation of soluble phenolic acids, flavone C-glycosides, and lignin (Tomé-Sánchez et al., 2020). Germinated sweet corn and oat grains also had much more phenolic compounds than un-germinated grains (Xiang et al., 2017). Germination substantially altered the profile of phenolic compounds in naked barley, and the level of phenolic compounds rose significantly when germination was extended up to 36 h (Ge et al., 2021). Phenolic acids are biosynthesized during sprouting, and the synthesized cell wall-degrading enzymes, such as cellulases and endoxylanases in combination with cinnamoyl esterases and feruloyl esterases, could hydrolyze phenolic compounds bound to cell walls. This increases the levels of free phenolic compounds, mainly ferulic acid, and the antioxidant capacity (Lemmens et al., 2019). Oats contain avenanthramides, a group of phenolic antioxidants that are unique to oats. The levels of these bioactive compounds in oats improved by the germination process and power ultrasound (25 kHz). The power ultrasound treatment for 5 min after soaking significantly enhanced the total avenanthramides (24 h), (48–96 h), total phenolic content (24 h), and GABA levels in the germinated oats (Ding et al., 2018).

In recent years, the development of γ-Aminobutyric acid (GABA) enriched foods has aroused increasing interest in the food industry because of the wide variety of recognized health benefits, such as blood pressure control and prevention of diabetes and cancer, for this compound (Aparicio-García et al., 2021a). GABA is a metabolic end product originated primarily in plants by glutamic acid decarboxylation catalyzed by glutamate decarboxylase (GAD) enzyme. During germination, proteins are partially hydrolyzed, increasing the availability of free amino acids, and the activity of GAD is also enhanced (Ding et al., 2018). Relatively high sprouting temperatures are needed to stimulate GABA synthesis. Acidic conditions during steeping can also potentiate GABA synthesis (Lemmens et al., 2019). Germination parameters (temperature and time) were optimized to produce sprouted oat flours with enhanced contents of protein, essential amino, riboflavin, minerals, polyunsaturated fatty acids, as well as phenolic compounds, GABA, and antioxidant activity. GABA levels of sprouted oat were over threefold higher than the control (Aparicio-García et al., 2021a). Also, elicitation can distinctly change the content of GABA in sprouts. In previous studies, more elicitors, such as glutamic acid, light, ultrasound, hypoxia and slightly acidic electrolyzed water, and composite elicitors, such as chitosan/glutamic acid, could enhance the accumulation of GABA in sprouts (Ding et al., 2018; Liu et al., 2019). For instance, the GABA content in the ultrasound treated rice after soaking for 12 h increased by 60.9% compared to the control (Ding et al., 2018). Compared with un-germinated wheat, GABA maximum enhanced by 88.73% and 50.70% in hypoxia- and normoxia-germinated wheat, respectively (Wang et al., 2018b). Ascorbic acid and light can enhance the accumulation of GABA but can also decrease the content of GABA in sprouts. Therefore, the influence of these elicitors on GABA depends on many factors, such as genetic characteristics, growth phase, and growing environment of sprouts, and the nature, dose, and the application way of the elicitor (Liu et al., 2019). It is noteworthy that GABA is very heat-stable and is not degraded during kilning/drying (Lemmens et al., 2019). However, Baranzelli et al. (2018) observed a 75–90% loss in GABA content in bread baked (175 °C) from sprouted wheat, which was attributed to the degradation of free amino acids that are used in the Maillard reaction (Lemmens et al., 2019).

The effect of germination on dietary fiber content of sprouted grains is often inconsistent and depends strictly on fiber fraction, germination time, and genotypes (Benincasa et al., 2019). No significant effects in dietary fiber fractions were recorded for barley grains sprouted for 72 h (Teixeira et al., 2016) and oat grains sprouted for 72 h (Aparicio-García et al., 2021a). The β-glucan levels decreased 40% during 72 h of oat germination (Aparicio-García et al., 2021a) and suffer a higher depletion (72–93%) in barley sprouted at 15 °C for 36–72 h (Marconi et al., 2014). However, germination of barley at 16 °C for 3.5 days retained 87% of the initial β-glucan content, and exhibited levels of phenolic compounds and GABA between 1.4-fold and 2.5-fold higher than the non-sprouted grain (Rico et al., 2020). Germinated grains have been used to develop several kinds of functional cereal-based foods including breakfast cereals (Jribi et al., 2019; Kapoor & Haripriya, 2020), gluten-free bread (Comettant-Rabanal et al., 2021), noodles with sprouted pulses (Sofi et al., 2020), Se-enriched breads (Guardado-Félix et al., 2020), non-dairy fermented probiotic drinks (Chavan et al., 2018), and symbiotic beverages with sprouted oat flour (Aparicio-García et al., 2021b). On the other hand, continuous processing of germination/extrusion cereals produces functional flour with better functional and sensory properties (Gong et al., 2018). As shown in Table 13.2, the addition of sprouted cereals to cereal-based products improves their nutraceutical properties and bioactive content.

Table 13.2 Germinated cereals to produce ingredients and functional cereal-based products

Product	Functional properties	Germination conditions	Reference
Gluten-free bread	Gluten-free product	Germination (soaking for 24 h germinating for 48 h at 30 °C, relative humidity of 90%. Dried in a fan oven at 30 °C/24 h)	Comettant-Rabanal et al. (2021)
Wheat breads added with Se-enriched Chickpea sprouted flours	Increased cellular antioxidant activity (CAA)	Chickpea grains after soaking with different sodium selenite (Na_2SeO_3) concentrations (0, 1 and 2 mg/100 g seeds) were germinated during four days at 24 °C.	Guardado-Félix et al. (2020)
Breakfast cereal based product (41.4% germinated dried wheat)	Increase of +32% in total phenol content	Durum wheat seeds germinated for 48 hours.	Jribi et al. (2019)
Gluten free noodle	Enhanced antioxidant activity, total phenolic acids and *in vitro* protein digestibility	Chickpea seeds soaked for 30 min in sodium hypochlorite solution, soaked for 12 h in distilled water and were germinated for 36 h at 30 °C.	Sofi et al. (2020)
Novel gluten-free ingredient from germinated oat	Sprouted oat presented better levels of GABA (54.9 mg/100 g), free phenolics (507.4 mg GAE/100 g) and antioxidant capacity (1744.3 mg TE/100 g)	Oat were sprouting at 18 °C for 96 h.	Aparicio-García et al. (2021a)
Gluten free fermented beverage based on sprouted oat flour	Sprouted oat fermented beverage was a good source of protein β-glucan (79 mg/100 mL), and phenolic compounds (61.4 mg GAE/100 mL), and had a high antioxidant potential (164.3 mg TE/100 mL).	Oat grains were germinated at 18 °C for 4 days. Sprouted oat flour was mixed with tap water (18% w/v), sucralose (0.2% w/v) and salt (0.1% w/v), pasteurized and cooled. Fermentation was carried out at 30 °C for 4 h.	Aparicio-García et al. (2021b)
Non-dairy fermented probiotic drink	DPPH assay and polyphenol content were increased as the concentration of drink mixture increased in three different milk and distilled water	Grains were germinated at 30 °C in different times (barely 48 h, ragi 36 h moth bean 24 h). Germinated grains were roasted, grinded and mixed in the ratio of 2.5:1.5:1 (barley: finger millet: moth bean) with sugar and cardamom.	Chavan et al. (2018)

GAE gallic acid equivalent, *GABA* gamma aminobutyric acid, *TE* trolox equivalent, *DPPH* 2,2-diphenyl-1-picrylhydrazyl

Extruded Grains

The extrusion process is a low-cost technology that uses high temperature and short time (HTST). Furthermore, the whole grains can be used in this process, maintaining all the anatomic components of the grains in the final products. Cereals are the most convenient grains for developing extruded snacks owing to their high popularity, low cost, and availability. Most bioactive and phenolic compounds are mainly concentrated in the outer layer of cereal grains; thus, consuming whole-grain products is considered the best solution to increase the health benefits of cereal products under conditions that favor quality retention, high productivity, and low cost.

Although food extruders were initially limited to mixing and forming macaroni and ready-to-eat cereal pellets, they are now considered a HTST bioreactor that transforms raw ingredients into several food products, including pasta, breakfast cereals, bread crumbs, biscuits, crackers, croutons, baby foods, snack foods, confectionery items, chewing gum, texturized vegetable protein, pet foods, dried soups, dry beverage mixes, instant flours to make tortillas, modified starches, and novel ingredients (Reyes-Moreno et al., 2018; Gómez-Favela et al., 2021).

The extrusion processing conditions (humidity, temperature, screw speed) have a crucial impact on the levels of chemical compounds (e.g., starch, proteins, lipids, phenolic compounds) by releasing and/or degrading them (Salazar Lopez et al., 2016; Reyes-Moreno et al., 2018). In general, the increase in phenolic during the extrusion process could be due to the destruction of cell walls, the consequent release of phenolic compounds, and the Maillard reaction products quantified as phenolic compounds. On the other hand, a reduction of the phenolic content caused by the thermal process resulted from the polymerization of these compounds and consequently less extractability (Espinoza-Moreno et al., 2016). In general, all phenolic compounds showed higher contents in the bound fraction than the free fraction in extruded maize, wheat, and sorghum (Félix-Medina et al., 2021; Gaxiola-Cuevas et al., 2017; Ramos-Enríquez et al., 2018). Ferulic acid was the main phenolic acid reported in maize and sorghum bran before and after the extrusion process (Aguayo-Rojas et al., 2012; Salazar Lopez et al., 2016). Highest extrusion temperature and lowest bean flour content in mixtures of common beans and corn showed the highest values of AOXA, TPC, and melanoidin content (Félix-Medina et al., 2020). In contrast, the extrusion process did not show a significant effect on AOXA of free, bound, and total phenolic fractions, but increased the phenolic content on free, bound, and total phenolic fractions, as well as the free and total flavonoid content in mixtures of maize and tepary bean (Gómez-Favela et al., 2021). In addition, a significant decrease in total anthocyanin in extruded blue Mexican maize flour was mainly attributed to the flavonoids' sensitivity to high temperatures (Aguayo-Rojas et al., 2012). Important losses of heat sensible compounds (lipids) are reported, due to the thermal effects and chemical changes that occur during lime extrusion to produce tortillas. In this regard, maize carotenoids and lipophilic AOXA (ORAC assay) were affected during lime extrusion of pigmented creole Mexican maize (yellow and red). This thermal process might induce carotenoid degradation and reactions, such as isomerization and oxidation, that decrease the AoxA of lipophilic extracts (Corrales-Bañuelos et al., 2016).

Applying the extrusion process at optimal conditions allows us to produce extruded grain flours with high AOXA and TPC, and specific bioactive compounds. The response surface methodology has been used to optimized extrusion conditions to maximizing the phenolic content and antioxidant capacity of wheat bran (Ramos-Enríquez et al., 2018), sorghum (Ortiz-Cruz et al., 2020), wheat bran, oat bran and dehydrated fruit pomace (Delgado-Nieblas et al., 2019), maize and common beans (Reyes-Moreno et al., 2012, 2014; Félix-Medina et al., 2020; Espinoza-Moreno et al., 2021), maize and chickpea (Rochín-Medina et al., 2015a), maize and tepary bean (Gómez-Favela et al., 2021), maize and amaranth (Gámez-Valdez et al., 2021), brown rice (Hyun et al., 2016); most of the extrudates were used to produce flours as functional cereal food (Table 13.3) ingredients, including porridge for older adults (Espinoza-Moreno et al., 2021), expanded snacks (Félix-Medina et al., 2020, 2021; Gómez-Favela et al., 2021), breakfast cereals (Delgado-Nieblas et al., 2019), beverages (Rochín-Medina et al., 2015a; Reyes-Moreno et al., 2012, 2014), and functional tortillas (Gámez-Valdez et al., 2021) and noodles (Hyun et al., 2016). The presence of bioactive compounds in the extrude cereals or products have been related *in vitro* and *in vivo* (Table 13.3) with antioxidant activity (Ortiz-Cruz et al., 2020), cellular antioxidant activity (Gaxiola-Cuevas et al., 2017), ACE inhibition (Reyes-Moreno et al., 2014), hypoglycemic potential (Gámez-Valdez et al., 2021), and anti-inflammatory activity (Salazar Lopez et al., 2016).

In addition, extrusion cooking is widely used to improve the functional properties of dietary fiber of food processing by-products. Moreover, extrusion cooking was also indicated to increase the nutritional properties of those by-products (Zhong et al., 2019).Extrusion cooking increased SDF content of a range of high IDF sources such as wheat bran (Abd Rashid et al., 2015), rice and rye bran (Andersson et al., 2017; Wu et al., 2021), along with beneficially modified physico-chemical properties (e.g., viscosity, solubility and water binding capacity). Brown rice noodles with extruded rice bran had the highest total dietary fiber phenolic profiles compared to white rice noodles and brown rice noodles with un-extruded rice bran (Wu et al., 2021). Also, enzyme treatment (xylanase) and extrusion have been used to improve the solubility of rice bran dietary fiber and other soluble components. Especially the sequential extrusion-enzyme treatment significantly increased the total soluble pentosan content compared to individual or simultaneous treatments (Dang & Vasanthan, 2019). Preparation of breakfast cereals using whole cereal grains has been recommended since they conserve the bran fraction, which refers to the external parts of the grain. Extruded mixtures of wheat-bran and oat-bran with different levels of dehydrated naranjita pomace (DNP) were used to produce breakfast cereal. When the extrusion temperature was elevated, the soluble dietary fiber increased, whereas when DNP was higher, the soluble dietary fiber, total phenolic compounds, and antioxidant activity of the breakfast cereals were improved (Delgado-Nieblas et al., 2019).

Several strategies have been implemented to increase pulse consumption, such as their incorporation in bakery products, breakfast cereals, and snacks. Pulses are best

Table 13.3 Functional cereal-based products developed by extrusion process

Functional cereal product	Extrusion conditions	Functional improvement	Reference
Beverage of extruded quality protein maize and fermented bean flours (proportion = 60%: 40%, respectively)	Optimized extrusion variables were: 85 °C y 240 rpm. Beverage was prepared by mixed 25 g of extruded flours with water 134 mL, 6.5 g of low-calorie sweetener and 1 g of vanillin powder.	The mixture had high AoxA, TPC, antihypertensive capacity and strong antimutagenic capacity. A 200 mL portion of the beverage contributed with 69–115% of the recommended daily intake of antioxidants.	Reyes-Moreno et al. (2014)
Beverage from extruded whole Maize (60%) and extruded chickpea flours (40%).	Optimized extrusion variables were: Maize BT = 109 °C/ SS = 158 rpm and chickpea BT = 127 °C/SS = 151 rpm.	The optimized mixture retained more than 91% the AoxA (ORAC).	Rochín-Medina et al. (2015a)
Sorghum bran	The sorghum bran was hydrate for 8 h (20% and 30%) and extruding with a single screw at 180 °C and SS rpm and the die diameter was 5 mm.	Extrusion increases phenol content and percentage of inhibition of nitric oxide (NO) production by RAW cells	Salazar Lopez et al. (2016)
Expanded snack of extruded maize and extruded common bean	Optimized extrusion process variables BT = 157 °C/ SS = 238 rpm.	Extrusion increased the AoxA (ORAC, ABT), TDF and TPC total phenolic content.	Espinoza-Moreno et al. (2016)
Breakfast cereals (BC) were elaborated by mixing citrus fruit pomace with wheat-bran or oat-bran	A simple extrusion screw using SS = 110 rpm compression ratio 2:1, BT = 88.79–131.21 °C.	Higher BT increased SDF, whereas when addition of citrus fruit pomace was higher SDF, TPC, and AoxA increased.	Delgado-Nieblas et al. (2019)
Wheat bran	The optimized extrusion conditions: feedmoisture (FM) = 30% and T = 140 °C.	Optimization of FM and BT improved BTPC and bound DPPH	Reyes-Moreno et al. (2018)
Snacks directly expanded (second-generation) developed from three mixtures of whole corn flour (WCF) and common bean flours (CBF) in relation 84:16, 70:30, and 60:40.	Optimized extrusion conditions to 100% WCF (ET = 170 °C/SS = 200 rpm and mixtures 84:16 (BT = 159 °C/SS = 214 rpm), 70:30 (BT = 164 °C/187 rpm), and 60:40 (BT = 163 °C/ SS = 165 rpm).	Optimized extrusion conditions increased AoxA (ABTS, DPPH, FRAP), TPC and melanoidins content. When higher bean flour content (30% and 40%) was used an increase in the SDF content occurred.	Félix-Medina et al. (2020)

(continued)

Table 13.3 (continued)

Functional cereal product	Extrusion conditions	Functional improvement	Reference
Extruded sorghum bran	The optimal extrusion conditions were FM = 30% and T = 160 °C	Increased free total phenolic compounds with a value of (7428.95 μg GAE/g) and AoxA (14.12 μmol TE/g)	Ortiz-Cruz et al. (2020)
Creole blue maize tortillas added with 30% extruded amaranth flour	Optimized extrusion conditions: Creole blue maize: ET = 85 °C/SS = 240 rpm Amaranth: ET = 130 °C/SS = 124 rpm.	Extrusion improved TDF, IVPD, C-PER and antihypertensive and hypoglycemic potentials.	Gámez-Valdez et al. (2021)
Gluten-free healthy snack (GFHS) developed from a mixture of protein quality maize/tepary bean (70:30)	The best combination to produce GFHS was FMC = 16.4%/ET = 137 °C/VT = 237 rpm.	GFHS (50 g) was compared with two commercial snacks showing better protein content, TDF and AoxA, as well as high acceptability and lower energy content than commercial snacks.	Gómez-Favela et al. (2021)
Ready-to-eat expanded snack (RES) produced using a maize/common bean (70/30%)	Optimized extrusion conditions: BT = 164 °C and SS = 187 rpm.	RES presented three phenolic acids (ferulic, diferulic, p-coumaric) and three flavonoids (naringenin, kaempferol, methyl isoflavone). Seven fatty acids were also identified and the two most abundant were unsaturated (oleic, linoleic).	Félix-Medina et al. (2021)
Functional food for older adults (FFOA) was prepare using extruded whole maize flour (EWMF 60%) and extruded whole common bean flour (EWBF 40%)	Optimized extruded conditions to EWMF: BT = 97 °C/SS = 236 rpm, and EWBF: ET = 152 °C/SS = 240 rpm.	Extrusion increased the content of phenolic compounds and AoxA.	Espinoza-Moreno et al. (2021)
Starch based extruded cereals with chokeberry	Extrusion temperature 100C, screw speed 200, 400, 600, 800 rpm	Dietary fiber was unaffected, anthocyanins were degraded 70%, while phenolics acids and flavonoids were fully retained	Schmid et al. (2022)

(continued)

Table 13.3 (continued)

BT barrel temperature, *SS* screw speed, *DPPH* 2,2-diphenyl-1-picrylhydrazyl, *FRAP* ferric reducing antioxidant power, *ABTS* 2,2′-azinobis (3-ethylbenzothiazoline-6-sulfonic acid), *ORAC* oxygen radical absorbance capacity, *FAE* ferulic acid equivalent, *GAE* gallic acid equivalent, *BTPC* bound total phenol content, *TDF* total dietary fiber, *SDF* soluble dietary fiber

consumed in combination with cereals, as they complete each other by balancing the limiting essential amino acids. Most cereals are deficient in lysine, but they are significantly higher in pulses, and similarly, cysteine and methionine, which are limiting in pulses, are present in cereals (Sharma et al., 2021). Incorporating pulses in cereal-based extruded products has shown promising results in improving the protein quality of the resultant products. Several authors have reported enhancement in the bioactive profile of extrudates with the addition of pulse flours (Espinoza-Moreno et al., 2016; Sharma et al., 2021; Félix-Medina et al., 2020, 2021; Gómez-Favela et al., 2021). Extruded snacks exhibited high protein and dietary fiber but low-fat content, contrary to the high fat content of deep fat–fried snacks. Gluten-free extruded snacks from rice flour, cowpea, and whey protein concentrate increased the *in vitro digestibility* and amino acid composition (Dilrukshi et al., 2022). The decrease in lipids content has been reported in extruded snacks (Espinoza-Moreno et al., 2016; Félix-Medina et al., 2020). Thus, the expanded snack could be a good alternative to the high-caloric snacks (cookies, fries and others) currently consumed (Félix-Medina et al., 2021; Gómez-Favela et al., 2021) (Table 13.4).

The incorporation of bioprocessed grains into extruded grain flours has not been sufficiently explored. In this sense, fermented common bean (tempeh) and extruded maize were used to develop a flour mixture with high antioxidant activity, total phenolic content, strong antimutagenic capacity, and antihypertensive capacity. This mixture was used to develop a functional beverage which contributed with 69–115% of the recommended daily intake of antioxidants (Reyes-Moreno et al., 2014).Germinated brown rice contains many nutritional components, such as dietary fiber, mineral, essential amino acid, and γ -aminobutyric acid (GABA), and can be used to enrich extruded cereal-based products. The substitution of corn grit at 20% (w/w) with nutrition rich by-products, defatted soybean meal, germinated brown rice meal, and mango peel fiber, into ready to eat snacks increased protein content (up to 2.65 folds), dietary fiber (more than 3 folds), total phenolic compounds (22.33–33.53 folds), and antioxidant activity (5.30–11.53 folds) (Korkerd et al., 2016).

Novel studies focusing on the optimization of thermal and nonthermal operations during extrusion and the use of bioprocessed ingredients have a vast potential for the food industry. Processing operations optimized for food safety may be combined with phytochemical studies to analyze both the nutritional and safety aspects.

Table 13.4 Effect of pulses and by product addition on techno-functional profile of cereal-based products

Cereal based product	Effect on quality	Reference
Tepache (traditional Mexican beverage)	The optimum fermentation conditions for higher acceptance of tepache were: temperature 22 °C, brown sugar 10%, an initial pH of 5, and 72 h of fermentation.	Corona-González et al. (2013)
Yogurt-like beverage containing 25% of oat flakes	Sensory analyses showed that fermented oat flakes beverage had the typical features of a yogurt-like beverage, enhancing the overall intensity of odor and flavor compared to the non-fermented control.	Luana et al. (2014)
Beverage of extruded blue maize and black common beans	SSB process increased the protein content, soluble, insoluble and total dietary fiber contents, resistant starch and C-PER	Rochín-Medina, Milán-Carrillo, et al. (2015a)
Bread gluten free	Addition of quinoa flour increased batter elasticity, improved dough structure, increased fiber content	Turkut et al. (2016)
Expanded snack developed	Expanded snack (maize/common bean) had good expansion ratio, bulk density and hardness, which are quality indicators for expanded snacks, as well as high antioxidant activity (AoxA).	Espinoza-Moreno et al. (2016)
Bread gluten free	Addition of Maize arabinoxylans improved WAC and batter rheology, higher specific volume, softer crumb.	Ayala-Soto et al. (2017)
Bread gluten-free	Addition of teff flour (fermented) increased specific volume, decreased crumb hardness	Martins et al. (2017)
Extruded corn starch/soy protein isolate	Soy protein isolate increased the expansion of the snacks, but bulk density significantly declined; water solubility index and hardness also reduced with increasing level of protein	Chen et al. (2017)
Vegan cookies	Incorporation of tempeh in relation: 1:7 (tempeh:wheat) resulted in higher sensory acceptance scores compared to regular cookies without tempeh	Budsabun et al. (2019)
Cereal bars	The aforementioned characteristics positively affected the chemical composition of CB added with tempeh flour (10, 15% and 20%), resulting on a higher protein and lipid (especially PUFA) content,	De Melo et al. (2020)
Tortillas added with extruded defatted chia flour	The tortillas elaborated with extruded defatted chia flour had higher IVPD, C-PER, TPC and AoxA, and better antihypertensive potential than elaborated with only nixtamalized flour.	León-López et al. (2019)
Snacks directly expanded	Snacks directly expanded ((maize/common bean) had acceptable values of expansion ratio (ER), hardness (H) and bulk density (BD).	Félix-Medina et al. (2020)
Bread	Incorporating 5% of freeze-dried tempeh resulted in sensorially accepted bread with increased protein content and polyunsaturated/saturated fatty acids ratio	De Melo et al. (2020)
Functional food for older adults (FFOA).	FFOA (maize/common bean)) prepared using a blend of whole maize and common bean flours had high values of sensory acceptability, nutritional properties	Espinoza-Moreno et al. (2021)

(continued)

Table 13.4 (continued)

Cereal based product	Effect on quality	Reference
Tortillas produced from extrudedcreole blue maize flour	The sensory evaluation of the tortillas using a hedonic scale of 9 points, where 1 corresponds to "dislike extremely", 5 to "neither like or dislike", and 9 to "Like extremely". The functional tortillas showed a general acceptability value of 8.01.	Gámez-Valdez et al. (2021)
Optimized Gluten-free healthy snack (OGFHS).	OGFHS (protein quality maize/tepary bean) showing better protein content, total dietary fiber and AoxA, as well as high acceptability and lower energy content than commercial snacks.	Gómez-Favela et al. (2021)
Expanded snack from a maize-bean mixture (70/30)	Snack have an acceptable nutritional balance, good antioxidant properties, and important bioactive compounds.	Félix-Medina et al. (2021)

ER expansion ratio, *SME* specific mechanical energy, *WAC* water absorption capacity, *WAI* water absorption index, *WSI* water solubility index, *C-PER* calculated protein efficiency ratio, *EAA* essential amino acids, *IVPD in vitro* protein digestibility, *TPC* total phenolic content, *AoxA* antioxidant activity

5 Quality Aspects

The successful delivery of a newly developed functional food on the market is not only affected by the carried health benefits but strictly depends on several other factors. Besides the typical quality markers (e.g., taste, convenience, price), these include market positioning and functionality communication (Alongi & Anese, 2021). From the reviewed literature, it can be observed that several strategies and processing technologies for the enhancement of bioactive compounds and dietary fiber contents of cereal and cereal-based foods have been proposed and applied in research (Schmid et al., 2022). However, the limited application of some processing technologies can be attributed to a number of challenges in sensory characteristics such as taste, color, and texture of cereal foods during processing and after supplementation with natural sources rich in bioactive compounds and dietary fiber (Saleh et al., 2019).

The processing at high temperature, such as baking, extrusion, and puffing with the presence of natural supplements such as fruit and vegetable by-products, plant ingredients and extracts, whole grain flour, and cereal by-products, may result in significant changes in taste and color of the end products compared with products prepared from the refined flour only. The changes in taste and color could be attributed to that these supplements contain high amounts of high-temperature sensitive compounds, such as pigments (carotenoids, stilbenes), phenolics (anthocyanins, tannins), sugars, free amino acids, and vitamins. Under high temperature, these compounds are degraded or involved in non-enzymatic browning reactions, such as oxidation and Maillard reaction, and produce reaction products and flavor active substances, which may contribute to dark color and bitter or astringent taste of the end product. Furthermore, overall acceptability of cereal foods supplemented with natural sources of bioactive compounds and dietary fiber was found to be influenced by the storage time depending on the type of supplemental ingredients. The special texture and volume of cereal foods, such as breads, mainly depend on formation of the wheat gluten network and the ability to hold the gas produced by yeast during fermentation (Heinio et al., 2016; Saleh et al., 2019). However, by-products (BP) functional ingredients incorporation on bakery items at 10% was frequently successful. Overall, BP functional ingredients improve the nutritional profile of bakery products; however, they can also impair some functional and sensorial properties. From a nutritional point of view, cakes and biscuits enriched with BP functional ingredients cannot compete with the healthier profile presented by breads. A balance of BP incorporation has to be found in order to obtain healthier bakery products that can compete in sensory characteristics with the traditional ones (Martins et al., 2017). In addition, certain antinutrient factors present in legumes, including tannins, phytates, and enzyme inhibitors, have been reported to reduce the digestibility of proteins and starch along with a restricted bioavailability of minerals due to the formation of complex compounds, thus slandering the potential of pulses in value-added functional foods. Extrusion cooking is a good strategy to reduce the anti-nutritional factors ascribing to high temperature short-time processing (Sharma et al., 2021). Extruded cereal-based products, such as snack (Espinoza-Moreno et al., 2016; Félix-Medina et al., 2020; Gómez-Favela et al., 2021; Dilrukshi et al., 2022), porridge (Espinoza-Moreno et al., 2021), tortillas (León-López et al., 2019; Gámez-Valdez et al., 2021), and beverages (Rochín-Medina et al., 2015a), improved their nutritional properties, including *in vitro* protein digestibility, protein content, C-PER and essential amino acid content after the extrusion process. Extrusion conditions can be optimized applying response surface methodology (RSM) and a desirability function to maximize sensorial and nutritional quality of breakfast cereal (Espinoza-Moreno et al., 2016; Delgado-Nieblas et al., 2019; Dos Santos et al., 2019), expanded snacks (Félix-Medina et al., 2020; Gómez-Favela et al., 2021), tortillas (León-López et al., 2019), and gluten-free extrudates based on germinated quinoa flour (Jan et al., 2017).

Gluten free (GF) grains and crops (corn, rice, millet, amaranth, teff, cassava, tapioca, quinoa, etc.) are being applied to many widely consumed products, including but not limited to bread, pasta, snacks, and beverages. GF foods have posed a challenge to product developers because they lack the viscoelastic property imparted by gluten. However, many solutions have been explored to improve organoleptic characteristics, functionality, and nutritional profiles to meet the growing demand for high quality GF products. Possible solutions to improve the physical and organoleptic properties of GF products include the use of functional ingredients (hydrocolloids, enzymes, starch, protein) as well as technological advances (instant controlled pressure drop, extrusion, hot air toasting, and so on (Woomer & Adedeji, 2021).

Gluten-free bread formulations with different levels of yogurt addition were evaluated using dough rheology measurements and baking quality parameters. Although the yogurt incorporation significantly reduced the rheology properties of doughs, such effects resulted in significant improvements in the overall quality of the corresponding breads. Good linear correlations between bread firmness, specific volume with flow behavior, and viscoelastic functions were found, supporting the results obtained (Graça et al., 2020). Optimized thermal process conditions and bioprocessed ingredients can be used to improve quality properties of the final products. Crumb hardness of gluten-free (GF) bread can be reduced with the addition of germinated grains. Additionally, this natural ingredient contributes to CO_2 gasification during the fermentation and the retardation of bread staling. Germinated millet (*Pennisetum glaucum* (L.) R. Br.) would allow technological benefits due to its high degree of germination ratio and considerable enzymatic activity compared to other grains. Also, thermoplastic extrusion improving pasting and rheological properties of GF bread with and without the incorporation of 5% germinated pearl millet (Comettant-Rabanal et al., 2021). Milling-enzyme treatment changed the physicochemical components and increased the pasting viscosities of the rice flour to prepare noodles (Geng et al., 2020). Biscuits with 6% defatted rice bran (obtained through *Trichodermaviride* fermentation) got the highest acceptability and texture score (Jia et al., 2020). In bread, incorporating 5% of freeze-dried tempeh resulted in sensorially accepted bread with increased protein content and polyunsaturated/saturated fatty acids ratio (De Melo et al., 2020). Partial and complete substitution of sucrose could be achieved, and appropriate concentration of resistant starch (RS) or green banana flour contributed to better dough and biscuit texture (Di Cairano et al., 2021). Sourdough fermentation improves dough workability, bread structure, and organoleptic and nutritional properties of raw flours. Furthermore, it increases the content of biogenic compounds and the uptake of minerals and decreases the level of anti-nutritional factors and the value of the glycemic response. The fully hydrolyzed wheat flour produced during sourdough fermentation generates gluten-free breads safe for consumption by individuals with celiac disease. However, these GF breads had technology behavior and properties similar to the naturally gluten-free flours (e.g., corn), and the use of structuring agents was needed for making baked goods (Gobbetti et al., 2014). In this sense, wheat bread with an intermediate content of gluten maintains the texture and sensory features of traditional breads (Rizzello et al., 2014). Also, the use of sprouted cereals for baking can also positively impact bread shelf-life and sensory properties. The impact of 48 h sprouted quinoa (SQ) was assessed in bread-making. The best bread-making performance (highest bread specific volume and lowest crumb firmness) was obtained at 20:80 (SQ: wheat flour) replacement level (Suárez-Estrella et al., 2020). The use of quinoa sourdough improved the chemical, textural, and sensory features of wheat bread, showing better performances compared to the use of quinoa flour (Rizzello et al., 2016). In the production of tortillas, sprouting can make wholemeal products more acceptable for consumers. Fully or partially substituting wheat flour with flour from wheat sprouted for 2 days at 25 °C significantly increased the size of tortillas (Zhu et al., 2017). Dough produced from sprouted wheat flour readily expands during extrusion resulting in puffy tortillas, which are valued by consumers. Indeed, such extruded products have an increased sensory score and overall good acceptability (Pawar et al., 2016). Gluten-free pregelatinized rice noodle enriched with germinated chickpea flour (20 g/100 g) showed improved nutritional and quality characteristics (Sofi et al., 2020).

Fermentation can improve the shelf life of beverages by increasing the acidity and thereby inhibiting certain spoilage and harmful microorganisms. Moreover, some LAB and molds can also produce bacteriocins and antibiotics, which can inhibit the growth of spoilage and pathogenic microorganisms. Fermentation also has been used to enhance the flavor, safety, nutritional value and health-promoting properties of cereal beverages (Xiong et al., 2022). The volatile compounds content in fermented foods significantly affects their aroma and has a considerable impact on all of their organoleptic characteristics. It was observed that, during the fermentation of barley and malt substrates cultured individually with *L. acidophilus*, *L. reuteri* and *L. plantarum*, the volatile content was unique for each of the formulations. Acetaldehyde, acetone and ethyl acetate values were significantly higher in *L. plantarum* fermented malt substrates. *L. reuteri* produced greater values of ethanol in the malt medium, and the three LAB formed diacetyl only in the malt medium. *L. plantarum* were preferred according to the acceptability test. Cereal beverages with higher acetaldehyde concentrations were well accepted in the sensorial tests (Salmeron, 2017). A functional beverage, fermented with *L. plantarum* LP09 containing 25% (wt/wt) of oat flakes, had the typical features of a yogurt-like beverage and was considered optimal on the basis of sensory and technological properties (Luana et al., 2014). Optimization applying RSM allowed to correlate the fermentation products with the degree of acceptance of a traditional Mexican beverage (tepache) (Corona-González et al., 2013). Further research could identify the key attributes and substrates/probiotics that are associated with premium flavor and aroma and develop standardized starter cultures to control the microbial profile and enhance food safety, while maintaining sensory quality and health benefits (Salmeron, 2017; Xiong et al., 2022).

6 Conclusion and Future Trends

Despite the importance of selecting the appropriate food-processing conditions to improve the nutritional characteristics and increase the amount of biocomponents of the final product, research in this area is limited. Also, there is still a lack of knowledge on the metabolic pathways as well as the relationship between specific metabolites and their corresponding bioactivity, bio accessibility and bioavailability, which thus demands more research in this field. Thus, novel studies focusing on the optimization of thermal and non-thermal operations have a vast potential for the food industry. More research should be conducted, particularly focusing on modern microbial and biotechnological techniques, as well as the adoption of advanced techniques, including, but not limited to, metabolomics, metagenomics, metatranscriptomics, proteomics and artificial intelligence models to better optimize, standardize and describe the bioprocess for an overall improved food quality, enhanced nutrition and health as well as other associated socioeconomic benefits. The use of bioprocesses and versatile thermal processes or their combination to produce functional cereal-based products at low cost is another key area of investigation for the coming researchers. Investigations focusing on the metabolic studies with clinical trials are limited, and systematic studies need to be done. Hence, future studies must consider higher levels of cereals for clinical data, and the effect on postprandial insulin and glycaemic response must be recorded.

References

Abd Rashid, N. Y., Abd Razak, D. L., Jamaluddin, A., Sharifuddin, S. A., & Long, K. (2015). Bioactive compounds and antioxidant activity of rice bran fermented with lactic acid bacteria. *Malaysian Journal of Microbiology, 11*(2), 156–162.

Abd Razak, D. L., Abd Rashid, N. Y., Jamaluddin, A., Sharifudin, S. A., & Long, K. (2015). Enhancement of phenolic acid content and antioxidant activity of rice bran fermented with Rhizopus oligosporus and Monascus purpureus. *Biocatalysis and Agricultural Biotechnology, 4*(1), 33–38.

Adebo, O. A., Kayitesi, E., & Njobeh, P. B. (2019). Reduction of mycotoxins during fermentation of whole grain sorghum to whole grain ting (a Southern African food). *Toxins, 11*(3), 180.

Adebo, J. A., Njobeh, P. B., Gbashi, S., Oyedeji, A. B., Ogundele, O. M., Oyeyinka, S. A., & Adebo, O. A. (2022). Fermentation of cereals and legumes: Impact on nutritional constituents and nutrient bioavailability. *Fermentation, 8*(2), 63.

Aguayo-Rojas, J., Mora-Rochín, S., Cuevas-Rodríguez, E. O., Serna-Saldivar, S. O., Gutierrez-Uribe, J. A., Reyes-Moreno, C., & Milán-Carrillo, J. (2012). Phytochemicals and antioxidant capacity of tortillas obtained after lime-cooking extrusion process of whole pigmented Mexican maize. *Plant Foods for Human Nutrition, 67*(2), 178–185.

Ahmad, A., & Khalid, N. (2018). Dietary fibers in modern food production: A special perspective with β-glucans. In A. M. Grumezescu & A. M. Holban (Eds.), *Biopolymers for food design* (pp. 125–156). Elsevier.

Ahnan-Winarno, A. D., Cordeiro, L., Winarno, F. G., Gibbons, J., & Xiao, H. (2021). Tempeh: A semicentennial review on its health benefits, fermentation, safety, processing, sustainability, and affordability. *Comprehensive Reviews in Food Science and Food Safety, 20*(2), 1717–1767.

Alongi, M., & Anese, M. (2021). Re-thinking functional food development through a holistic approach. *Journal of Functional Foods, 81*, 104466.

Alongi, M., Melchior, S., & Anese, M. (2019). Reducing the glycemic index of short dough biscuits by using apple pomace as a functional ingredient. *LWT, 100*, 300–305.

Andersson, A. A., Andersson, R., Jonsäll, A., Andersson, J., & Fredriksson, H. (2017). Effect of different extrusion parameters on dietary fiber in wheat bran and rye bran. *Journal of Food Science, 82*(6), 1344–1350.

Aoe, S., Nakamura, F., & Fujiwara, S. (2018). Effect of wheat bran on fecal butyrate-producing bacteria and wheat bran combined with barley on Bacteroides abundance in Japanese healthy adults. *Nutrients, 10*(12), 1980.

Aparicio-García, N., Martínez-Villaluenga, C., Frias, J., & Peñas, E. (2021a). Sprouted oat as a potential gluten-free ingredient with enhanced nutritional and bioactive properties. *Food Chemistry, 338*, 127972.

Aparicio-García, N., Martínez-Villaluenga, C., Frias, J., & Peñas, E. (2021b). Production and characterization of a novel gluten-free fermented beverage based on sprouted oat flour. *Foods, 10*(1), 139.

Augustin, L. S., Aas, A. M., Astrup, A., Atkinson, F. S., Baer-Sinnott, S., Barclay, A. W., & Jenkins, D. J. (2020). Dietary fiber consensus from the international carbohydrate quality consortium (ICQC). *Nutrients, 12*(9), 2553.

Ayala-Soto, F. E., Serna-Saldivar, S. O., & Welti-Chanes, J. (2017). Effect of arabinoxylans and laccase on batter rheology and quality of yeast-leavened gluten free breads. *Journal of Cereal Science, 73*, 10–17.

Baranzelli, J., Kringel, D. H., Colussi, R., Paiva, F. F., Aranha, B. C., Miranda, M. Z. D., et al. (2018). Changes in enzymatic activity, technological quality and gamma-aminobutyric acid (GABA) content of wheat flour as affected by germination. *Food Science and Technology, 90*, 483–490.

Bartkiene, E., Bartkevics, V., Krungleviciute, V., Juodeikiene, G., Zadeike, D., Baliukoniene, V., Bakutis, B., Zelvyte, R., Santini, A., & Cizeikiene, D. (2018). Application of hydrolases and probiotic *Pediococcusacidilactici* BaltBio01 strain for cereal by-products conversion to bio-product for food/feed. *International Journal of Food Sciences and Nutrition, 69*, 165–175.

Basinskiene, L., & Cizeikiene, D. (2020). *Cereal-based nonalcoholic beverages. Trends in non-alcoholic beverages* (pp. 63–99). Academic Press.

Bei, Q., Chen, G., Lu, F., Wu, S., & Wu, Z. (2018). Enzymatic action mechanism of phenolic mobilization in oats (*Avena sativa* L.) during solid-state fermentation with *Monascusanka*. *Food Chemistry, 245*, 297–304.

Benincasa, P., Falcinelli, B., Lutts, S., Stagnari, F., & Galieni, A. (2019). Sprouted grains: A comprehensive review. *Nutrients, 11*(2), 421.

Budsabun T, Panphut W Chansukh KK (2019) Development vegan cookies by using fermented soybean (tempeh). International Academic Research Conference in Amsterdam, pp 43–46

Cai, S., Gao, F., Zhang, X., Wang, O., Wu, W., Zhu, S., Zhang, D., Zhou, F., & Ji, B. (2014). Evaluation of γ-aminobutyric acid, phytate and antioxidant activity of tempeh-like fermented oats (*Avena sativa* L.) prepared with different filamentous fungi. *Journal of Food Science and Technology, 51*(10), 2544–2551.

Călinoiu, L. F., Cătoi, A. F., & Vodnar, D. C. (2019). Solid-state yeast fermented wheat and oat bran as a route for delivery of antioxidants. *Antioxidants, 8*(9), 372.

Chavan, M., Gatb, Y., Harmalkara, M., & Waghmarea, R. (2018). Development of non-dairy fermented probiotic drink based on germinated and un-germinated cereals and legume. *LWT, 91*, 339–344.

Chávez-Santoscoy, R. A., Gutiérrez-Uribe, J. A., Serna-Saldivar, S. O., & Perez-Carrillo, E. (2016). Production of maize tortillas and cookies from nixtamalized flour enriched with anthocyanins, flavonoids and saponins extracted from black bean (Phaseolus vulgaris) seed coats. *Food Chemistry, 192*, 90–97.

Chen, B., Yu, C., Liu, J., et al. (2017). Physical properties and chemical forces of extruded corn starch fortified with soy protein isolate. *International Journal of Food Science & Technology, 52*, 2604–2613.

Chen, Y., Ma, Y., Dong, L., Jia, X., Liu, L., Huang, F., Chi, J., Xiao, J., Zhang, M., & Zhang, R. (2019). Extrusion and fungal fermentation change the profile and antioxidant activity of free and bound phenolics in rice bran together with the phenolic bioaccessibility. *LWT, 115*, 108461.

Chen, G., Liu, Y., Zeng, J., Tian, X., Bei, Q., & Wu, Z. (2020). Enhancing three phenolic fractions of oats (*Avena sativa* L.) and their antioxidant activities by solid-state fermentation with *Monascusanka* and *Bacillus subtilis*. *Journal of Cereal Science, 93*, 102940.

Coda, R., Rizzello, C. G., & Gobbetti, M. (2010). Use of sourdough fermentation and pseudo-cereals and leguminous flours for the making of a functional bread enriched of gamma-aminobutyric acid (GABA). *International Journal of Food Microbiology, 137*(2–3), 236–245.

Coda, R., Rizzello, C. G., Pinto, D., & Gobbetti, M. (2012). Selected lactic acid bacteria synthesize antioxidant peptides during sourdough fermentation of cereal flours. *Applied and Environmental Microbiology, 78*, 1087–1096.

Comettant-Rabanal, R., Carvalho, C. W. P., Ascheri, J. L. R., Chavez, D. W. H., & Germani, R. (2021). Extruded whole grain flours and sprout millet as functional ingredients for gluten-free bread. *LWT, 150*, 112042.

Corona-González, R. I., Ramos-Ibarra, J. R., Gutiérrez-González, P., Pelayo-Ortiz, C., Guatemala-Morales, G. M., & Arriola-Guevara, E. (2013). The use of response surface methodology to evaluate the fermentation conditions in the production of tepache. *Revista Mexicana de Ingenieria Quimica, 12*(1), 19–28.

Corrales-Bañuelos, A. B., Cuevas-Rodríguez, E. O., Gutiérrez-Uribe, J. A., Milán-Noris, E. M., Reyes-Moreno, C., Milán-Carrillo, J., & Mora-Rochín, S. (2016). Carotenoid composition and antioxidant activity of tortillas elaborated from pigmented maize landrace by traditional nixtamalization or lime cooking extrusion process. *Journal of Cereal Science, 69*, 64–70.

Czubaszek, A., Czaja, A., Sokół-Łętowska, A., Kolniak-Ostek, J., & Kucharska, A. Z. (2021). Changes in antioxidant properties and amounts of bioactive compounds during simulated *in vitro* digestion of wheat bread enriched with plant extracts. *Molecules (Basel, Switzerland), 26*(20), 6292.

Dang, T. T., & Vasanthan, T. (2019). Modification of rice bran dietary fiber concentrates using enzyme and extrusion cooking. *Food Hydrocolloids, 89*, 773–782.

De Melo, P. F., Kalschne, D. L., da Silva-Buzanello, R. A., Amaral, J. S., Torquato, A. S., Corso, M. P., et al. (2020). Cereal bars functionalised with tempeh: Nutritional composition, isoflavone content and consumer acceptance. *International Journal of Food Science & Technology, 55*(1), 397–405.

De Villa, R., Roasa, J., Mine, Y., & Tsao, R. (2021). Impact of solid-state fermentation on factors and mechanisms influencing the bioactive compounds of grains and processing by-products. *Critical Reviews in Food Science and Nutrition, 63*(21), 5388–5413.

Delgado-Nieblas, C., Ruiz-Beltrán, K., Sánchez-Lizárraga, J., Zazueta-Morales, J. D. J., Aguilar-Palazuelos, E., Carrillo-López, A., et al. (2019). Effect of extrusion on physicochemical, nutritional and antioxidant properties of breakfast cereals produced from bran and dehydrated naranjita pomace. *CyTA – Journal of Food, 17*(1), 240–250.

Dey, T. B., & Kuhad, R. C. (2014). Enhanced production and extraction of phenolic compounds from wheat by solid-state fermentation with *Rhizopus oryzae* RCK2012. *Biotechnology Reports, 4*, 120–127.

Di Cairano, M., Caruso, M. C., Galgano, F., Favati, F., Ekere, N., & Tchuenbou-Magaia, F. (2021). Effect of sucrose replacement and resistant starch addition on textural properties of gluten-free doughs and biscuits. *European Food Research and Technology, 247*(3), 707–718.

Dilrukshi, H. N., Torrico, D. D., Brennan, M. A., & Brennan, C. S. (2022). Effects of extrusion processing on the bioactive constituents, in vitro digestibility, amino acid composition, and antioxidant potential of novel gluten-free extruded snacks fortified with cowpea and whey protein concentrate. *Food Chemistry, 389*, 133107.

Ding, J., Ulanov, A. V., Dong, M., Yang, T., Nemzer, B. V., et al. (2018). Enhancement of γ -aminobutyric acid (GABA) and other health-related metabolites in germinated red rice (*Oryza sativa* L.) by ultrasonication. *Ultrasonics Sonochemistry, 40*, 791–797.

Dos Santos, P. A., Caliari, M., Júnior, M. S. S., Silva, K. S., Viana, L. F., Garcia, L. G. C., & de Lima, M. S. (2019). Use of agricultural by-products in extruded gluten-free breakfast cereals. *Food Chemistry, 297*, 124956.

Dziki, D., Różyło, R., Gawlik-Dziki, U., & Świeca, M. (2014). Current trends in the enhancement of antioxidant activity of wheat bread by the addition of plant materials rich in phenolic compounds. *Trends in Food Science and Technology, 40*(1), 48–61.

Esfandi, R., Walters, M. E., & Tsopmo, A. (2019). Antioxidant properties and potential mechanisms of hydrolyzed proteins and peptides from cereals. *Heliyon, 5*(4), e01538.

Espinoza-Moreno, R. J., Reyes-Moreno, C., Milán-Carrillo, J., Lopez-Valenzuela, J. A., Paredes-López, O., & Gutiérrez-Dorado, R. (2016). Healthy ready-to-eat expanded snack with high nutritional and antioxidant value produced from whole amarantin transgenic maize and black common bean. *Plant Foods for Human Nutrition, 71*(2), 218–224.

Espinoza-Moreno, R. J., Reyes-Moreno, C., Gutiérrez-Dorado, R., Milán-Carrillo, J., Cuevas-Rodríguez, E. O., Mora-Rochín, S., Gómez-Favela, M. A., & Perales-Sánchez, J. X. K. (2021). Functional food for older adults produced by extrusion from whole grains of maize/common bean. *Acta Universitatis, 31*, 1–18.

Feizollahi, E., Hadian, Z., & Honarvar, Z. (2018). Food fortification with omega-3 fatty acids; microencapsulation as an addition method. *Current Nutrition & Food Science, 14*(2), 90–103.
Félix-Medina, J. V., Montes-Ávila, J., Reyes-Moreno, C., Perales-Sánchez, J. X. K., Gomez-Favela, M. A., Aguilar-Palazuelos, E., et al. (2020). Second-generation snacks with high nutritional and antioxidant value produced by an optimized extrusion process from corn/common bean flours mixtures. *LWT, 124*, 109172.
Félix-Medina, J. V., Gutiérrez-Dorado, R., López-Valenzuela, J. A., López-Ángulo, G., Quintero-Soto, M. F., Perales-Sánchez, J. X. K., & Montes-Ávila, J. (2021). Nutritional, antioxidant and phytochemical characterization of healthy ready-to-eat expanded snack produced from maize/common bean mixture by extrusion. *LWT, 142*, 111053.
Freire, A. L., Ramos, C. L., da Costa Souza, P. N., Cardoso, M. G. B., & Schwan, R. F. (2017). Nondairy beverage produced by controlled fermentation with potential probiotic starter cultures of lactic acid bacteria and yeast. *International Journal of Food Microbiology, 248*, 39–46.
Funck, G. D., Marques, J. D. L., Cruxen, C. E. D. S., Sehn, C. P., Haubert, L., Dannenberg, G. D. S., et al. (2019). Probiotic potential of *Lactobacillus curvatus* P99 and viability in fermented oat dairy beverage. *Journal of Food Processing and Preservation, 43*(12), e14286.
Galli, V., Mazzoli, L., Luti, S., Venturi, M., Guerrini, S., Paoli, P., Vincenzini, M., Granchi, L., & Pazzagli, L. (2018). Effect of selected strains of lactobacilli on the antioxidant and anti-inflammatory properties of sourdough. *International Journal of Food Microbiology, 286*, 55–65.
Gámez-Valdez, L. C., Gutiérrez-Dorado, R., Gómez-Aldapa, C. A., Perales-Sánchez, J. X. K., Milán-Carrillo, J., Cuevas-Rodríguez, E. O., et al. (2021). Effect of the extruded amaranth flour addition on the nutritional, nutraceutical and sensory quality of tortillas produced from extruded creole blue maize flour. *Biotecnia, 23*(2), 103–112.
Gaxiola-Cuevas, N., Mora-Rochín, S., Cuevas-Rodriguez, E. O., León-López, L., Reyes-Moreno, C., Montoya-Rodríguez, A., & Milán-Carrillo, J. (2017). Phenolic acids profiles and cellular antioxidant activity in tortillas produced from Mexican maize landrace processed by nixtamalization and lime extrusion cooking. *Plant Foods for Human Nutrition, 72*(3), 314–320.
Ge, X., Saleh, A. S. M., Jing, L., Zhao, K., Su, C., Zhang, B., et al. (2021). Germination and drying induced changes in the composition and content of phenolic compounds in naked barley. *Journal of Food Composition and Analysis, 95*, 103594.
Geng, D. H., Zhou, S., Wang, L., Zhou, X., Liu, L., Lin, Z., Qin, W., Liu, L., & Tong, L. (2020). Effects of slight milling combined with cellulase enzymatic treatment on the textural and nutritional properties of brown rice noodles. *LWT, 128*, 109520.
Giaretta, D., Lima, V. A., & Carpes, S. T. (2018). Improvement of fatty acid profile in breads supplemented with Kinako flour and chia seed. *IFSET, 49*, 211–214.
Gobbetti, M., Rizzello, C. G., Di Cagno, R., & De Angelis, M. (2014). How the sourdough may affect the functional features of leavened baked goods. *Food Microbiology, 37*, 30–40.
Gómez-Favela, M. A., Reyes-Moreno, C., Milán-Carrillo, J., Partida-Preciado, R. A., Espinoza-Moreno, R. J., Preciado-Ortiz, R., & Gutiérrez-Dorado, R. (2021). Gluten-free healthy snack with high nutritional and nutraceutical value elaborated from a mixture of extruded underutilized grains (quality protein maize/tepary bean). *Acta Universitatis, 31*, 1–18.
Gong, K., Chen, L., Li, X., Sun, L., & Liu, K. (2018). Effects of germination combined with extrusion on the nutritional composition, functional properties and polyphenol profile and related *in vitro* hypoglycemic effect of whole grain corn. *Journal of Cereal Science, 83*, 1–8.
Graça, C., Raymundo, A., & Sousa, I. (2020). Yogurt as an alternative ingredient to improve the functional and nutritional properties of gluten-free breads. *Foods, 9*(2), 111.
Granato, D., Barba, F. J., BursaćKovačević, D., Lorenzo, J. M., Cruz, A. G., & Putnik, P. (2020). Functional foods: Product development, technological trends, efficacy testing, and safety. *Annual Review of Food Science and Technology, 11*, 93–118.
Guardado-Félix, D., Lazo-Vélez, M. A., & Serna-Saldivar, S. O. (2019). Protein-Selenized enriched breads. In *Flour and breads and their fortification in health and disease prevention* (pp. 307–317). Academic Press.

Guardado-Félix, D., Lazo-Vélez, M. A., Pérez-Carrillo, E., Panata-Saquicili, D. E., & Serna-Saldívar, S. O. (2020). Effect of partial replacement of wheat flour with sprouted chickpea flours with or without selenium on physicochemical, sensory, antioxidant and protein quality of yeast-leavened breads. *LWT, 129*, 109517.

Heinio, R. L., Noort, M. W. J., Katina, K., Alam, S. A., Sozer, N., De Kock, H. L., et al. (2016). Sensory characteristics of wholegrain and bran-rich cereal foods–A review. *Trends in Food Science and Technology, 47*, 25–38.

Horvat, D., Šimić, G., Drezner, G., Lalić, A., Ledenčan, T., Tucak, M., Plavšić, H., Andrić, L., & Zdunić, Z. (2020). Phenolic acid profiles and antioxidant activity of major cereal crops. *Antioxidants (Basel, Switzerland), 9*(6), 527.

Hyun, J. H., Choi, H. W., Seo, D. H., Park, J. D., Kum, J. S., Lee, H., Kim, B. Y., & Baik, M. Y. (2016). Optimization of thermal processing conditions for brown rice noodles. *Applied Biological Chemistry, 59*, 517–524.

IDF (2021) International Diabetes Federation. IDF Diabetes Atlas. 10th ed. [Internet]. Brussels, Belgium, International Diabetes Federation

Ilowefah, M., Bakar, J., Ghazali, H. M., & Muhammad, K. (2017). Effect of yeast fermented Brown Rice flour substitution on nutritional, rheological and textural properties of steamed Brown Rice bread. In A. Amanullah & S. Fahad (Eds.), *Rice – technology and production*. IntechOpen. https://doi.org/10.5772/66880

Jan, R., Saxena, D. C., & Singh, S. (2017). Effect of storage conditions and packaging materials on the quality attributes of gluten-free extrudates and cookies made from germinated *chenopodium* (*Chenopodium album*) flour. *Journal of Food Measurement and Characterization, 11*(3), 1071–1080.

Janarny, G., & Gunathilake, K. D. P. P. (2020). Changes in rice bran bioactives, their bioactivity, bioaccessibility and bioavailability with solid-state fermentation by *Rhizopus oryzae*. *Biocatalysis and Agricultural Biotechnology, 23*, 101510.

Jia, M., Chen, J., Liu, X., Xie, M., Nie, S., Chen, Y., et al. (2019). Structural characteristics and functional properties of soluble dietary fiber from defatted rice bran obtained through *Trichodermaviride* fermentation. *Food Hydrocolloids, 94*, 468–474.

Jia, M., Yu, Q., Chen, J., He, Z., Chen, Y., Xie, J., Nie, S., & Xie, M. (2020). Physical quality and *in vitro* starch digestibility of biscuits as affected by addition of soluble dietary fiber from defatted rice bran. *Food Hydrocolloids, 99*, 105349.

Jin, W. J., Kim, M. J., & Kim, K. S. (2013). Utilization of barley or wheat bran to bioconvert glutamate to γ-aminobutyric acid (GABA). *Journal of Food Science, 78*(9), C1376–C1382.

Jribi, S., Sassi, K., & Debbabi, H. (2019). Incorporation of sprouted wheat seeds as functional ingredient in a Mediterranean breakfast cereal based product, Bsissa: A mixture design approach. *IJIAAR, 3*(3), 429–441.

Kapoor, T., & Haripriya, A. (2020). Formulation and evaluation of functional, nutritional and sensory properties of inulin incorporated ready to eat multi-grain breakfast cereal. *Indian Journal of Nutrition and Dietetics, 57*(4), 422.

Kasote, D., Tiozon, R. N., Jr., Sartagoda, K. J. D., Itagi, H., Roy, P., Kohli, A., Regina, A., & Sreenivasulu, N. (2021). Food processing technologies to develop functional foods with enriched bioactive phenolic compounds in cereals. *Frontiers in Plant Science, 30*(12), 771276.

Khan, S. A., Zhang, M., Liu, L., Dong, L., Ma, Y., Wei, Z., Chi, J., & Zhang, R. (2020). Coculture submerged fermentation by lactobacillus and yeast more effectively improved the profiles and bioaccessibility of phenolics in extruded brown rice than single-culture fermentation. *Food Chemistry, 326*, 126985.

Koistinen, V. M., Katina, K., Nordlund, E., Poutanen, K., & Hanhineva, K. (2016). Changes in the phytochemical profile of rye bran induced by enzymatic bioprocessing and sourdough fermentation. *Food Research International, 89*, 1106–1115.

Koistinen, V. M., Mattila, O., Katina, K., Poutanen, K., Aura, A. M., & Hanhineva, K. (2018). Metabolic profiling of sourdough fermented wheat and rye bread. *Scientific Reports, 8*(1), 1–11.

Korkerd, S., Wanlapa, S., Puttanlek, C., Uttapap, D., & Rungsardthong, V. (2016). Expansion and functional properties of extruded snacks enriched with nutrition sources from food processing by-products. *Journal of Food Science and Technology, 53*(1), 561–570.

Lee, D. P. S., Peng, A., Taniasuri, F., Tan, D., & Kim, J. E. (2022). Impact of fiber-fortified food consumption on anthropometric measurements and cardiometabolic outcomes: A systematic review, meta-analyses, and meta-regressions of randomized controlled trials. *Critical Reviews in Food Science and Nutrition*, 1–19. https://doi.org/10.1080/10408398.2022.2053658

Leite, R., Carrão-Panizzi, M. C., Curti, J. M., Dias, I. P., & Seibel, N. F. (2013). Tempeh flour as a substitute for soybean flour in coconut cookies. *Food Science and Technology, 33*(4), 796–800.

Lemmens, E., Moroni, A. V., Pagand, J., Heirbaut, P., Ritala, A., Karlen, Y., et al. (2019). Impact of cereal seed sprouting on its nutritional and technological properties: A critical review. *Comprehensive Reviews in Food Science and Food Safety, 18*(1), 305–328.

León-López, L., Bañuelos-Piña, A. M., Reyes-Moreno, C., Milán-Carrillo, J., Contreras-Andrade, I., Sánchez-Magaña, L. M., & Cuevas-Rodríguez, E. O. (2019). Optimisation of temperature and time for the dark germination bioprocess of Moringa oleifera seeds to boost nutritional value, total phenolic content and antioxidant activity. *International Food Research Journal, 26*(3), 831–839.

Liu, H., Kang, Y., Zhao, X., Liu, Y., Zhang, X., & Zhang, S. (2019). Effects of elicitation on bioactive compounds and biological activities of sprouts. *Journal of Functional Foods, 53*, 136–145.

Longoria-García, S., Cruz-Hernández, M. A., Flores-Verástegui, M. I. M., Contreras-Esquivel, J. C., Montañez-Sáenz, J. C., & Belmares-Cerda, R. E. (2018). Potential functional bakery products as delivery systems for prebiotics and probiotics health enhancers. *Journal of Food Science and Technology, 55*(3), 833–845.

Luana, N., Rossana, C., Curiel, J. A., Kaisa, P., Marco, G., & Rizzello, C. G. (2014). Manufacture and characterization of a yogurt-like beverage made with oat flakes fermented by selected lactic acid bacteria. *International Journal of Food Microbiology, 185*, 17–26.

Majerska, J., Michalska, A., & Figiel, A. (2019). A review of new directions in managing fruit and vegetable processing by-products. *Trends in Food Science and Technology, 88*, 207–219.

Manini, F., Brasca, M., Plumed-Ferrer, C., Morandi, S., Erba, D., & Casiraghi, M. C. (2014). Study of the chemical changes and evolution of microbiota during sourdoughlike fermentation of wheat bran. *Cereal Chemistry, 91*(4), 342–349.

Marconi, O., Tomasi, I., Dionisio, L., Perretti, G., & Fantozzi, P. (2014). Effects of malting on molecular weight distribution and content of water-extractable β-glucans in barley. *Food Research International, 64*, 677–682.

Martins, Z. E., Pinho, O., & Ferreira, I. M. P. L. V. O. (2017). Food industry by-products used as functional ingredients of bakery products. *Trends in Food Science and Technology, 67*, 106–128.

Menezes, A. G. T., Ramos, C. L., Dias, D. R., & Schwan, R. F. (2018). Combination of probiotic yeast and lactic acid bacteria as starter culture to produce maize-based beverages. *Food Research International, 111*, 187–197.

Menis-Henrique, M. E. C., Scarton, M., Piran, M. V. F., & Clerici, M. T. P. S. (2020). Cereal fiber: Extrusion modifications for food industry. *Current Opinion in Food Science, 33*, 141–148.

Milán-Carrillo, J., Gutiérrez-Dorado, R., Cuevas-Rodríguez, E. O., Sánchez-Magaña, L. M., Rochín-Medina, J. J., & Reyes-Moreno, C. (2017). Bebida funcional con potencial antidiabético y antihipertensivo elaborada con maíz azul y frijol negro bioprocesados. *Revista Fitotecnia Mexicana, 40*(4), 451–459.

Min, M., Bunt, C. R., Mason, S. L., & Hussain, M. A. (2019). Non-dairy probiotic food products: An emerging group of functional foods. *Critical Reviews in Food Science and Nutrition, 59*(16), 2626–2641.

Mori, H. (2019). Fermented Brown Rice and Rice Bran with *Aspergillus Oryzae* as a new promising agent for cancer prevention. *American Journal of Biomedical Science and Research, 6*(2), 232–233.

Murugkar, D. A., Zanwar, A. A., & Shrivastava, A. (2021). Effect of nano-encapsulation of flaxseed oil on the stability, characterization and incorporation on the quality of eggless cake. *Applied Food Research, 1*(2), 100025.

Ortiz-Cruz, R. A., Ramírez-Wong, B., Ledesma-Osuna, A. I., Torres-Chávez, P. I., Sánchez-Machado, D. I., Montaño-Leyva, B., et al. (2020). Effect of extrusion processing conditions on the phenolic compound content and antioxidant capacity of sorghum (*Sorghum bicolor* (L.) Moench) bran. *Plant Foods for Human Nutrition, 2*, 252–257.

Palla, M., Blandino, M., Grassi, A., Giordano, D., Sgherri, C., Quartacci, M. F., et al. (2020). Characterization and selection of functional yeast strains during sourdough fermentation of different cereal wholegrain flours. *Scientific Reports, 10*(1), 1–15.

Pasqualone, A., & Summo, C. (2021). Qualitative and nutritional improvement of cereal-based foods and beverages. *Foods, 10*(2), 338.

Patel, A., Desai, S. S., Mane, V. K., Enman, J., Rova, U., Christakopoulos, P., & Matsakas, L. (2022). Futuristic food fortification with a balanced ratio of dietary ω-3/ω-6 omega fatty acids for the prevention of lifestyle diseases. *Trends in Food Science and Technology, 120*, 140–153.

Pawar, V. S., Shere, D. M., & Khapre, A. P. (2016). Effect of sprouting treatment on quality characteristics of cereal-legume based extruded product. *International Journal of Food and Fermentation Technology, 6*, 35–40.

Petrovna, L. G., Vladimirovna, S. E., Dashievna, Z. S., & Sergeevna, K. T. (2022). The functional properties of bread enriched with essential fatty acids. *KnE Life Sciences*, 2022, 236–250.

Qu, T., He, S., Ni, C., Wu, Y., Xu, Z., Chen, M. L., et al. (2022). In vitro anti-inflammatory activity of three peptides derived from the byproduct of rice processing. *Plant Foods for Human Nutrition, 77*(2), 172–180.

Quiles, A., Campbell, G. M., Struck, S., Rohm, H., & Hernando, I. (2016). Fiber from fruit pomace: A review of applications in cereal-based products. *Food Reviews International, 34*(2), 162–181.

Ramos-Enríquez, J. R., Ramírez-Wong, B., Robles-Sánchez, R. M., Robles-Zepeda, R. E., González-Aguilar, G. A., & Gutiérrez-Dorado, R. (2018). Effect of extrusion conditions and the optimization of phenolic compound content and antioxidant activity of wheat bran using response surface methodology. *Plant Foods for Human Nutrition, 73*, 228–234.

Rawat, N., & Indrani, D. (2015). Functional ingredients of wheat-based bakery, traditional, pasta, and other food products. *Food Reviews International, 31*(2), 125–146.

Reyes-Moreno, C., Argüelles-López, O. D., Rochín-Medina, J. J., Milán-Carrillo, J., Rivera-López, J., Valdez-Ortiz, A., López-Valenzuela, J. A., & Gutiérrez-Dorado, R. (2012). High antioxidant activity mixture of extruded whole quality protein maize and common bean flours for production of a nutraceutical beverage elaborated with a traditional Mexican formulation. *Plant Foods for Human Nutrition, 67*(4), 450–456.

Reyes-Moreno, C., Milán-Carrillo, J., Valdez-Ortiz, A., Delgado-Vargas, F., Sánchez-Magaña, L. M., & Gutiérrez-Dorado, R. (2014). Bebida con potencial nutracéutico elaborada con maíz QPM extrudido y frijol fermentado. *Revista Iberoamericana de Ciencia, 1*(3), 61–77.

Reyes-Moreno, C., Reyes-Fernández, P. C., Cuevas-Rodríguez, E. O., Milán-Carrillo, J., & Mora-Rochín, S. (2018). Changes in nutritional properties and bioactive compounds in cereals during extrusion cooking. In *Extrusion of metals, polymers, and food products* (pp. 104–124). Intech.

Rico, D., Peñas, E., García, M. D. C., Martínez-Villaluenga, C., Rai, D. K., Birsan, R. I., & Martín-Diana, A. B. (2020). Sprouted barley flour as a nutritious and functional ingredient. *Foods, 9*(3), 296.

Ripari, V., Bai, Y., & Ganzle, M. G. (2019). Metabolism of phenolic acids in whole wheat and rye malt sourdoughs. *Food Microbiology, 77*, 43–51.

Rizzello, C. G., Calasso, M., Campanella, D., De Angelis, M., & Gobbetti, M. (2014). Use of sourdough fermentation and mixture of wheat, chickpea, lentil and bean flours for enhancing the nutritional, texture and sensory characteristics of white bread. *International Journal of Food Microbiology, 180*, 78–87.

Rizzello, C. G., Lorusso, A., Montemurro, M., & Gobbetti, M. (2016). Use of sourdough made with quinoa (Chenopodium quinoa) flour and autochthonous selected lactic acid bacteria for enhancing the nutritional, textural and sensory features of white bread. *Food Microbiology, 56*, 1–13.

Robledo-Márquez, K., Ramírez, V., González-Córdova, A. F., Ramírez-Rodríguez, Y., García-Ortega, L., & Trujillo, J. (2021). Research opportunities: Traditional fermented beverages in Mexico. Cultural, microbiological, chemical, and functional aspects. *Food Research International, 147*, 110482.

Rochín-Medina, J. J., Milán-Carrillo, J., Gutiérrez-Dorado, R., Cuevas-Rodríguez, E. O., Mora-Rochín, S., Valdez-Ortiz, A., & Reyes-Moreno, C. (2015a). Bebida funcional de valor nutricional/nutracéutico alto elaborada a partir de una mezcla de granos integrales (maíz+ garbanzo) extrudidos. *Revista Iberoamericana de Ciencia, 2*(1), 51–65.

Rochín-Medina, J. J., Gutiérrez-Dorado, R., Sánchez-Magaña, L. M., Milán-Carrillo, J., & Cuevas-Rodríguez, E. O. (2015b). Solid state bioconversion for producing functional flours from whole quality protein maize and common beans with enhanced nutritional value, antioxidant and antihypertensive potential. *Journal of Nutritional Health & Food Engineering, 2*(3), 00057.

Salar, R. K., Purewal, S. S., & Bhatti, M. S. (2016). Optimization of extraction conditions and enhancement of phenolic content and antioxidant activity of pearl millet fermented with *Aspergillus awamori* MTCC-548. *Resource-Efficient Technologies, 2*, 148–157.

Salar, R. K., Purewal, S. S., & Sandhu, K. S. (2017). Fermented pearl millet (*Pennisetum glaucum*) with *in vitro* DNA damage protection activity, bioactive compounds and antioxidant potential. *Food Research International, 100*, 204–210.

Salazar Lopez, N. J., Loarca-Piña, G., Campos-Vega, R., Gaytán Martínez, M., Morales Sánchez, E., Esquerra-Brauer, J. M., & Robles Sánchez, M. (2016). The extrusion process as an alternative for improving the biological potential of sorghum bran: Phenolic compounds and antiradical and anti-inflammatory capacity. *Evidence-Based Complementary and Alternative Medicine, 2016*, 8387975.

Saleh, A. S., Wang, P., Wang, N., Yang, S., & Xiao, Z. (2019). Technologies for enhancement of bioactive components and potential health benefits of cereal and cereal-based foods: Research advances and application challenges. *Critical Reviews in Food Science and Nutrition, 59*(2), 207–227.

Salmeron, I. (2017). Fermented cereal beverages: From probiotic, prebiotic and synbiotic towards nanoscience designed healthy drinks. *Letters in Applied Microbiology, 65*(2), 114–124.

Sánchez-Magaña, L. M., Reyes-Moreno, C., Milán-Carrillo, J., Mora-Rochin, S., León-López, L., Gutierrez-Dorado, R., & Cuevas-Rodríguez, E. O. (2019). Influence of solid-state bioconversion by *Rhizopus oligosporus* on antioxidant activity and phenolic compounds of maize (*Zea mays* L.). *Agrociencia, 53*(1), 45–57.

Sandhu, K. S., & Punia, S. (2017). Enhancement of bioactive compounds in barley cultivars by solid substrate fermentation. *Journal of Food Measurement and Characterization, 11*(3), 1355–1361.

Sandhu, K. S., Punia, S., & Kaur, M. (2016). Effect of duration of solid state fermentation by *Aspergillus awamorinakazawa* on antioxidant properties of wheat cultivars. *LWT, 71*, 323–328.

Santhanam, A. K., Lekshmi, M., Chouksey, M. K., Tripathi, G., & Gudipati, V. (2015). Delivery of omega-3 fatty acids into cake through emulsification of fish oil-in-milk and encapsulation by spray drying with added polymers. *Drying Technology, 33*(1), 83–91.

Schmid, V., Mayer-Miebach, E., Behsnilian, D., Briviba, K., Karbstein, H. P., & Emin, M. A. (2022). Enrichment of starch-based extruded cereals with chokeberry (Aronia melanocarpa) pomace: Influence of processing conditions on techno-functional and sensory related properties, dietary fibre and polyphenol content as well as in vitro digestibility. *LWT, 154*, 112610.

Seyedain-Ardabili, M., Sharifan, A., & Ghiassi Tarzi, B. (2016). The production of synbiotic bread by microencapsulation. *Food Technology and Biotechnology, 54*(1), 52–59.

Sharma, R., Dar, B. N., Sharma, S., & Singh, B. (2021). *In vitro* digestibility, cooking quality, biofunctional composition, and sensory properties of pasta incorporated with potato and pigeon pea flour. *International Journal of Gastronomy and Food Science, 23*, 100300.

Silva, E. M. M., Ascheri, J. L. R., & Ascheri, D. P. R. (2016). Quality assessment of gluten-free pasta prepared with a brown rice and corn meal blend via thermoplastic extrusion. *LWT, 68*, 698–706.

Silva, M. S., Ramos, C. L., González-Avila, M., Gschaedler, A., Arrizon, J., Schwan, R. F., & Dias, D. R. (2017). Probiotic properties of *Weissella cibaria* and *Leuconostoc citreum* isolated from tejuino – A typical Mexican beverage. *LWT, 86*, 227–232.

Socas-Rodríguez, B., Álvarez-Rivera, G., Valdés, A., Ibáñez, E., & Cifuentes, A. (2021). Food by-products and food wastes: Are they safe enough for their valorization? *Trends in Food Science and Technology, 114*, 133–147.

Sofi, S. A., Singh, J., Mir, S. A., & Dar, B. N. (2020). *In vitro starch* digestibility, cooking quality, rheology and sensory properties of gluten-free pregelatinized rice noodle enriched with germinated chickpea flour. *LWT, 133*, 110090.

Starzyńska-Janiszewska, A., Stodolak, B., Socha, R., Mickowska, B., & Wywrocka-Gurgul, A. (2019). Spelt wheat Tempe as a value-added whole-grain food product. *LWT, 113*, 108250.

Suárez-Estrella, D., Cardone, G., Buratti, S., Pagani, M. A., & Marti, A. (2020). Sprouting as a pre-processing for producing quinoa-enriched bread. *Journal of Cereal Science, 96*, 103111.

Subiria-Cueto, R., Coria-Oliveros, A. J., Wall-Medrano, A., Rodrigo-García, J., González-Aguilar, G. A., Martinez-Ruiz, N. D. R., & Alvarez-Parrilla, E. (2021). Antioxidant dietary fiber-based bakery products: A new alternative for using plant-by-products. *Food Science and Technology*. https://doi.org/10.1590/fst.57520

Svensson, L., Sekwati-Monang, B., Lutz, D. L., Schieber, A., & Ganzle, M. G. (2010). Phenolic acids and flavonoids in nonfermented and fermented red sorghum (*Sorghum bicolor* (L.) Moench). *Journal of Agricultural and Food Chemistry, 58*(16), 9214–9220.

Teixeira, C., Nyman, M., Andersson, R., & Alminger, M. (2016). Effects of variety and steeping conditions on some barley components associated with colonic health. *Journal of the Science of Food and Agriculture, 96*, 4821–4827.

Tomasik, P., & Tomasik, P. (2020). Probiotics, non-dairy prebiotics and postbiotics in nutrition. *Applied Sciences, 10*(4), 1470.

Tomé-Sánchez, I., Martín-Diana, A. B., Peñas, E., Bautista-Expósito, S., Frias, J., Rico, D., et al. (2020). Soluble phenolic composition tailored by germination conditions accompany antioxidant and anti-inflammatory properties of wheat. *Antioxidants, 9*(5), 426.

Tu, J., Zhao, J., Liu, G., Tang, C., Han, Y., Cao, X., Jia, J., Ji, G., & Xiao, H. (2020). Solid state fermentation by *Fomitopsispinicola* improves physicochemical and functional properties of wheat bran and the bran-containing products. *Food Chemistry, 328*, 127046.

Turkut, G. M., Cakmak, H., Kumcuoglu, S., & Tavman, S. (2016). Effect of quinoa flour on gluten-free bread batter rheology and bread quality. *Journal of Cereal Science, 69*, 174–181.

Verni, M., Verardo, V., & Rizzello, C. G. (2019). How fermentation affects the antioxidant properties of cereals and legumes. *Foods, 8*(9), 362.

Von Mollendorff, J. W., Vaz-Velho, M., & Todorov, S. D. (2016). Boza, a traditional cereal-based fermented beverage: A rich source of probiotics and bacteriocin-producing lactic acid bacteria. In *Functional properties of traditional foods* (pp. 157–188). Springer.

Wang, C., Liang, S., Wang, H., & Guo, M. (2018a). Physiochemical properties and probiotic survivability of symbiotic oat-based beverage. *Food Science and Biotechnology, 27*, 735–743.

Wang, P., Liu, K., Gu, Z., & Yang, R. (2018b). Enhanced γ-aminobutyric acid accumulation, alleviated componential deterioration and technofunctionality loss of germinated wheat by hypoxia stress. *Food Chemistry, 269*, 473–479.

Williams, B. A., Mikkelsen, D., Flanagan, B. M., & Gidley, M. J. (2019). "Dietary fibre": Moving beyond the "soluble/insoluble" classification for monogastric nutrition, with an emphasis on humans and pigs. *Journal of Animal Science and Biotechnology, 10*(1), 1–12.

Woomer, J. S., & Adedeji, A. A. (2021). Current applications of gluten-free grains–a review. *Critical Reviews in Food Science and Nutrition, 61*(1), 14–24.

Wu, H., Rui, X., Li, W., Xiao, Y., Zhou, J., & Dong, M. (2018). Whole-grain oats (*Avena sativa* L.) as a carrier of lactic acid bacteria and supplement rich in angiotensin I-converting enzyme inhibitory peptides through solid-state fermentation. *Food and Function, 9*(4), 2270–2281.

Wu, N. N., Ma, Z. Q., Li, H. H., Tian, X. H., Fang, Y., & Tan, B. (2021). Nutritional and cooking quality improvement of brown rice noodles prepared with extruded rice bran. *Cereal Chemistry, 98*(2), 346–354.

Xiang, N., Guo, X., Liu, F., Li, Q., Hu, J., & Brennan, C. S. (2017). Molecular sciences effect of light-and dark-germination on the phenolic biosynthesis, phytochemical profiles, and antioxidant activities in sweet corn (*Zea mays* L.) sprouts. *International Journal of Molecular Sciences, 2017*, 1806146.

Xiao, Y., Rui, X., Xing, G., Wu, H., Li, W., Chen, X., Jiang, M., & Dong, M. (2015). Solid state fermentation with *Cordycepsmilitaris* SN-18 enhanced antioxidant capacity and DNA damage protective effect of oats (*Avena sativa* L.). *Journal of Functional Foods, 16*, 58–73.

Xiong, Y., Zhang, P., Warner, R. D., Shen, S., & Fang, Z. (2022). Cereal grain-based functional beverages: From cereal grain bioactive phytochemicals to beverage processing technologies, health benefits and product features. *Critical Reviews in Food Science and Nutrition, 62*(9), 2404–2431.

Xu, K., Debelo, H., Roman, L., Guo, M., Ferruzzi, M. G., & Martinez, M. M. (2021). Co-extruded wheat/okra composite blends result in soft, cohesive and resilient crumbs rich in health-promoting compounds. *Food Chemistry, 364*, 130395.

Yadav, P., Chauhan, A. K., & Al-Sebaeai, M. A. (2020). Omega-3 fatty acid from plant sources and its application in food technology. In *Innovations in food technology* (pp. 41–52). Springer.

Zhang, B., Xu, Y., Zhao, C., Zhang, Y., Lv, H., Ji, X. M., et al. (2022). Protective effects of bioactive peptides in foxtail millet protein hydrolysates against experimental colitis in mice. *Food and Function, 13*, 2594–2605.

Zhong, L., Fang, Z., Wahlqvist, M. L., Hodgson, J. M., & Johnson, S. K. (2019). Extrusion cooking increases soluble dietary fibre of lupin seed coat. *LWT, 99*, 547–554.

Zhu, L., Adedeji, A. A., & Alavi, S. (2017). Effect of germination and extrusion on physicochemical properties and nutritional qualities of extrudates and tortilla from wheat. *J Food Sci, 82*(8), 1867–1875.

Zlateva, D., Stefanova, D., & Ninova-Nicolova, N. (2017). Study on the development of lactic acid bacteria in bread dough enriched with zinc and selenium. *J Food Sci Eng, 7*, 497–504.

Chapter 14
3D Food Printing

Plachikkattu Parambil Akhila, Basheer Aaliya, Muhammed Navaf, Kundukulangara Pulissery Sudheer, Shabir Ahmad Mir, and Kappat Valiyapeediyekkal Sunooj

1 Introduction

The three-dimensional (3D) printing technique is a revolutionary method which has created rapid advancements in conceptual designs, feasibility analysis, and prototype creation. 3D printing is a process that operates on an 'additive' principle by depositing the food ink layer- by layer based on a pre-designed model (Hao et al., 2010). This technology is also known as additive manufacturing, rapid prototyping, or food-layered manufacturing (Hussain et al., 2022). Various industries, academia, and the public have taken a keen interest in this technology due to its potential applicability in industry and research. Many sectors have adopted 3D printing, including automobiles, prototype machine design, biomechanics, pharmaceuticals, artwork, fashion technology, and architecture. This technique has recently been used in aerospace, tissue engineering, and food processing (Zheng et al., 2021).

The introduction of 3D printing for food manufacturing processes has demonstrated great promise for producing structures with structural complexity enabling mass customization with environmental sustainability (Kim et al., 2017). The technologically advanced contemporary society is expected to change food production by expanding food innovation with automated proficiency, which drives the need for 3D printing in the food business (Zheng et al., 2019). The food research scientific community has investigated 3D printing over the last ten years from a variety of angles, including the printing characteristics of food formulas, the effects of

P. P. Akhila · B. Aaliya · M. Navaf · K. V. Sunooj (✉)
Department of Food Science and Technology, Pondicherry University, Puducherry, India

K. P. Sudheer
Department of Agricultural Engineering, Kerala Agricultural University, Thrissur, India

S. A. Mir
Department of Food Science and Technology, Government College for Women, Jammu, India

M. A. Shah et al. (eds.), *Cereal-Based Food Products*,
https://doi.org/10.1007/978-3-031-40308-8_14

Table 14.1 Principles, advantages, and food ink suitable in different 3D printing technology for the printing of food

Printing technology	Principle	Food ink materials	Advantages
Extrusion	Extrusion and deposition	Dough, meat puree, chocolate	More material choices
Inkjet	Drop on demand /continuous deposition of food ink	Sugar icing, chocolate, meat paste, liquid dough, jams, cheese	Complex 3D food fabrication, varying textures
Binder jet	Binders are deposited on the powdered ink and bind each other	Chocolate, sugar, flour, starch	Complex 3D food fabrication, varying textures
Selective sintering	Heat source (laser or IR) that melt and fuse the specific region of the powdered formulation.	Sugar	More material choices, fast fabrication, better printing quality,

printing variables, the improvement of nutritional composition, and the impact of 3D products' texture, color, and flavor (Derossi et al., 2021; Shi et al., 2021).

In recent years, research and development on 3D food printing have been carried out using different printing technologies like extrusion, inkjet, binder jet, and selective sintering with a variety of food formulation ingredients, including chocolate, cheese, ground meat, and dough (Table 14.1) (Kim et al., 2017). 3D printing is easily manageable with cereal-based dough due to its homogeneous viscosity and can be applied to make pizza, cookies, and pasta (Çakmak & Gümüş, 2020). Extrusion is the most popular technique used in 3D food printing, which involves supplying food ink through a nozzle and depositing them in a predetermined pattern onto a target board. It is imperative to comprehend the rheology and printability of food materials before fabricating complex structures using 3D printing, and maintaining the shape of 3D-printed products is challenging (Liu et al., 2019; Vancauwenberghe et al., 2018). The current chapter focuses on the history, recent research, developments, and challenges in 3D food printing, cereal-based products and material characteristics, and opportunities of 4D/5D/6D printing in food sectors.

2 History of 3D Printing Technology

The history of 3D printing started in the mid-1980s with the introduction of additive technologies that work by building objects from the bottom up, bit by bit, layer by layer, rather than from top to bottom. Presently, 3D printing is employed in almost all disciplines and sectors for a wide range of purposes, from the creation of basic models, food, toys, high-end clothing, and accessories to sophisticated aerospace and orbital transport components (Jakus, 2019). Charles Hull worked on UV-based photo printing for fabricating plastic devices from the University of Colorado and obtained a stereo lithography (SLA) patent in 1986 (Gross et al., 2014). He developed the .stl file format, which contributes the easy communication from software

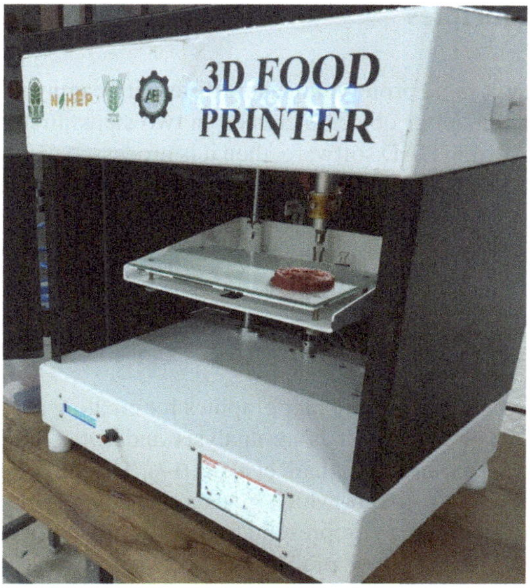

Fig. 14.1 Image of a 3D food printer

to printing devices (Gross et al., 2014). The first 3D printer, the SLA-250, was commercialized by Hull (Jakus, 2019). Meanwhile, Carl Deckard, an undergraduate at the University of Texas, invented the selective laser sintering (SLS) process. Through the use of a laser beam, SLS selectively solidified powder. The first SLS printer was manufactured by Desktop Manufacturing Corporation (DTM Corp) in 1992 and was introduced into the market as Sinter station 2000 by Deckard in 1993 (Su & Al'Aref, 2018). Fused deposition modeling (FDM) was invented by S. Scott and Lisa Crump in 1989. FDM is the extrusion-based printing technology in which melted plastic or metal wire were extruded through the nozzle under computer-guided deposition (Su & Al'Aref, 2018). By the second half of the twentieth century, computer software's CAD and CAM (computer-aided manufacturing) had advanced, revolutionizing the design, fabrication, and prototyping of materials (Savini & Savini, 2022). 3D printers were limited and costly before 2000 and a low-cost and nonproprietary printer was initiated around the year 2005. A variety of additive technologies have been proposed over the past decade, ranging from photo polymerization to fusion deposition to recent advances in bioprinting. Meanwhile, several open-source computer programs have been developed to aid the 3D printing process. These days, nearly anyone can afford the price of a desktop 3D printer (Savini & Savini, 2022). Figure 14.1 represents the image of a 3D food printer. From the earlier stage till the year 2012, much of the attention was on the advancements in software and hardware of the 3D printer. At present, researchers are concentrating on the invention of the novel, advanced, and functional3D printable materials in the fields of industry, government, medicine, and academia (Jakus, 2019).

3 General Principles

3D printing facilitates rapid prototyping using different imaging techniques (Hangge et al., 2018). A 3D model or object is created in two steps: first, it is reconstructed into a virtual object using CAD software, then it is processed and printed (Fig. 14.2). Following these stages, 3D-printed food products either undergo post-processing or are ready for consumption.

Modeling

3D modeling is a technique in computer graphics for creating a 3D digital representation of any object or surface. It is an analysis and collection of data about an object's shape and appearance (Gokhare et al., 2017). The 3D printing process relies heavily on programming software which gives a command to the hardware for the printing of objects. A schematic representation of themodeling and 3D printing of food is given in Fig. 14.3. The use of CAD software systems has been prevalent for many years in the engineering and architecture domains, but this sophisticated software usually requires substantial expertise to use (Hurst, 2016). The process of modeling 3D objects using CAD software results in a representation of the three-dimensional shape on a two-dimensional computer screen that can be saved as .stl-files (Standard Tessellation Language). Currently, there are numerous choices for

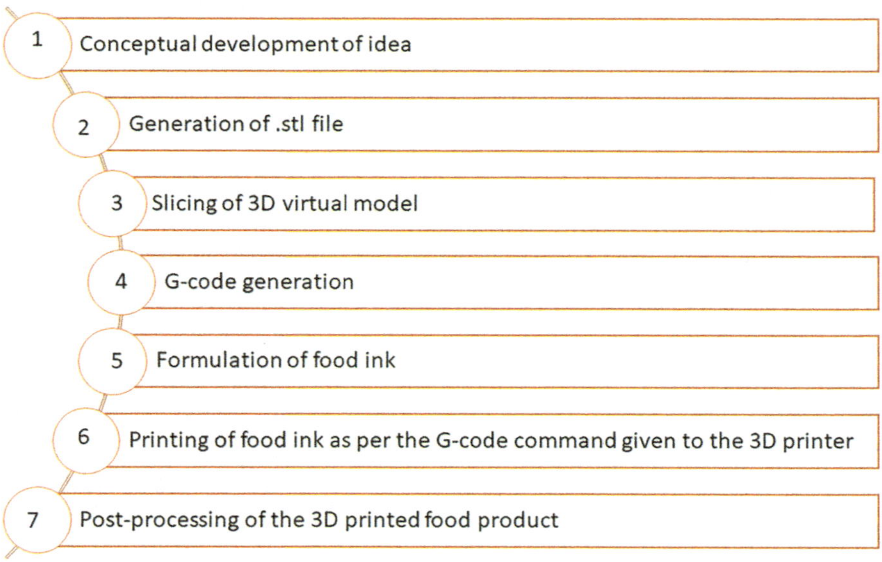

Fig. 14.2 Steps in the 3D food printing process

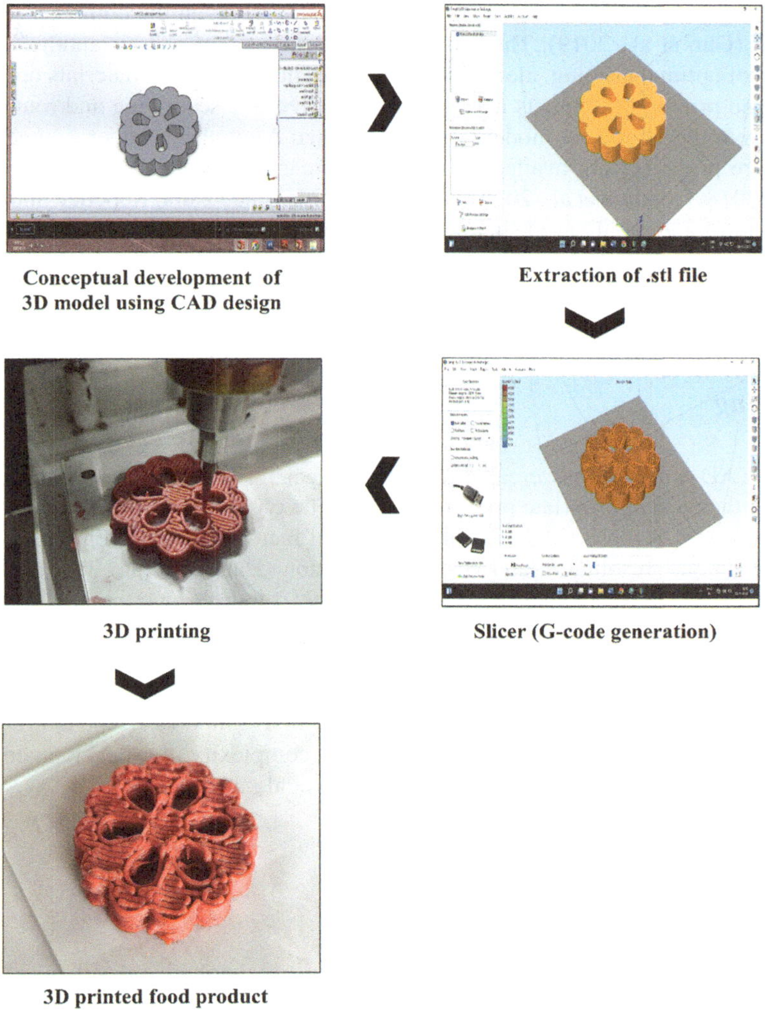

Fig. 14.3 Schematic representation of steps in modeling and 3D printing of food

3D modeling software, many of which are open-source and offers training (Hurst, 2016). The most popular design (CAD) programs, are AutoCAD, AutoDesk, SolidWorks, SketchUp, or Creo Parametric (Gross et al., 2014). Moreover, computerized tomography (CT) and magnetic resonance imaging (MRI) can be used to make a virtual 3D object (Hurst, 2016).

After creating the 3D virtual image by the CAD program, it is saved as a .stl file. The next step in the modeling is slicing of the 3D image. Slicing software converts 3D models (.stlfiles) into a series of thin layers and produces a G-code file. G-code is nothing but a type of printer command that the printer can recognize and print accordingly. G-code act as an interactive language between the computer and the printer (Guo et al., 2019). The use of the right slicing software is another crucial aspect of optimal printing, along with an optimal model design, materials optimization, and printer type. It acts as a middleman driver for sectioning and route planning from the 3D object model to the 3D printer. A number of open-source slicer software programs are available, including Slic3r, Cura, and KISSlicer (Hangge et al., 2018; Gokhare et al., 2017). Studies have identified Cura, Repetier, and Slic3r as the most commonly used slicing software for food 3D printers. It was found that the software was widely applicable to computer systems attached to FDM 3D printers (Guo et al., 2019).

Printing

Once CAD software has created a virtual object for 3D printing, the user must choose the most appropriate printing method. The type of 3D printing method suitable for a material or model rely on a number of factors, such as properties of food ink, technique, resolution, and cost of production (Hangge et al., 2018). The 3D printer prints continuous layers of powder, liquid, or sheet material as directed by G-code in order to build the model. In this process, layers are formed according to digital cross-sections of the CAD model. These printed layers joined or adhere one over the other forming the final product. A primary advantage of this technique is the ability to generate different shapes or geometric models. Depending on the approach employed, the model's size, and its complexity, the development of a model could take many hours to days (Gokhare et al., 2017).

4 Types of 3D Printing

In the last few decades, there have been various technologies developed for 3D printing depending on the functions. The International organization for standardization recently categorized printing technology into seven, including directed energy deposition, sheet lamination, material jetting, binder jetting, powder bed fusion, vat polymerization, and material extrusion. A number of industries use 3D printing technologies, including pharmaceuticals, medical, aerospace, engineering, and automobiles (Caporizzi et al., 2019).

The first application of 3D printing in food science was for food customization. It combines digital prototyping with 3D printing technology, allowing people to tailor food based on their specific needs in terms of shape, structure, and flavor, as opposed to conventional mass-produced foods (Chen et al., 2022b). The food industry uses 3D printing techniques in a variety of ways, but not all foods are compatible with each technique. Therefore, it is imperative to choose the technology appropriately based on the property of food formulation and the end product. Selective sintering printing, extrusion-based printing, inkjet printing, and adhesive printing are the four types of 3D food printing employed in the food industry (Fig. 14.4) (He et al., 2020; Chen et al., 2022b). Because each is aimed at a specific application, there is no debate about which is better or more efficient (Shahrubudin et al., 2019).

Extrusion Based Printing

The extrusion-based 3D printing gained much attention in the food industry due to its easy handling and usage. Extrusion processes have been applied to3D food printing from the FDM technique (Godoi et al., 2016). An automated 3D model directs the cylinder at predetermined locations for sequential layer deposition (Le-Bail et al., 2020). The structure maintenance of the printed product is significantly influenced by food ink (formulated food ingredients). In extrusion-based printing, the food ink is loaded in a cylindrical portion and is extruded through a printing head, where the nozzle with the cylinder moves towards the x-y plane along the surface where the material layer is extruded through a nozzle (Liu et al., 2017). According to Lipton (2017), the ink must have the ability to be extruded and must have the ability to support itself when it is printed on a surface. Cereals-based formulations greatly prefer extrusion-based 3D printing due to its pseudoplastic and thixotropic nature (Fahmy et al., 2020).

The rheology of the food ink influences the extrusion ability of the material. The ability of a material to flow is indicated by the shear modulus parameters (G*), which include the shear storage modulus (G′) and shear loss modulus (G″). IfG″

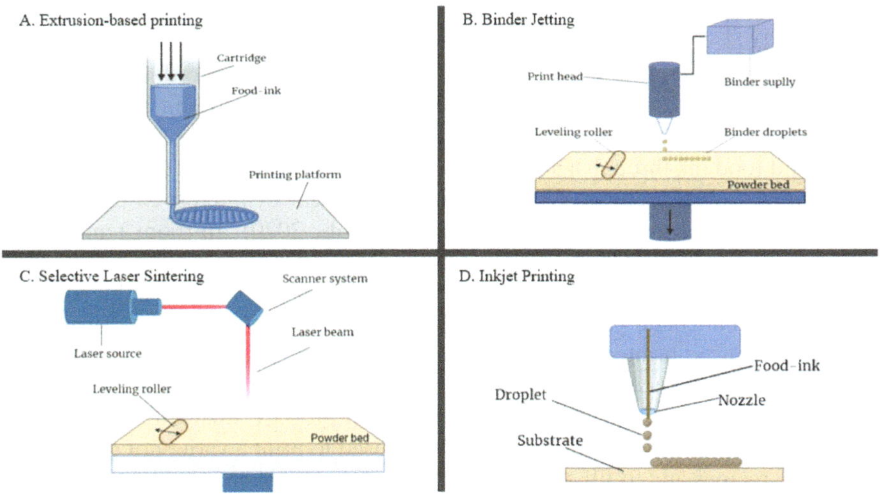

Fig. 14.4 Schematic illustration of different 3D printing technology for the printing of food (Varvara et al., 2021)

exceeds G′, the material can flow and be extruded (Le-Bail et al., 2020). The printing accuracy is influenced by the variables such as the properties of the material (mechanical, rheological, and thermal characteristics) to be printed, the process parameters (movement rate of the nozzle and printer, nozzle diameter, printing height), and the post-processing steps (drying, baking, and cooking) (Baiano, 2022). In addition to its versatility and simplicity of equipment, this technology offers the advantage of printing on many types of materials. A major disadvantage is its incapability to produce complex structures.

Extrusion of food ink is carried out in 3D food printing by three different methods: screw-, syringe-, and air pressure-based extrusion (Liu et al., 2017). The screw-based extrusion involves mixing food ingredients separately and fed into the cartridge. Screw tubes give the movement of food ink to the nozzle tip. Howbeit, a highly viscous food ink formulation cannot print by the screw-based extrusion. Because the printed samples lack the requisite mechanical strength to hold the successive layers and result in compression deformation of the printed product (Liu et al., 2017). Syringe-based extrusion involves putting printable materials into a syringe tube with a plunger inside. Stepper motors are attached to the plunger and help the plunger to move linearly and the food ink moves towards the nozzle tips by the impulse of the stepper motor. The syringe-based extrusion unit is capable of printing food materials with high viscosities and high mechanical strengths, which could be used to fabricate complex 3D objects (Liu et al., 2017). Furthermore, semi-solid materials, such as chocolate or dough, can be printed using this syringe-based extrusion printer (Hussain et al., 2022). While continuous feeding of food materials is not possible with air pressure-based and syringe-based extrusion (Liu et al., 2017). Air pressure-based extrusion process uses a compressor to produce air pressure and food ink is fed into the cartridge tube. Compressed air pushes printable materials forward through the nozzle tip of the printer to the printer bed (Hussain et al., 2022).

Extrusion-based printing can also be further categorized according to temperature, including hot-melt, room temperature, and hydrogel-forming extrusion (Liu et al., 2017). The technique of hot-melt extrusion involves extruding molten or semi-solid food ink through a nozzle to the printer bed, which directly solidifies and bonds with the preceding layer. There is a great deal of interest in creating 3D chocolate products using this technique (He et al., 2020). In-room temperature extrusion, natively printable materials are extruded smoothly at room temperature without the need for temperature control or phase transitions. Dough blends with starchy gels, chocolate paste, hummus, jelly, cheese, peanut butter, and fruits and vegetables can be printed by this extrusion technology (Le-Bail et al., 2020; He et al., 2020). The process of room temperature extrusion is being used to fabricate complex confectioneries, which were difficult to produce manually, with a high degree of repeatability. Also, extrusion can be used to print a variety of pasta based on classic recipes and to fill cookies, pizzas, and graphical decorations with a variety of fillers at room temperature (Le-Bail et al., 2020). In contrast to the previously mentioned extrusion techniques, hydrogel-forming extrusion involves printing hydrocolloid dispersions into a gel setting bath or polymer using syringe pipettes, jet cutters, and vibrating nozzles. This extrusion technology is limited to the food formulation which has viscoelastic characteristics during extrusion and should form a self-supporting gel after printing. Mashed potatoes, cheese, vegetables, and fruits can be used as food ink in hydrogel extrusion printing (Wilms et al., 2021).

Selective Sintering

Selective sintering is a technique that facilitates the formation of 3D objects in a short interval of time by using powdered materials (Mantihal et al., 2020). Selective sintering uses a laser or hot air source to generate energy that allows selective fusion of areas of a powder bed (Baiano, 2022). Once the first layer has finished sintering, another layer of powder is distributed by a roller on top of the previous one. The material is constructed layer by layer of sintering over the powder bed and once the printing is over it is separated from the powder bed (Wang & Yang, 2021). This technology is flexible enough to make multi-layered material food products by the layering of different food material components (Baiano, 2022). Essentially, selective binding uses powder binders like melted fats, sugars, and starch. After spreading a layer of powdered food formulation, a sintering agent, such as lasers or hot air, moves between the X and Y axes to fuse the powder particles (Navaf et al., 2022). Printing precision of selective sintering is determined by a variety of factors, including glass transition temperature, particle size, flowability, melting temperature, processing factors (spot diameter, thickness, and types of laser energy), and post-processing treatments (baking, cooking, boiling, or removal of excess parts) (Baiano, 2022). A homemade 3D printer (CandyFab 4000) based on a special version of selective sintering was used for the first time to create 3D-printed sugar. The technology behind CandyFab was hot air sintering, which melts particles together using a focused heat source to create 'caramelized' sculptures (Liu et al., 2017). A major benefit of this technique is that complex food items can be quickly constructed. Despite, the fabrication operation is complex and is limited to fat and sugar-based materials with a relatively low melting point (Le-Bail et al., 2020; Sun et al., 2015b).

Inkjet Printing

An inkjet printer's fundamental principle involves accumulating droplets of material layer by layer under controlled circumstances (Nachal et al., 2019). Inkjet printers mostly function using piezoelectric or thermal heads. An acoustic wave generated by the piezoelectric crystal inside the printer head separates the liquid food ink into a continuous droplet at even intervals in piezoelectric inkjet printers. Printer heads in thermal inkjet printers are heated electrically to create pulses of pressure which propel droplets through the nozzles (Godoi et al., 2016). A commercialized inkjet food printing technology called FoodJet printing was developed by Grood et al. (2013). It's working based on the drop-on-demand ejection of ink from the printer head with the help of pressure (Godoi et al., 2016). Droplets ejected from the printer are carried by gravity, impact the substrate, and are dried by solvent evaporation (Sun et al., 2015a). Although drop-on-demand printing takes longer than continuous printing, it produces more precise and high-resolution prints. Compared to continuous printing, drop-on-demand printing is time-consuming and produces more precise and high-resolution prints. The factors that affect printing precision, include the temperature, rheological properties of the ink, its surface properties, the printing height and rate, and the diameter of the nozzle. Materials with low viscosities, such as liquid dough, chocolate, meat paste, icing, and jams can be printed using this method (Baiano, 2022). The FoodJet printer deposit material drops on cupcakes, pizza bases, and biscuits, using pneumatic membrane nozzles (Sun et al., 2015a). Ink jet printing technology is more commonly used to draw print on flat products or surface filling than to construct complex food products (Sun et al., 2015b). Confectionery and decorations have benefited mainly from this technique (Le-Bail et al., 2020).

Bioprinting is a fascinating application of 3D printing. It is an advanced form of inkjet printing and is patented by Forgacs. This technique was initially used to build tissue scaffolds but is now being used to print meat analogs in food sectors. Technology such as inkjet, micro-extrusion, and laser-assisted printing is utilized for the deposition of biological materials and the culture of living cells. In bioprinting, cells self-assemble to produce tissue. An inkjet nozzle deposits droplets of a multicellular aggregate bio-ink onto an agarose rod (supporting agent). After printing the maturation take place in a bioreactor (Baiano, 2022).

Binder Jetting

The process of binder jetting involves selectively depositing a liquid binding agent onto powder food ink to create layers and form a final 3D object (Shahrubudin et al., 2019). A counter-rotating roller is typically operated to spread powder onto each layer. The unfused powder supports the fused components during manufacturing and is eventually removed and recycled after the completion of printing (Baiano, 2022). Printing rates, layer thickness, nozzle diameters of the printer, particle size, wettability, and binder viscosity of the food ink, and postprocessing procedures, will affect printing accuracy. The advantages of binder jetting include the capacity to print on a range of materials, high printing speeds, low material cost and can produce complex structures (Baiano, 2022). The printed objects may have rough surfaces and sometimes need dehydration or further processing to improve the final structure and strength (Sun et al., 2015b). Printing with binder jetting is possible on powdered materials, like starch, chocolate, protein, and sugar (Baiano, 2022).

5 3D Printing Technologies for Cereal-Based Formulations

The majority of nations around the world consume cereal-based foods as part of their daily diet. 3D printing is easier with dough due to its homogeneous viscosity and can be applied to prepare pizza, cookies, and pasta. Studies related to 3D food printing in cereal-based formulations and their major findings are given in Table 14.2. Material extrusion is the most common 3D printing technology employed in cereal-based products. In the extrusion process, a force from a hydraulic piston, an auger screw, or pressure from a compressor is utilized to propel a molten or semi-solid food filament through the movable nozzle. Printing is progressed by the movement of nozzle tips in the X and Y plane coordinated and the printer bed along the Z axis. This controlled movement allows building of objects as same as the virtual model designed by CAD (Caporizzi et al., 2019). The rheological characteristics of the formulation, hydrogel formation during extrusion, and solidification upon cooling support the binding between the layers and thereby construction of 3D products. Extrusion-based 3D food printing has several advantages, such as digitization of the food manufacturing process, low maintenance costs, personalization of dimensions and geometry of the product, improved process efficiency, and the ability to use novel and alternative ingredients. However, it is time-consuming and only a few materials can be printed using this technology. The traditional food extrusion process is fully digitalized by 3D printing, allowing real-time manipulation of food manufacturing. Most of the cereal-based 3D printed products require post-processing before consumption. In order to provide the desired texture of the final product, formulation, and processing, as well as post-processing suitability, must be considered equally (Dick et al., 2021). Yang et al. (2018) explored the effect of raw material composition on the physicochemical and functional characteristics of 3D-printed dough. Rheological characteristics were used to calculate the quality of the products. It was observed that the extrusion ability of dough depends on the viscous modulus and flow behavior, meanwhile, the structural support of the 3D material and maintenance of the shape of the printed product is determined by the gel strength, elastic modulus, adhesiveness, and rupture strength of the dough. According to the rheological properties, the cereal-based dough can be effortlessly extruded due to its pseudoplastic nature. The linear viscoelastic study showed a significantly higher value of G'' compared to G', this suggests that an elastic gel or gel-like network could be formed (Jiang et al., 2019). Martínez et al. (2013) studied the impact of adding extruded wheat flour on dough that has been tested using a rapid viscosity analyzer. Extruded flours were shown to reduce dough extensibility while increasing dough plasticity. These attributes make it possible to create 3D-printed products with precise geometric shapes (Martínez et al., 2013). Addition of fat influences the extrusion, flow, and deposition of dough in 3D printing (Zhang et al., 2022). A study by Yang et al. (2018) showed good printing advantages and demonstrates improved extrudability, enhanced gel strength, and elasticity, and decreased ductility upon the addition of butter to the dough. Another study by Liu et al. (2019)showed incorporation of olive oil improved the fluidity and printability

Table 14.2 Studies on 3D printing in cereal-based food

Technique	Composition of food ink	Post-processing	Major findings	Reference
Extrusion	Wheat flour, dried mealworm powder,	Baked in a convection oven	Snacks that were printed accurately reproduced the geometric design of the virtual 3D image. Enrichment of insect in the dough resulted in improvement of the amino acid profile of snacks.	Severini et al. (2018)
Extrusion	Rice flour, jaggery, and water	Steaming at 100 °C for 15 min	85.95 g, 33.04 g, and 100 g of rice flour, jaggery, and water showed best printability food ink combination.	Thangalakshmi et al. (2021)
Extrusion	Wheat flour, powdered sugar, butter, xanthan gum, milk, and antioxidants extracted from *A. platensis*	Baked at 150 °C for 25 min	Antioxidant encapsulation improved the antioxidant and color stability of the cookie	Vieira et al. (2020)
Extrusion	Low-gluten flour, icing sugar, eggs, butter, and water	Low temperature refrigeration (−65 °C, for 0, 5, 10 min), baked in an oven at 190 °C for 10 min	The process parameter such as filament diameter (2.30 mm), nozzle movement speed (25 mm/s), nozzle diameter (2.0 mm) and nozzle height (2.40 mm) showed best printing stability and resolution. Cooling at −65 °C for a period of 10 min improved the baking quality.	Yang et al. (2019)
Extrusion	Whole milk powder	Baking at 150 °C, 5 min	Milk powder gave volume and glossiness to the baked samples, whereas rye flour seemed to increase the baking stability and rigidity.	Lille et al. (2020)

(continued)

Table 14.2 (continued)

Technique	Composition of food ink	Post-processing	Major findings	Reference
Extrusion	Sodium alginate, waxy rice, indica rice, and japonica rice	Steaming	The best printability and highest precision of 3D printed products are in the order of japonica rice > waxy rice > indica rice	Liu et al. (2020a, b)
Extrusion	Wheat, calcium caseinate strain of *Lactobacillus plantarum*	145 °C for 16 min, and at 175 or 205 °C for 8 min	The probiotics were viable in the baked product (duration of 6 min at 145 °C)	L. Zhang et al. (2018)
Extrusion	Whole meal, wheat starch wheat flour	No post-processing	Printability of the material is in the order of whole meal > Wheat starch > Wheat flour	Zheng et al. (2021)

of 3D-printed dough while also enhancing the sensory qualities of the printed objects. Furthermore, the amount of fat in the mixture influences particle deposition, water-holding capacity, and the integrity and maintenance of the 3D-printed geometry.

Powder-based printing (PBP) technologies, such as SLS (selective laser sintering) offer potential applications in food production. The working principle of SLS is based on fusing powder particles at predetermined locations through the sintering of laser or infrared beam or by using liquid binders to build complex 3D printed designs. The formulating ingredients must satisfy the unique requirements of these technologies. The advantage of using powder in 3D printing is that it serves as a foundation for following layers, allowing for the creation of 3D structures with highly porous microstructures and the leftover powder can be cleaned up later. Initially, 2D honeycomb cell constructions with different cell sizes were made in order to provide different fracture behaviors. An extremely porous and delicate texture was produced during the SLS printing technology. The samples could alternatively undergo postprocessing via a steaming or baking stage and can provide better textural characteristics (Noort et al., 2017). Aregawi et al. (2015) prepared cookies with various flour types, such as soft wheat flour, semolina, and a combination of flour and starch, using the SLS technology. Additionally, they employed various binder and structure ratios when the traditional flour/starch blend was replaced by semolina and soft wheat flour for the preparation of cookies, the prepared cookies have thicker walls. The structure:binder ratio has no impact on wall thickness, but a binder composition of 90:10 resulted in noticeably thicker walls.

The capacity to produce cereal-based items using various 3D methods may be of considerable interest in the future. Flowabilities and non-Newtonian fluid properties of the composite cereal dough made extrusion-based 3D printing easier in various products (Kewuyemi et al., 2021). Thangalakshmi et al. (2021) investigated various combinations of food ink formulation (rice flour, jaggery, and water) in extrusion-based food printing. The study reported that rice flour of 85.95 g, jaggery of 33.04 g, and water of 114.93 g concentration in the food ink maintained the best dimensional accuracy and 3D structure after the post-processing. Many researchers have also studied the 3D printing fortification of cereal-based products. Researchers successfully infused probiotics into a dough made from wheat flour, water, and calcium caseinate. A significant amount of probiotics survived and the shape of the product was maintained during baking (Çakmak & Gümüş, 2020). A study by Letras et al. (2022) reported an inventive replacement for existing commercially available gluten-free snacks by the incorporation of spirulina in 3D-printed gluten-free cereal snacks. Snacks with 5% spirulina had the highest nutritional value, with increased antioxidant activity, phenolic content, and better coloring properties.

Material Properties of Cereal-Based 3D Printed Food Products

Printability is described as 'the property of a material which can be printable and self-supporting and retains the structure' (Caporizzi et al., 2019). The impact of rheological, textural structural, and morphological attributes of cereal-based food materials on 3D printed structures is discussed in the following sections.

Rheology

The rheological characteristics of the printable material are crucial to achieve the best printing stability and support of successive layers during printing. These include G*, G′, G″, loss tangent, flow stress, shear-thinning behavior, and yield stress of the materials (Hussain et al., 2022). The overall rheology of the food ink is significantly altered by the ingredients used for the formulation, their interaction, and the effect of the binding agent (Caporizzi et al., 2019). The flour with low viscosity and low starch content showed poor printability and it can be improved by the incorporation of additives into the ink formulation (Wang et al., 2022). Lipid is a good additive in cereal-based food ink formulation for improving the printability of the material.

The viscous modulus and gel strength were improved with the addition of butter to the dough. Yang et al. (2018) discovered that dough with 6% butter had an erratic shape upon 3D printing. Here butter limits the gluten formation of the dough and increases the elastic modulus and adhesiveness of the material thereby better formation of the gel network and helping maintenance of shape. Increments in the shear rate and angular frequency were observed with the addition of butter, sucrose, and flour in the food ink. Pavičić et al. (2021) investigated the3D printing accuracy and repeatability of dough formulated with olive oil as a substitution with butter in two different flour (whole grain flour and carob flour). The study observed that dough formulated with olive oil showed a superior consistency coefficient and viscosity than those with butter. The study concluded that the lowest n values and maximum consistency coefficient were found in the dough made from oat, rye, and carob combined with olive oil or butter. In addition, to meet the mouth-feel and dietary demands of consumers, fats are essential for the production of quality bakeries by providing the correct rheological qualities and improving printability (Pavičić et al., 2021).

Water plays another important role in the cereal-based formulation which helps in reducing the loss tangent and increase the printability of the dough (Hussain et al., 2022). Liu et al. (2020b) examined the effect of material formulation on the quality of 3D-printed rice-based foods. The findings demonstrated that the formulation of rice doughs affected its rheological characteristics. It was found that waxy rice dough had a greater tan δ than japonica and indica rice and that tan δ increased as the rice flour/water ratio increased. The best shape, the highest level of accuracy, and most compact microstructure were displayed by the 3D-printed waxy rice product, which had a flour-to-water ratio of 100:90 (Liu et al., 2020b). An interesting review by Godoi et al. (2016) explained the influence of food ingredients on the rheological characteristics and the printability of the ink. A study by Correa et al. (2012) showed that dough hardness and strength remarkably increased with the addition of salt. Salt in the formulation cross-linked with the protein and forming a more stable network structure and thereby improving the gel hardness and self-stability of the 3D structure. A study by Wang et al. (2016) compared the impact of different salt such as NaF, NaCl, and K_2SO_4, on the strength of the gel formed by mashed potatoes. It was found that the addition of salt accelerated the gel structure formation.

Hydrocolloids are typically added for improving the printability of materials such as meat, rice, vegetables, and fruits. For example, the gum isolated from flaxseed, basil seed, or xanthan gum, and guar gum has the capacity to improve the viscosity of the material and the G′ is much greater than the loss modulus, which can be exploited in 3D printing (Jiang et al., 2019). Additionally, the printable material should exhibit pseudoplastic and non-Newtonian shear thinning behavior to obtain self-stable printed products. The effects of using hydrocolloids in the post-processing of 3D-printed cookie dough were examined by Kim et al. (2019). The results show that cookie dough that contains 0.5% xanthan gum prints smoothly and maintains 3D shapes after processing. The 3D printable dough for cookie preparation was formulated from wheat dough blended with xanthan gum and methylcellulose at different proportions. In this study, the dough material containing less methylcellulose demonstrated lower G′ and G″ than the dough containing xanthan gum. Formulation of dough with xanthan gum increased the yield stress of the material and the viscosity of the printed product (Kim et al., 2019). Furthermore, Liu et al. (2020a) investigated the viability of printing pastes made from different varieties of rice and the impact of sodium alginate on their structural and physicochemical properties. Sodium alginate-containing rice pastes showed shear-thinning characteristics, indicating their applicability for 3D printing. Japanica rice variety with sodium alginate (0.50%) had the best printability and accuracy among others.

Textural Properties of Cereal-Based Printable Materials

A revolutionary structure with a customized nutritional profile, flavor, and texture can be made via 3D printing (Pereira et al., 2021). Food texture involves all aspects of food that can be perceived by the human senses, especially by the mouth and hands. It is often possible to assess a sample's texture either using texture profile analysis or through sensory evaluation. The advantage of instrumental tools is that texture can be quantified and standardized to ensure food quality control during the production process. The most commonly studied textural properties are hardness, springiness, chewiness, gumminess, cohesiveness, and adhesiveness (Pereira et al., 2021). A material's textural properties are strongly influenced by the constituents of the printable ink, additives, instrumental parameters, and infill percentage (Liu et al., 2018). A study demonstrated that cookies with varying filling densities (30% or 50% infill) and varied design shapes (cubic or circular-based prisms) can maintain structural stability after 3D printing (Pulatsu et al., 2022). The print ink is extremely sensitive to processing variables like extrusion rate, heating, melting, and steaming.

The formulation of wheat-based cookie dough, including the concentration of butter, sugar, and yolk has a substantial impact on extrusion, deposition, and shape stability after baking. In fact, the amount of leavening agent used as well as mixing and kneading conditions can alter the microporosity (1–100 μm) of the dough (Dick et al., 2021). The effect of dough characteristic on the addition of two different lipids (sunflower oil and bakery fat) were compared by Jacob and Leelavathi (2007). The study observed that sunflower oil-added dough showed a relatively harder texture due to lower entrapment of air and bakery fat-formulated dough exhibited a soft texture. On contrary, the dough mixed with bakery fat provided a printable product with a soft texture. Further, the viscosity of the material and adhesive force is reduced when fats like butter, olive oil, or egg white are added to the material. Adding egg yolk to cereal-based formulation enhances the hardness of materials by forming multifaceted networks between egg yolks and starches (Hussain et al., 2022).

Derossi et al. (2021) explored the possibility of regulating the texture of a cereal-based snack employing 3D printing technology. It is necessary to add flow enhancers to the cereal-based food ink formulations to improve their printing abilities and textural properties. Many investigations use hydrocolloids to provide and improve viscoelastic properties for the printing of dough. In printing materials, xanthan gum and methylcellulose increase hardness because hydrogen bonds are formed between molecules as a result of the addition of these ingredients (Hussain et al., 2022). There should be enough care in using the type and concentration of hydrocolloids in the formulations as it can make the food ink harder, inhibit extrusion, and cause poor printing accuracy. Cohen et al. (2009) studied the changes in the textural characteristics of food formulation and the mouthfeel of the printed product with the addition of varying concentrations of hydrocolloid composites.

Hardness measures the capacity of a food product to maintain its shape (Pereira et al., 2021). Nozzle diameter and printing speed play an important role in the hardness and fracturability of printed material. The hardness of the printed products increases with an increase in the nozzle diameter. The internal factors of printed food, such as pattern, void fraction, and level of infill, have a significant impact on its textural and structural characteristics (Pereira et al., 2021). Infill levels and perimeter strongly influence the hardness, firmness, gumminess, and Young's modulus of the materials and showed a positive correlation. Whereas the decreased value of gumminess and harness was observed with an increase in the void fraction of a material (Liu et al., 2018). A study by Huang et al. (2019) investigated the effects of the printing parameters (infill levels, nozzle diameter, and perimeter) on the textural characteristics and printing precision of brown rice-formulated 3D printed products. Printing precision was improved by reducing the diameter of the nozzle and widening the perimeter of the print bed. The study noticed a significant shift from the soft texture by increasing the nozzle diameter in brown rice-based printed products (Huang et al., 2019). Severini et al. (2018) 3D printed wheat flour-based snacks with different edible insect substitution levels and compared their hardness based on insect level. Insect enrichment resulted in a gradual hardening of the products and 20% insect-enriched product showed a hardness value of 52 N. There was no linear relationship between baking temperature elevation and the hardness of the material. The lowest hardness was observed at intermediate baking temperatures, and the highest hardness had been observed at the highest baking temperatures.

There are some pre-and/or post-treatments recommended for printed food products in order to maintain their structural stability. Usually, pre-treatment involves kneading at the time of preparations of powdered materials, doughs, or gels, whereas post-treatments typically involve advanced cooking and rapid chilling technologies carried out after the printing of products. Research is currently underway to improve the rheological and textural properties of materials by applying innovative pre- and post-treatment technologies. In one experiment, the stability and printing precision of 3D printed products were enhanced by preheating a wheat starch-papaya paste with a combination of ultrasonic and microwave energy. The authors found that non-uniform microwave heating promoted gelatinization, which increased strength, viscoelasticity, and compactness, giving internal structural integrity to 3D printed products, while ultrasound improved starch granule swelling capacity, making them improved water absorption capacity (Xu et al., 2020).

Structural and Morphological Properties of Cereal-Based Printable Food Materials

Morphological characteristics of 3D printed products rely on the food ink and processing parameters. A sample's internal structure and 3D modeling may be affected by the distribution of different ingredients in the food ink. It is possible to observe changes in the morphological characteristics as well as the distribution of various components in the dough by scanning electron microscopy, laser scanning confocal microscopy, and low-frequency nuclear magnetic resonance.

The study Fahmy et al. (2020) quantitatively analyzed 3D-printed cereal morphologies using onboard imaging and analysis. This method of analysis helps to determine the infill pattern and output internal structure such as density, pore structure, and distribution (Fahmy et al., 2020). A study by Derossi et al. (2020) examined how 3D printing affects the microstructure and mechanical characteristics of snacks made from cereals. According to the study, fewer pores could be seen in 3D-printed structures than in conventional products. The reduction of pores and subsequent deposition of a more dense food formula filament were caused by the natural compression of the food ink in both the piston chamber and the screw-extrusion system. The pores in 3D products are due to the mismatch between layer height and extrusion rate. These variations in sample microstructures had a major impact on the samples' mechanical characteristics. The 3D-printed products exhibited noticeably higher levels of hardness, chewiness, and cohesiveness than conventional products. Water plays an important role in maintaining the structure of 3D printed objects, in which the bound water binds strongly with the macromolecule, and the free water is responsible for the liquefaction of the product. Additionally, protein has a high capacity for absorbing water and a high intention to combine with water to form a complex gluten network; as a result, the interior structure of the 3D model becomes more robust with less void space (Kim et al., 2019; Yang et al., 2018). A cereal-based material's morphology is greatly affected by its infill percentage, mixing time, particle size, and total solid content. Further, proper mixing of the formulation decreases the particle size of the mixture, and the smaller particles produce a dense product (Severini et al., 2016). 3D structure and internal filling texture were negatively affected by increasing the flour concentration (Liu et al., 2019). According to the study, the cereal contains small particles of micronutrients that are dispersed in a macromolecular network of gluten made suitable for best 3D printed structures.

6 Development of Cereal-Based Functional Food

An ongoing research area in the field of nutrition is the incorporation of novel components (probiotics or edible insects) to enhance the nutritional value of cereal-based 3D-printed products. Zhang et al. (2018) tested the robustness of probiotic-infused dough under various printing structures and baking conditions. Probiotic-infused dough formulations with varying compositions of wheat flour, water, and calcium caseinate were tested, followed by baking (post-processing) at temperatures of 145, 175, and 205 °C. The study successfully formed 3D printed honeycomb structure probiotic food with a viable probiotic bacterium above 10^6 CFU/g after baking at a specific temperature and time (145 °C for 6 min).

As part of the recent 3D printed product development efforts, Severini et al. (2018) developed wheat snacks that contain wheat flour enriched with insect powder (yellow mealworm larvae), providing greater levels of essential amino acids and protein. A 3D-printed snack with high fiber and protein content has been developed by Krishnaraj et al. (2019). Study state that regular consumption of this formulated snack complements a healthy diet daily. Hence, 3D food printing as incipient technology shows promise as a means to produce customized, nutritious food. The dough was infused with an extract of *Arthrospiraplatensis* to create antioxidant-active cookies (Vieira et al., 2020). Encapsulated antioxidant extract by alginate microbeads demonstrated better antioxidant activity and long-term color stability to the finished product in comparison to dry biomass or freeze-dried antioxidant extract added to the cookies. Varghese et al. (2020) have created high-quality nutrient-dense 3D-printed cookies by employing underutilized foods (millet and pineapple seed meal) as basic ingredients (Demei et al., 2022). Studies like these open the enormous potentiality of nutrient-enriched 3D-printed foods.

7 Post-processing of 3D-Printed Food Products

3D printing may result in some surface defects on the product, It may be necessary to use postprocessing procedures to increase printing precision and shape stability. The postprocessing processes for 3D-printed non-food products are sufficiently developed and include support removal, polishing, model repair, and surface painting. However, the post-processing of food items created with 3D printing is very different and mostly involves baking, drying, frying, boiling, cooling, freezing, cooking, and 4D printing (He et al., 2020). During the post-processing, the finished product may have an enhanced sensory aspect. Baking, frying, and steaming decreased the moisture content of the product, improving its hardness and reducing its stickiness (Hussain et al., 2022). The high-temperature cooking stimulates a number of chemical reactions, including Maillard reactions, evaporation of water, denaturation of proteins, change in volume, and nutritional content (Dankar et al., 2018). Baking is a common post-processing process that is often used in studies dealing with 3D-printed cereal-based products (Demei et al., 2022). In certain cases, the complex structure of 3D-printed food products results in non-uniform heat distribution or difficulty in maintaining the desired structure can be seen during the post-processing time (Çakmak & Gümüş, 2020). Additionally, the layer height and infill density of the printer can affect the printing accuracy and texture of cereal-based snacks (wheat) after post-processing (baking). A large object with a variety of textures can be made by increasing the void fraction and decreasing the solid content of the final product (Dick et al., 2021). Packages and storage conditions are crucial for ensuring the structure and stability of 3D-printed foods (Çakmak & Gümüş, 2020). Severini et al. (2016) analyzed the changes in size, shape, texture, and microstructure of wheat-based foods after the post-processing, and found that the diameter of the samples decreased after boiling due to the removal of water. An innovative 3D printing method was developed for the preparation of collapse-resistant biscuits in various shapes by Pulatsu et al. (2022). Krishnaraj et al. (2019) examined the post-processing acceptability of 3D-printed products in terms of their nutritional, color, texture, and sensory properties. The products were post-processed by hot-air drying, deep frying, and microwave drying. In comparison to the other two post-processing techniques, microwave drying is able to retain better amounts of nutrients while assuring minimum changes in color and textural features. The study showed that the sensory qualities of all post-processed samples were satisfactory, and the produced snack has commercial potential (Krishnaraj et al., 2019).

An experiment was conducted by Lipton et al. (2010) to determine shape retention after baking a modified cookie recipe. The inside geometry of the baked cookie was effectively retained even though the exterior slumped. Another study conducted by Lille et al. (2020) looked into the application of extrusion-based 3D printing for the manufacture of nutrient-dense snack meals derived from whole wheat rye flour and whole milk powder. The experiment involved preparing a water-based paste by adjusting the raw materials in different proportions, printing the food ink at room temperature, and baking the printed samples in an oven at 150 °C. All of the formulae were found to have excellent printability and structural stability after printing. Moreover, in the traditional baking process, case hardening or expansion deforestation problem was resolved by adding rye flour to the milk formula. The effect of post-processing by two drying techniques in a 3D printed product formulated with varying concentrations of starch, protein, and fiber was investigated by Lille et al. (2018). The study concluded that freeze drying can more effectively retain the printed structure compared to oven drying at 100 °C (Lille et al., 2018). A material dehydrated in a frozen state maintains its shape better due to its solid structure. Since oven drying decreased the viscosity of the material upon heating and made the printed samples slightly more spreadable (Dankar et al., 2018). A 3D-printed pizza with an ultra-thin crust created a new sensory experience for mouthfeel, swallowing, and chewing after baking. It initially tasted like a normal pizza, but after a few minutes, it became more like crackers (Garfield, 2016). Additionally, several researchers have suggested different post-processing techniques as an alternative to cooking. Sun et al. (2018) suggested using direct infrared radiation rather than baking. Howbeit, this has not yet been accomplished. Furthermore, altering the composition of printing materials and creating new post-processing techniques clearly work to address the negative changes that result from the post-processing of dough, and they offer important research insights (Wang et al., 2022).

8 Challenges and Prospects

People are always learning and developing the aspects that influence the growth of 3D food printing through continuous research. Globally, significant efforts have been made over the past 10 years to adopt 3D printing to the food sector, with particular attention paid to the use of new food sources, the creation of personalized food, the optimization of printing parameters, and the design of printing equipment (He et al., 2020). However, there are still several obstacles in the way of the widespread use of 3D food printing, some of which are covered here. The advantages and disadvantages of 3D printing in food are represented in Fig. 14.5. A major drawback

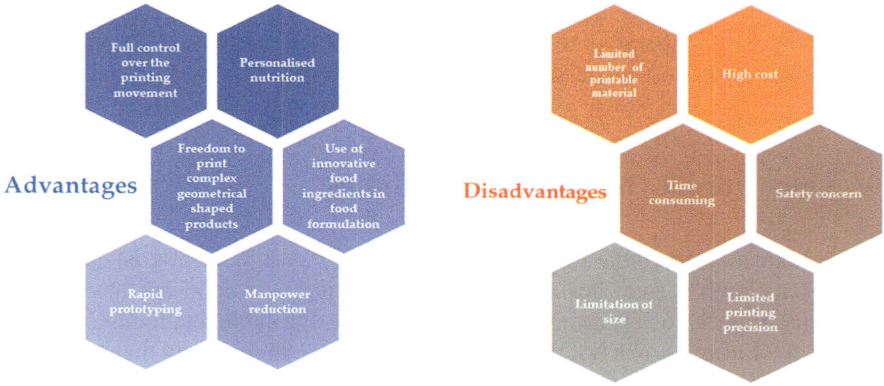

Fig. 14.5 Advantages and disadvantages of 3D printing

of 3D printing is that it involves the time-consuming task of constantly optimizing printing conditions for the materials utilized and the designs being printed (Pereira et al., 2021). Another hurdle for 3D printing technology is to tackle the geometrical stability and precision of printed products, especially for a cereal-based formulation that has post-processing as a necessary step (Wang et al., 2022). A better understanding of the relationship between the qualities of food components and their printability is necessary for formulating food ink with accurate printing (Wang et al., 2022). The majority of food items produced with 3D printing need to be post-processed, including baking, cooking, frying, cooling, and drying. This post-processing sometimes made advantages as well as disadvantages to the sensory properties of printed foods. For example, the ultra-thin crust of the pizza created by the Bee Hex 3D printer after baking made it taste more like crackers after a few minutes, providing customers with a novel flavor experience (Sun et al., 2018). However, sometimes it is not desirable, such as when nutrients are lost, unwanted flavors are formed, or the color and shape of the product change (Dankar et al., 2018). The limitations might be overcome by combining 3D food printing with other cutting-edge technologies (Wang et al., 2022). Post-processing via baking of cereal-based 3D printed products at high temperatures may result in shape instability or significant deformation of printing patterns as a result of modifications on the micro- and macromolecular levels (Lille et al., 2020). Yang et al. (2019) tackled this difficulty by introducing rapid freeze technology in between the 3D printing process. The printed cookie dough was treated at −65 °C for 10 minutes before baking at 190 °C, which helped to improve the structure and shape of the baked cookie. In the current state, 3D-printed food usually satisfies only one requirement, such as nutrition, color, flavor, or texture. 3D food printing will likely be able to expand product functionality and create customized nutrition products in the future (Demei et al., 2022). Further, there are concerns about food safety as well as consumer acceptance of 3D-printed foods, but research on these topics is scarce (Varghese et al., 2020). In conclusion, 3D printing technology has a wide range of potential applications. In order to enhance the qualities of food manufacturing and processing technology, it is imperative to continue strengthening research on 3D printing in combination with post-processing technology.

9 Opportunities for 4D/5D/6D Printing

The introduction of edible ink printing technology has significantly advanced the food sector with a great deal of innovation. Printing with food ink has evolved gradually from 2D to 4D printing as decoration of food to intelligent food. Currently, 3D/4D food printing is a hot topic and will encourage digital food manufacturing and advance the intellectualization of the food business (Cheng et al., 2022). 4D printing refers to the fusion of 3D printing technology and smart materials or stimulus-responsive materials, i.e., employing stimulus-responsive materials as the target for 3D printing with respect to time (Fig. 14.6) (He et al., 2020). Material having the ability to change shape or color by exposure to different environments is termed as smart materials. They are self-sensing, responsive, decision-making, self-adapting, shape memory, self-repairing, and multi-functionality in nature (Chen et al., 2022a). The development of 4D printing technology adds a "space-time axis" to the 3D coordinate system, extending the capabilities of 3D printing (Cheng et al.,

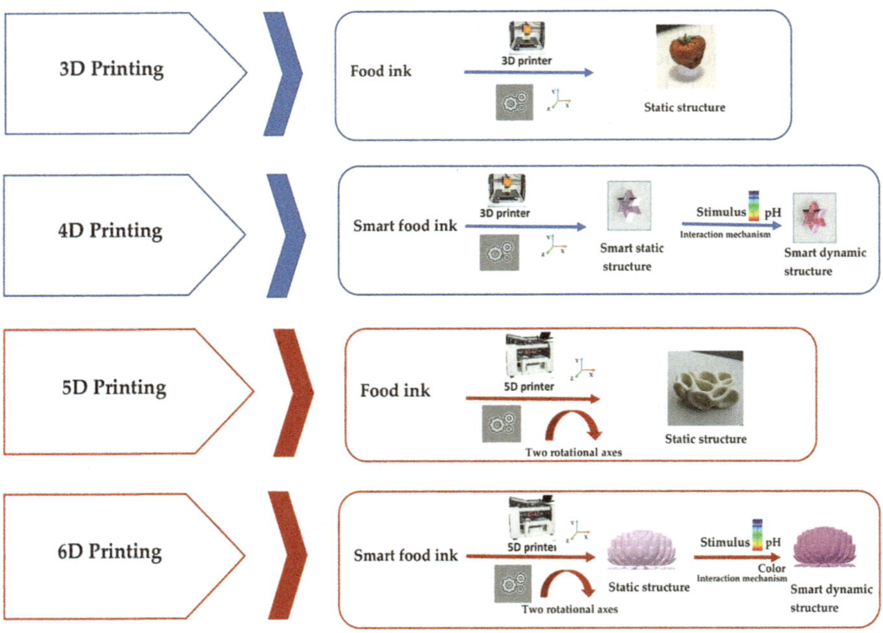

Fig. 14.6 Schematic illustration of 3D, 4D, 5D, and 6D food printing

2022). The concept of 4-D printing in food came from the report by Wen Wang et al. (2017). They created edible 2D films using ordinary food constituents (protein, starch, and cellulose) that on processing (cooking) result in twisted, curled, and self-transformed 3D structures.

There are several types of deformation behaviors inherent to the 4D food printing process, including bending, deformation, crimping of surfaces, and linear and non-linear expansion/contraction (Chen et al., 2022a). Several fields, including materials science, bioengineering, manufacturing, and medicine, have greatly benefited from the use of 4D printing (He et al., 2020). A study by Guo et al. (2021), explored the suitability of applying microwave heating as a stimulus and gelatin-gum Arabic-oil complex coacervates microcapsules as a stimulus-response material to achieve color and aroma alteration in 3D-printed buckwheat dough incorporated with peaches. The study opens up a way to produce 4D food printing with selective flavor and color changes by combining microwave heating and3Dprinting. Wen Wang et al. (2017) successfully applied 4D printing in food and printed deformable pasta. The study exploited the discrete water absorption capacities of ethylcellulose and gelatin on printed bilayer structures under water-induced stimulation. A combination of printing paper and pumpkin exhibited deformation abilities due to distinct thermal responses (Chen et al., 2021). Several factors affected the controlled shape change, including the pumpkin's water content, the design of the model, and the level of adhesion between the pumpkin and the printing paper (Chen et al., 2021).

5D/6D printing is the latest innovation in additive manufacturing, a technology that can create complex and intricately shaped products with a reduced amount of material (Cheng et al., 2022). American universities introduced the concept of 5D printing, a technology derived from 3D printing. A team led by senior principal research scientist William Yerazunis used this novel technology for the first time at Mitsubishi electric research laboratories (Ghazal et al., 2022). 5D printing enables the printing of objects through 5 axes motion, three axes (X,Y and Z) by the printer head and two rotation axes by the print bed (Fig. 14.6). The freedom of movement of the print head and print bed allows layers, planes, and curves to be printed, resulting in a more appealing product (Ghazal et al., 2022). The latest advancement in technology enables the creation of intricately shaped implants, devices, and products with better physical properties than 3D printing. The printing of food materials with various infill types and densities allows the product to retain its shape after printing, which is beneficial for fabricating food products with various textures to meet the needs of a wide range of consumers (Ghazal et al., 2022). The higher expense for the two more axes of bed movement is the primary difficulty and restriction of 5D printing technology. Additionally, highly skilled human resources are also required for the development, maintenance, and use of 5D printers. Similar to 3D and 4D printing, 5D printing has the power to create incredible advancements and improvements in the near future in a variety of fields, including engineering, dentistry, orthopedics, medicine, and food (Ghazal et al., 2022).

The combination of 4D and5D printing techniques has led to a new additive manufacturing technology known as 6D (Fig. 14.6). Georgantzinos and Giannopoulos (2021) introduced the concept for the first time. 6D printing is defined as the process of printing food ink with smart material using a 5-axis printer. Combining the flexibility of 4D printing to alter the shape, color, or other features over time in response to stimulation with the ability of 5D printing to produce complex structures with enhanced strength and less material is one of the main advantages of 6D printing. The inherent flexibility of 6D printing technology can also reduce processing times and material usage simultaneously. Until now, 6D printing has not been used in any field.

10 Conclusion

3D printing technology is a promising tool for creating novel food products from various food formulations with improved sustainability. In addition to providing freedom for geometric design, 3D printing makes it feasible to alter products' mechanical, textural, and nutritional properties in ways that are not possible with conventional subtractive manufacture. As of now, the production of 3D-printed objects on a large scale is limited by cost, time, and production volume. Adaptation of the 4D concept in food increases the possibility of printing stimulus response products for the research field. Furthermore, 5D and 6D printing appeared to be a promising technology in various fields due to their dynamic structure with less material, but further research is needed to adopt the technology into food in the future.

Acknowledgments The authors would like to express their gratitude to Dr. Shijin A, veterinary surgeon, Government of Kerala and Arjun P P, Student, National Institute of Technology, Calicut, Kerala.

References

Aregawi, W., Verboven, P., Vancauwenberghe, V., Bongaers, E., Van Den Eijnden, E., Van Bommel, K., Diaz, J., & Nicolai, B. (2015). *Structural-mechanical analysis of cookies produced by conventional and 3D printing techniques. Proceedings of the 29th EFFoSt International, Volume I* (pp. 415–419).

Baiano, A. (2022). 3D printed foods: a comprehensive review on technologies, nutritional value, safety, consumer attitude, regulatory framework, and economic and sustainability issues. *Food Reviews International, 38*(5) Taylor & Francis. https://doi.org/10.1080/87559129.2020.1762091

Çakmak, H., & Gümüş, C. E. (2020). 3D food printing with improved functional properties: A review. *International Journal of 3D Printing Technologies and Digital Industry, 4*(2), 178–192. https://doi.org/10.46519/ij3dptdi.746389

Caporizzi, R., Derossi, A., & Severini, C. (2019). Cereal-based and insect-enriched printable food: from formulation to postprocessing treatments. Status and perspectives. In *Fundamentals of 3D food printing and applications* (pp. 93–116). Academic Press. https://doi.org/10.1016/B978-0-12-814564-7.00004-3

Chen, F., Zhang, M., Liu, Z., & Bhandari, B. (2021). 4D deformation based on double-layer structure of the pumpkin/paper. *Food Structure, 27*, 100168. https://doi.org/10.1016/j.foostr.2020.100168
Chen, X., Zhang, M., Teng, X., & Mujumdar, A. S. (2022a). Recent progress in modeling 3D/4D printing of foods. *Food Engineering Reviews, 14*(1), 120–133. https://doi.org/10.1007/s12393-021-09297-6
Chen, Y., Zhang, M., Sun, Y., & Phuhongsung, P. (2022b). Improving 3D/4D printing characteristics of natural food gels by novel additives: A review. *Food Hydrocolloids, 123*, 107160. https://doi.org/10.1016/j.foodhyd.2021.107160
Cheng, Y., Fu, Y., Ma, L., Lay, P., Losic, D., Wang, H., & Zhang, Y. (2022). Food hydrocolloids rheology of edible food inks from 2D / 3D / 4D printing, and its role in future. *Food Hydrocolloids, 132*, 107855. https://doi.org/10.1016/j.foodhyd.2022.107855
Cohen, D. L., Lipton, J. I., Cutler, M., Coulter, D., Vesco, A., & Lipson, H. (2009). *Hydrocolloid printing: a novel platform for customized food production. International solid freeform fabrication symposium* (pp. 807–817). University of Texas at Austin.
Correa, M. J., Pérez, G. T., & Ferrero, C. (2012). Pectins as Breadmaking additives: Effect on dough rheology and bread quality. *Food and Bioprocess Technology, 5*(7), 2889–2898. https://doi.org/10.1007/s11947-011-0631-6
Dankar, I., Haddarah, A., Omar, F. E. L., Sepulcre, F., & Pujolà, M. (2018). 3D printing technology: The new era for food customization and elaboration. *Trends in Food Science and Technology, 75*, 231–242. https://doi.org/10.1016/j.tifs.2018.03.018
Demei, K., Zhang, M., Phuhongsung, P., & Mujumdar, A. S. (2022). 3D food printing: Controlling characteristics and improving technological effect during food processing. *Food Research International, 156*(March), 111120. https://doi.org/10.1016/j.foodres.2022.111120
Derossi, A., Caporizzi, R., Oral, M. O., & Severini, C. (2020). Analyzing the effects of 3D printing process per se on the microstructure and mechanical properties of cereal food products. *Innovative Food Science and Emerging Technologies, 66*, 102531. https://doi.org/10.1016/j.ifset.2020.102531
Derossi, A., Caporizzi, R., Paolillo, M., & Severini, C. (2021). Programmable texture properties of cereal-based snack mediated by 3D printing technology. *Journal of Food Engineering, 289*, 110160. https://doi.org/10.1016/j.jfoodeng.2020.110160
Dick, A., Prakash, S., & Bhandari, B. (2021). 3D printing. In *Food formulation: Novel ingredients and processing techniques* (pp. 101–119) http://www.globalview.gr/2016/06/30/62949/
Fahmy, A. R., Becker, T., & Jekle, M. (2020). 3D printing and additive manufacturing of cereal-based materials: Quality analysis of starch-based systems using a camera-based morphological approach. *Innovative Food Science and Emerging Technologies, 63*, 102384. https://doi.org/10.1016/j.ifset.2020.102384
Garfield, L. (2016). This robot can 3D-print a pizza in under five minutes. Tech Insider, 282016. https://doi.org/https://www.businessinsider.in/tech/this-robot-can-3d-print-a-pizza-in-under-five-minutes/articleshow/52733227.cms
Georgantzinos, S. K., & Giannopoulos, G. I. (2021). Additive manufacturing for effective smart structures: The idea of 6D printing. *Journal of Composites Science, 5*(5), 119.
Ghazal, A. F., Zhang, M., Mujumdar, A. S., & Ghamry, M. (2022). Progress in 4D / 5D / 6D printing of foods: Applications and R & D opportunities. *Critical Reviews in Food Science and Nutrition*, 1–24. https://doi.org/10.1080/10408398.2022.2045896
Godoi, F. C., Prakash, S., & Bhandari, B. R. (2016). 3d printing technologies applied for food design: Status and prospects. *Journal of Food Engineering, 179*, 44–54. https://doi.org/10.1016/j.jfoodeng.2016.01.025
Gokhare, V. G., Raut, D. N., & Shinde, D. K. (2017). A review paper on 3D-printing aspects and various processes used in the 3D-printing. *IJERT International Journal of Engineering Research and Technology, 6*(6), 953–958. www.ijert.org
Grood, D., Wilhelmus, J. P., & De Grood, P. J. (2013). *Method and device for analysis of a liquid sample* (Patent No. U.S. 8,556,392).

Gross, B. C., Erkal, J. L., Lockwood, S. Y., Chen, C., & Spence, D. M. (2014). Evaluation of 3D printing and its potential impact on biotechnology and the chemical sciences. *Analytical Chemistry, 86*(7), 3240–3253. https://doi.org/10.1021/ac403397r

Guo, C., Zhang, M., & Bhandari, B. (2019). Model building and slicing in food 3D printing processes: A review. *Comprehensive Reviews in Food Science and Food Safety, 18*(4), 1052–1069. https://doi.org/10.1111/1541-4337.12443

Guo, C., Zhang, M., & Devahastin, S. (2021). Color/aroma changes of 3D-printed buckwheat dough with yellow flesh peach as triggered by microwave heating of gelatin-gum Arabic complex coacervates. *Food Hydrocolloids, 112*, 106358. https://doi.org/10.1016/j.foodhyd.2020.106358

Hangge, P., Pershad, Y., Witting, A. A., Albadawi, H., & Oklu, R. (2018). Three-dimensional (3D) printing and its applications for aortic diseases. *Cardiovascular Diagnosis and Therapy, 8*, S19–S25. https://doi.org/10.21037/cdt.2017.10.02

Hao, L., Mellor, S., Seaman, O., Henderson, J., Sewell, N., & Sloan, M. (2010). Material characterisation and process development for chocolate additive layer manufacturing. *Virtual and Physical Prototyping, 5*(2), 57–64. https://doi.org/10.1080/17452751003753212

He, C., Zhang, M., & Fang, Z. (2020). 3D printing of food: Pretreatment and post-treatment of materials. *Critical Reviews in Food Science and Nutrition, 60*(14), 2379–2392. https://doi.org/10.1080/10408398.2019.1641065

Huang, M.-S., Zhang, M., & Bhandari, B. (2019). Assessing the 3D printing precision and texture properties of Brown Rice induced by infill levels and printing variables. *Food and Bioprocess Technology, 12*(7), 1185–1196. https://doi.org/10.1007/s11947-019-02287-x

Hurst, E. J. (2016). 3D printing in healthcare: Emerging applications. *Journal of Hospital Librarianship, 16*(3), 255–267. https://doi.org/10.1080/15323269.2016.1188042

Hussain, S., Malakar, S., & Arora, V. K. (2022). Extrusion-based 3D food printing: Technological approaches, material characteristics, printing stability, and post-processing. *Food Engineering Reviews, 14*(1), 100–119. https://doi.org/10.1007/s12393-021-09293-w

Jacob, J., & Leelavathi, K. (2007). Effect of fat-type on cookie dough and cookie quality. *Journal of Food Engineering, 79*(1), 299–305. https://doi.org/10.1016/j.jfoodeng.2006.01.058

Jakus, A. E. (2019). An introduction to 3D printing-past, present, and future promise. In *3D printing in orthopaedic surgery* (pp. 1–15). Elsevier Inc. https://doi.org/10.1016/B978-0-323-58118-9.00001-4

Jiang, H., Zheng, L., Zou, Y., Tong, Z., Han, S., & Wang, S. (2019). 3D food printing: Main components selection by considering rheological properties. *Critical Reviews in Food Science and Nutrition, 59*(14), 2335–2347. https://doi.org/10.1080/10408398.2018.1514363

Kewuyemi, Y. O., Kesa, H., & Adebo, O. A. (2021). Trends in functional food development with three-dimensional (3D) food printing technology: Prospects for value-added traditionally processed food products. *Critical Reviews in Food Science and Nutrition, 62*(28), 7866–7904. https://doi.org/10.1080/10408398.2021.1920569

Kim, H. W., Bae, H., & Park, H. J. (2017). Classification of the printability of selected food for 3D printing: Development of an assessment method using hydrocolloids as reference material. *Journal of Food Engineering, 215*, 23–32. https://doi.org/10.1016/j.jfoodeng.2017.07.017

Kim, H. W., Lee, I. J., Park, S. M., Lee, J. H., Nguyen, M. H., & Park, H. J. (2019). Effect of hydrocolloid addition on dimensional stability in post-processing of 3D printable cookie dough. *LWT, 101*, 69–75. https://doi.org/10.1016/j.lwt.2018.11.019

Krishnaraj, P., Anukiruthika, T., Choudhary, P., Moses, J. A., & Anandharamakrishnan, C. (2019). 3D extrusion printing and post-processing of fibre-rich snack from indigenous composite flour. *Food and Bioprocess Technology, 12*(10), 1776–1786. https://doi.org/10.1007/s11947-019-02336-5

Le-Bail, A., Maniglia, B. C., & Le-Bail, P. (2020). Recent advances and future perspective in additive manufacturing of foods based on 3D printing. *Current Opinion in Food Science, 35*, 54–64. https://doi.org/10.1016/j.cofs.2020.01.009

Letras, P., Oliveira, S., Varela, J., Nunes, M. C., & Raymundo, A. (2022). 3D printed gluten-free cereal snack with incorporation of Spirulina (Arthrospira platensis) and/or Chlorella vulgaris. *Algal Research, 68*, 102863. https://doi.org/10.1016/j.algal.2022.102863

Lille, M., Nurmela, A., Nordlund, E., Metsä-Kortelainen, S., & Sozer, N. (2018). Applicability of protein and fiber-rich food materials in extrusion-based 3D printing. *Journal of Food Engineering, 220*, 20–27. https://doi.org/10.1016/j.jfoodeng.2017.04.034

Lille, M., Kortekangas, A., Heiniö, R.-L., & Sozer, N. (2020). Structural and textural characteristics of 3D-printed protein- and dietary fibre-rich snacks made of milk powder and wholegrain Rye flour. *Foods, 9*(11), 1527. https://doi.org/10.3390/foods9111527

Lipton, J. I. (2017). Printable food: The technology and its application in human health. *Current Opinion in Biotechnology, 44*, 198–201. https://doi.org/10.1016/j.copbio.2016.11.015

Lipton, J., Arnold, D., Nigl, F., Lopez, N., Cohen, D., Norén, N., & Lipson, H. (2010). Mutli-material food printing with complex internal structure suitable for conventional post-processing. *21st Annual International Solid Freeform Fabrication Symposium – An Additive Manufacturing Conference* (pp. 809–815). SFF.

Liu, Z., Zhang, M., Bhandari, B., & Wang, Y. (2017). 3D printing: Printing precision and application in food sector. *Trends in Food Science and Technology, 69*, 83–94. https://doi.org/10.1016/j.tifs.2017.08.018

Liu, Z., Bhandari, B., Prakash, S., & Zhang, M. (2018). Creation of internal structure of mashed potato construct by 3D printing and its textural properties. *Food Research International, 111*, 534–543. https://doi.org/10.1016/j.foodres.2018.05.075

Liu, Y., Liang, X., Saeed, A., Lan, W., & Qin, W. (2019). Properties of 3D printed dough and optimization of printing parameters. *Innovative Food Science and Emerging Technologies, 54*, 9–18. https://doi.org/10.1016/j.ifset.2019.03.008

Liu, Y., Tang, T., Duan, S., Qin, Z., Li, C., Zhang, Z., Liu, A., Wu, D., Chen, H., Han, G., Lin, B., He, J., & Wu, W. (2020a). Effects of sodium alginate and rice variety on the physicochemical characteristics and 3D printing feasibility of rice paste. *LWT – Food Science and Technology, 127*, 109360. https://doi.org/10.1016/j.lwt.2020.109360

Liu, Y., Tang, T., Duan, S., Qin, Z., Zhao, H., Wang, M., Li, C., Zhang, Z., Liu, A., Han, G., Wu, D., Chen, H., Shen, L., Wu, W., & Dong, Y. (2020b). Applicability of rice doughs as promising food materials in extrusion-based 3D printing. *Food and Bioprocess Technology, 13*(3), 548–563. https://doi.org/10.1007/s11947-020-02415-y

Mantihal, S., Kobun, R., & Lee, B. B. (2020). 3D food printing of as the new way of preparing food: A review. *International Journal of Gastronomy and Food Science, 22*, 100260. https://doi.org/10.1016/j.ijgfs.2020.100260

Martínez, M., Oliete, B., & Gómez, M. (2013). Effect of the addition of extruded wheat flours on dough rheology and bread quality. *Journal of Cereal Science, 57*(3), 424–429. https://doi.org/10.1016/j.jcs.2013.01.007

Nachal, N., Moses, J. A., Karthik, P., & Anandharamakrishnan, C. (2019). Applications of 3D printing in food processing. *Food Engineering Reviews, 11*(3), 123–141. https://doi.org/10.1007/s12393-019-09199-8

Navaf, M., Sunooj, K. V., Aaliya, B., Akhila, P. P., Sudheesh, C., Mir, S. A., & George, J. (2022). 4D printing: A new approach for food printing; effect of various stimuli on 4D printed food properties. A comprehensive review. *Applied Food Research, 2*(2), 100150. https://doi.org/10.1016/j.afres.2022.100150

Noort, M., Van Bommel, K., & Renzetti, S. (2017). 3D-printed cereal foods. *Cereal Foods World, 62*(6), 272–277. https://doi.org/10.1094/CFW-62-6-0272

Pavičić, T. V., Grgić, T., Ivanov, M., Novotni, D., & Herceg, Z. (2021). Influence of flour and fat type on dough rheology and technological characteristics of 3d-printed cookies. *Foods, 10*(1). https://doi.org/10.3390/foods10010193

Pereira, T., Barroso, S., & Gil, M. M. (2021). Food texture design by 3d printing: A review. *Foods, 10*(2), 1–26. https://doi.org/10.3390/foods10020320

Pulatsu, E., Su, J. W., Kenderes, S. M., Lin, J., Vardhanabhuti, B., & Lin, M. (2022). Restructuring cookie dough with 3D printing: Relationships between the mechanical properties, baking conditions, and structural changes. *Journal of Food Engineering, 319*, 110911. https://doi.org/10.1016/j.jfoodeng.2021.110911

Savini, A., & Savini, G. G. (2022). A short history of 3D printing, a technological revolution just started. *ICOHTEC/IEEE International History of High-Technologies and Their Socio-Cultural Contexts Conference (HISTELCON)* (pp. 1–8). https://doi.org/10.1109/HISTELCON.2015.7307314

Severini, C., Derossi, A., & Azzollini, D. (2016). Variables affecting the printability of foods: Preliminary tests on cereal-based products. *Innovative Food Science and Emerging Technologies, 38*, 281–291. https://doi.org/10.1016/j.ifset.2016.10.001

Severini, C., Azzollini, D., Albenzio, M., & Derossi, A. (2018). On printability, quality and nutritional properties of 3D printed cereal based snacks enriched with edible insects. *Food Research International, 106*, 666–676. https://doi.org/10.1016/j.foodres.2018.01.034

Shahrubudin, N., Lee, T. C., & Ramlan, R. (2019). An overview on 3D printing technology: Technological, materials, and applications. *Procedia Manufacturing, 35*, 1286–1296. https://doi.org/10.1016/j.promfg.2019.06.089

Shi, Y., Zhang, M., & Bhandari, B. (2021). Effect of addition of beeswax based oleogel on 3D printing of potato starch-protein system. *Food Structure, 27*, 100176. https://doi.org/10.1016/j.foostr.2021.100176

Su, A., & Al'Aref, S. J. (2018). History of 3D printing. In *3D printing applications in cardiovascular medicine* (pp. 1–10). Elsevier Inc. https://doi.org/10.1016/B978-0-12-803917-5.00001-8

Sun, J., Peng, Z., Zhou, W., Fuh, J. Y. H., Hong, G. S., & Chiu, A. (2015a). A review on 3D printing for customized food fabrication. *Procedia Manufacturing, 1*, 308–319. https://doi.org/10.1016/j.promfg.2015.09.057

Sun, J., Zhou, W., Huang, D., Fuh, J. Y. H., & Hong, G. S. (2015b). An overview of 3D printing technologies for food fabrication. *Food and Bioprocess Technology, 8*(8), 1605–1615. https://doi.org/10.1007/s11947-015-1528-6

Sun, J., Zhou, W., Yan, L., Huang, D., & Lin, L. ya. (2018). Extrusion-based food printing for digitalized food design and nutrition control. *Journal of Food Engineering, 220*, 1–11. https://doi.org/10.1016/j.jfoodeng.2017.02.028

Thangalakshmi, S., Arora, V. K., Kaur, B. P., & Malakar, S. (2021). Investigation on rice flour and jaggery paste as food material for extrusion-based 3D printing. *Journal of Food Processing and Preservation, 45*(4), 1–12. https://doi.org/10.1111/jfpp.15375

Vancauwenberghe, V., Delele, M. A., Vanbiervliet, J., Aregawi, W., Verboven, P., Lammertyn, J., & Nicolaï, B. (2018). Model-based design and validation of food texture of 3D printed pectin-based food simulants. *Journal of Food Engineering, 231*, 72–82. https://doi.org/10.1016/j.jfoodeng.2018.03.010

Varghese, C., Wolodko, J., Chen, L., Doschak, M., Srivastav, P. P., & Roopesh, M. S. (2020). Influence of selected product and process parameters on microstructure, rheological, and textural properties of 3D printed cookies. *Foods, 9*(7). https://doi.org/10.3390/foods9070907

Varvara, R. A., Szabo, K., & Vodnar, D. C. (2021). 3D food printing: Principles of obtaining digitally-designed nourishment. *Nutrients, 13*(10), 3617. https://doi.org/10.3390/nu13103617

Vieira, M. V., Oliveira, S. M., Amado, I. R., Fasolin, L. H., Vicente, A. A., Pastrana, L. M., & Fuciños, P. (2020). 3D printed functional cookies fortified with Arthrospira platensis: Evaluation of its antioxidant potential and physical-chemical characterization. *Food Hydrocolloids, 107*. https://doi.org/10.1016/j.foodhyd.2020.105893

Wang, Z., & Yang, Y. (2021). Application of 3D printing in implantable medical devices. *BioMed Research International, 2021*. https://doi.org/10.1155/2021/6653967

Wang, W., Zhou, H., Yang, H., & Cui, M. (2016). Effects of salts on the freeze–thaw stability, gel strength and rheological properties of potato starch. *Journal of Food Science and Technology, 53*(9), 3624–3631. https://doi.org/10.1007/s13197-016-2350-5

Wang, W., Yao, L., Zhang, T., Cheng, C. Y., Levine, D., & Ishii, H. (2017). Transformative appetite: Shape-changing food transforms from 2D to 3D by water interaction through cooking. *Proceedings of the 2017 CHI Conference on Human Factors in Computing Systems* (pp. 6123–6132). https://doi.org/10.1145/3025453.3026019

Wang, M., Li, D., Zang, Z., Sun, X., Tan, H., Si, X., Tian, J., Teng, W., Wang, J., Liang, Q., Bao, Y., Li, B., & Liu, R. (2022). 3D food printing: Applications of plant-based materials in extrusion-based food printing. *Critical Reviews in Food Science and Nutrition, 62*(26), 7184–7198. https://doi.org/10.1080/10408398.2021.1911929

Wilms, P., Daffner, K., Kern, C., Gras, S. L., Schutyser, M. A. I., & Kohlus, R. (2021). Formulation engineering of food systems for 3D-printing applications – A review. *Food Research International, 148*, 110585. https://doi.org/10.1016/j.foodres.2021.110585

Xu, K., Zhang, M., & Bhandari, B. (2020). Effect of novel ultrasonic- microwave combined pre-treatment on the quality of 3D printed wheat starch-papaya system. *Food Biophysics, 15*(2), 249–260. https://doi.org/10.1007/s11483-019-09615-w

Yang, F., Zhang, M., Prakash, S., & Liu, Y. (2018). Physical properties of 3D printed baking dough as affected by different compositions. *Innovative Food Science and Emerging Technologies, 49*, 202–210. https://doi.org/10.1016/j.ifset.2018.01.001

Yang, F., Zhang, M., Fang, Z., & Liu, Y. (2019). Impact of processing parameters and post-treatment on the shape accuracy of 3D-printed baking dough. *International Journal of Food Science and Technology, 54*(1), 68–74. https://doi.org/10.1111/ijfs.13904

Zhang, L., Lou, Y., & Schutyser, M. A. I. (2018). 3D printing of cereal-based food structures containing probiotics. *Food Structure, 18*, 14–22. https://doi.org/10.1016/j.foostr.2018.10.002

Zhang, J., Li, Y., Cai, Y., Ahmad, I., Zhang, A., Ding, Y., Qiu, Y., Zhang, G., Tang, W., & Lyu, F. (2022). Hot extrusion 3D printing technologies based on starchy food: A review. *Carbohydrate Polymers, 294*, 119763. https://doi.org/10.1016/j.carbpol.2022.119763

Zheng, L., Yu, Y., Tong, Z., Zou, Q., Han, S., & Jiang, H. (2019). The characteristics of starch gels molded by 3D printing. *Journal of Food Processing and Preservation, 43*(7), 1–11. https://doi.org/10.1111/jfpp.13993

Zheng, L., Liu, J., Liu, R., Xing, Y., & Jiang, H. (2021). 3D printing performance of gels from wheat starch, flour and whole meal. *Food Chemistry, 356*, 129546. https://doi.org/10.1016/j.foodchem.2021.129546

Chapter 15
Packaging of Cereal Products

Monjurul Hoque, Mahmoud Said Rashed, Animesh Sengar, and Shivani Pathania

1 Introduction

Cereal and cereal-based products are the essential constituents of a balanced diet consumed daily by people worldwide. In industrialized countries, about 50.0% of the standard requirement of carbohydrates is consumed through bread. Furthermore, these cereals are also a key source of proteins, minerals, vitamins, or other trace elements (Belitz et al., 2009). The cereal industry includes finished products such as fresh bread and bread rolls, pasta, extruded breakfast cereals, and principal ingredients including different varieties of functional flours such as all-purpose flour, baking flour, hard wheat flour, and semolina, couscous, and bulgur. Cereal products, such as cakes, pastries, cookies, crackers, or pretzels, can be found in the confectionery and snacks segment. In 2022, the bread and bakery products market generated \$1085 billion and is expected to grow 7.7% annually (Statisa, 2022).

In general, bakery products like bread and cakes, without preservatives, show a shelf life of about 3–5 days at room temperature. Their shelf life is primarily limited by staling and microbial growth. Fresh bakery products are highly perishable due to physical, chemical, and microbiological spoilage and are therefore highly sensitive

M. Hoque
Food Industry Development Department, Teagasc Food Research Centre, Dublin, Ireland

School of Food and Nutritional Sciences, University College Cork, Cork, Ireland

M. S. Rashed · A. Sengar
Food Industry Development Department, Teagasc Food Research Centre, Dublin, Ireland

Environmental Sustainability and Health Institute, Technological University Dublin, Dublin, Ireland

S. Pathania (✉)
Food Industry Development Department, Teagasc Food Research Centre, Dublin, Ireland
e-mail: Shivani.Pathania@teagasc.ie

M. A. Shah et al. (eds.), *Cereal-Based Food Products*,
https://doi.org/10.1007/978-3-031-40308-8_15

to the preservation, storage, and distribution practices. Undesirable physical and chemical modifications of the bakery products cause a loss in their freshness, texture, and taste, while microbiological spoilage results in their unpleasant appearance and food safety risks. Bakery products with low moisture ($a_w < 0.6$) and intermediate moisture content (a_w of 0.6–0.85) are most susceptible to physical spoilage (moisture loss, staling) and chemical spoilage (rancidity), while products with high moisture content ($a_w > 0.85$, and generally in the range of 0.95–0.99) undergo microbial spoilage (yeast, mold, bacterial growth) (Qian et al., 2021; Smith et al., 2004).

Some molds like *Penicillium spp.*, *Aspergillus niger*, *Cladosporium*, *Mucor*, and *Neurospora,* and bacteria such as *Bacillus subtilis* and *Bacillus licheniformis* and their colonies are the common causes of microbial deterioration of bakery items. Yeasts like *Hyphopichia burtonii* and *Saccharomycopsis fibuligera* give rise to visible physical spoilage -also referred to as the 'chalk mold defect,' one of bread's early stage microbial defects (Deschuyffeleer et al., 2011). The hyaline mycelium of certain fungi releases secondary metabolites such as mycotoxin and allergenic compounds even before fungal outgrowth becomes visible (Qian et al., 2021; Rüegg et al., 2022). Such microbial spoilage causes the off-flavors as well as unacceptable appearance while significantly threatening public health through food safety hazards, resulting in a high economic loss for the bakery industry and consumers.

Contamination of bakery products with microbes (yeast and molds) primarily occurs post baking, cooling and subsequent packaging. Therefore, chemical preservatives and physical techniques are applied to prevent microbial contamination. Chemical preservatives include potassium acetate, sodium acetate, ethanol, and organic acids such as propionic acid, benzoic acid, sorbic acid, and acetic acid andtheir salts (Qian et al., 2021). Various physical preservation techniques applied to bakery products includes ultraviolet light (UV), infrared (IFR), microwave heating (MWH), and ultra-high pressure (UHP) treatments. While consumers are reluctant to consume synthetic chemical preservatives added to bakery products directly, physical preservation techniques are limited by low penetration capacity of UV, condensation in MWH, and high cost for IFR. Therefore, research is needed to develop innovative preservative techniques to enhance bakery products' quality, safety, and shelf life, ensuring consumer acceptability and economic viability.

Currently several studies are being carried out that focus on developing innovative packaging systems that enhance bakery products' shelf life. Packaging is a fundamental part of the food processing system. It performs several functions like a barrier for food against external contaminations, containment, protection, transportation, distribution, and information display to the consumers. A suitable packaging system can extend bakery products' shelf life, retain their quality, and ensure safety besides minimizing food loss. Therefore this chapter aims to highlight the types and properties of the commonly used packaging material for bakery products. This chapter also highlights the factors affecting cereal and cereal-based products' shelf-life and thereby intends to present a comprehensive review of the product-specific packaging materials. Furthermore, the advanced packaging system used for the extension of the shelf life of these products and the challenges and future trends in the expansion of advanced packaging in the bakery industry are also discussed.

2 Packaging Functions

The primary functions of traditional packaging in the food sector include containment, prevention, preservation, displaying information (for communication and marketing), and convenience (Salgado et al., 2021; Bauer et al., 2022), as shown in Fig. 15.1. Suitable packaging prevents bakery products from spoilage and damage caused by external factors such as microorganisms, water vapor, gases (O_2, CO_2, and N_2), light, heat, odour, dirt, and dust. The packaging facilitates the containment of the products in different shapes and sizes and enhances logistic efficiency. Another critical function of the packaging material is communication and marketing. The manufacturers display vital information like ingredients list, nutritional facts, manufacturing and expiry dates, allergen warnings, brand logos, product storage instructions, and end use of the packaging material, such as recyclability. In recent times, packaging facilitates the modern trend of retail practices and caters to the changing lifestyle of the consumers such as convenient handling options like easy opening and closing, as well as being suitable for microwave heating, refrigeration, or freezing (Qian et al., 2021; Salgado et al., 2021).

Depending on the specific position in the supply chain, the packaging system can be categorized into three levels, namely primary, secondary and tertiary packaging, as shown in Fig. 15.2. Primary packaging is the first level of packaging that comes in direct contact with the products. Generally, primary packaging materials include boxes, cups, cartons, trays, bottles, tubs, cans, sacks, drums, and barrels. The primary level of packaging item not only protects the food materials but also satisfies

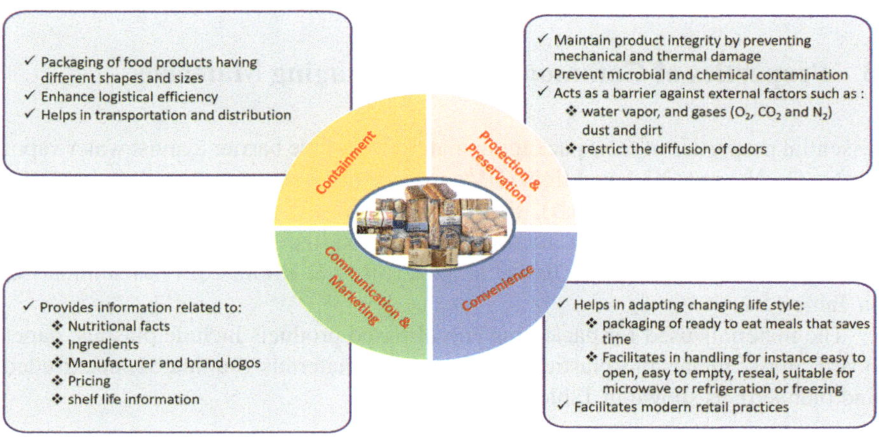

Fig. 15.1 General functions of food packaging

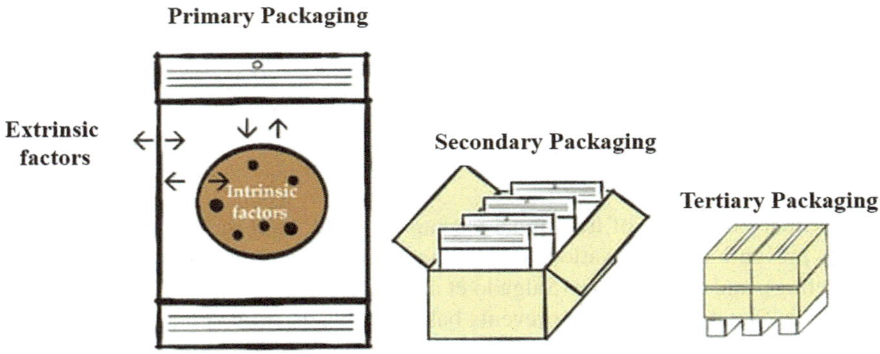

Fig. 15.2 Levels of cereal product packaging. (Reprinted from Bauer et al. (2022))

consumer preferences (product uses and features), handling, and apportionment of the product (Garcia-Arca et al., 2021; Ahmad et al., 2022).

Secondary level packaging consolidates multiple primary packages in a bundle containing the same or different products. Secondary packaging items include cardboard boxes, totes, cages, and cases. This level of packaging mainly protects the primary packaging and ease of handling at different stages, such as at the warehouse, during transportation and distribution, and in retail for sale (Garcia-Arca et al., 2021).

Tertiary packaging, also known as 'transport' or 'distribution' or 'unit load' packaging, typically contains primary or secondary level packages depending on the types of products. Generally, tertiary-level packaging materials include pallets (full, half quarter), crates, roll cages, bulkbags, drums, barrels, and shipping containers. This level of packaging facilitates the handling, storage, and transportation of a large volume of products (Garcia-Arca et al., 2021; Pålsson & Hellström, 2016).

3 Properties of Commonly Used Packaging Materials

Essential properties of food packaging materials include barrier against water vapor, gases (O_2, N_2, and CO_2), and light (UV), mechanical properties (tensile strength-TS and elongation at break-EAB), thermal properties, headspace gas composition, migration, protection against microbes, and an oxidizing agent (i.e., antimicrobial and antioxidant properties) (Qian et al., 2021; Roy & Rhim, 2021) as summarised in Table 15.1.

The materials used for packaging cereal-based products include plastics, paper/board, glass, metal, bio-plastic, and composite materials (laminated, coextruded, and blended), as shown in Table 15.2.

Table 15.1 Important packaging properties and their importance in food packaging

Properties	Importance	R
Mechanical	Mechanical properties such as static and dynamic stress and EAB play an important role in packing, storage, and transportation: Static stress arises when products are stacked or subjected to pressure, such as vacuum packaging (VP) or modified atmospheric packaging (MAP), as well as in the case of products having angular or pointed shapes Dynamic stress may arise during filling, forming, or printing as well due to vibration in transportation EAB denotes the amount a material plastically and elastically deforms before fracture, and it is important parameter to be considered when materials are subjected to extreme cold or heat and higher loading	Bauer et al. (2022); Hoque et al. (2021)
Barrier	Water vapor and gas (O_2, N_2, and CO_2) barrier properties influence the product quality and shelf life. Depending on the types of product being packed, the appropriate permeation property of packaging material is essential because: Water vapor can result in the variation of structural properties such as hardening, agglomeration, or softening of products Water vapor can facilitate microbial growth O_2 can cause lipid oxidation and degrade quality-determining constituents such as vitamins O_2 can also promote pathogenic and microbial spoilage N_2 and CO_2 used in the MAP and aroma component used in packaging are critical Light barrier property is also important as it can accelerate oxidative or other chemical reactions, as well it may change the structure of the product	Bauer et al. (2022); Hoque et al. (2022a)
Migration	Chemicals like plasticizers, antioxidants, light stabilizers, heat stabilizers, lubricants, slip compounds, antistatic agents, and monomers are found as migrant Migration of chemical compounds from packaging material into food may cause chemical contamination Some chemicals can be carcinogenic and may cause harm to the human life Migration from food to packaging material (sorption or scalping) may result in alteration of products, such as flavor loss and reduced reusability of packaging Migration of chemical compounds can be subject to various features such as material, Physico-chemical properties of the migrant, food (e.g., fat content),storage temperature, storage time, relative humidity, and size of the packaging material to the volume of food being packed	Bauer et al. (2022); Food-Packaging-Forum (2013)

(continued)

Table 15.1 (continued)

Properties	Importance	R
Thermal stability	In general thermal stability of the polymeric materials is analyzed using thermogravimetric analysis-TGA and differential scanning calorimetry-DSC Thermal stability of polymers influences the properties, such as strength, toughness, or elasticity at a given temperature Packaging material with high thermal stability protects food components from thermal damage, preventing itself from deformation and degradation	Ncube et al. (2020); Hoque et al. (2022b)
Sealing	Sealing can be either heat sealing or cold sealing Heat-sealed packages can withstand temperature, humidity changes, and rough handling Enhanced design flexibility, and it is cheap Heat sealing can enhance the functionality of the product, recyclability, and temperature resistance and imparts an attractive appearance while branding Cold sealing is self-sealing used for candy and chocolates, ice cream, and cereal products Adhesives can be used in cold sealing	Ilhan et al. (2021; Bostik-Smart-Adhesive (2022)
Antioxidant and antimicrobial	Antimicrobial packaging can inhibit pathogenic contamination in foods, thereby enhancing the shelf-life of foods Antioxidants in food packaging systems act as a scavenger of undesirable substances (e.g., oxygen, metal ions, or radical oxidative species) and thus enhances the shelf life of the products	Agarwal et al. (2020; Gómez-Estaca et al. (2014)

4 Categorization of Cereal products and Factors Affecting their Quality and Shelf Life

Cereals and cereals-based products comprise a wide range of products. Categorizing these products assists in grouping them based on their characteristics that govern their shelf life and susceptibility to various spoilages. This further enables the efficient design of suitable packaging systems. Different bakery items are prepared using cereals like wheat, rye, rice, barley, millet, oats, and corn. Bakery products include various sizes, shapes, colors, textures, and flavors.

Based on sugar content, bakery items can be classified as:

- unsweetened goods (bread, rolls, buns, crumpets, muffins, and bagels)
- sweet goods (pancakes, doughnuts, buns, wafers, waffles, cookies, biscuits)
- filled goods (fruit and meat pies, fruits tart and jams, sausage rolls, pastries, sandwiches, cream cakes, pizza, and quiche)" (Smith et al., 2004; Bauer et al., 2022; Giannou et al., 2014).

Based on the pH of the products, the classification of bakery products includes:

- high acid products (pH < 4.6, such as sourdough bread and apple pie)

Table 15.2 Properties of the commonly used packaging materials in food packaging

Packaging material	Colour	Barrier properties: Water vapor	Barrier properties: Gases	Barrier properties: Light	Mechanical properties	Thermal properties	Heat-sealability	Physical and chemical properties	Applications	References
LDPE	Natural milky color, translucent	High	High	High	Crystallinity is about 50–60%, and thus high flexibility; demonstrates relatively high impact strength in a cold environment; maximum permissible stress is 6–17 MPa at 20 °C	MP ~115 °C; density decreases significantly when subjected to past 20 °C	Yes	Resistive towards most alcohols, acids, and alkalis; demonstrate mild resistance towards oxidizing agents and selected hydrocarbons, and shows resistance to grease	Bags, flexible lids, and bottles	Bauer et al. (2022); Global-Plastic-Sheeting (2013); Giannou et al. (2014)
HDPE	Natural milky white, semi-translucent based on density	Extremely high	Extremely high	Extremely high	Crystallinity is greater than 90%, and thus inflexible and more rigid; high tensile and specific strength, maximum permissible stress is 14–32 MPa at 20 °C	Can tolerate heat and temperatures above 100 °C; MP ~ 135 °C	Yes	Highly resistant to solvents, alcohols, acids, and alkalis; mild resistant to hydrocarbons, and resistant to grease	Bottles, cardboard liners, tubs, bags	Bauer et al. (2022); Global-Plastic-Sheeting (2013); Giannou et al. (2014)

(continued)

Table 15.2 (continued)

Packaging material	Colour	Barrier properties			Mechanical properties	Thermal properties	Heat-sealability	Physical and chemical properties	Applications	References
		Water vapor	Gases	Light						
LLDPE	Natural milky color, translucent	Good	Poor	Low	Ultimate tensile strength is in the range of 7.93–45.5 MPa; film tensile strength at yield and film elongation at break at machine direction are 1.31–12.5 MPa and 460–1000%, respectively	Melting point 75.0–128 °C	Yes	Resistance to grease	(Strech) wrap	Bauer et al. (2022); Global-Plastic-Sheeting (2013)
PET		Good	Good	Low	Tensile strength 152 MPa, flexural strength 221 MPa	Melting temperature 260 °C	Yes	PET does not react with water or food; PET is lightweight, resistant to grease and oil,	Bottles, jars, tubs, trays, blisters, films (bags and wrappers	Bauer et al. (2022); Marsh and Bugusu (2007)

PP	Opaque, natural grayish yellow in natural form.	High	Low	Low	Tensile strength 31–45 MPa, elongation at break 50–145%	Melting temperature 160–163 °C	Yes	PP is highly resistant to chemicals (e.g. acids) but UV sensitive, possesses a low humidity absorption coefficient and is insoluble in water and organic solvents.	Bottles, cardboard liners, tubs, microwavable packaging, bags	Bauer et al. (2022); Global-Plastic-Sheeting (2013); Giannou et al. (2014)

(continued)

Table 15.2 (continued)

Packaging material	Colour	Barrier properties			Mechanical properties	Thermal properties	Heat-sealability	Physical and chemical properties	Applications	References
		Water vapor	Gases	Light						
Paper and board	May vary	Extremely low	Extremely low	Extremely low	Mechanical stability	Low thermal stability	No	Application of paper as primary packaging materials, i.e., when in direct contact with food materials, papers are coated or impregnated with waxes, resins, or lacquers to enhance the functionally	Corrugated box, folding cartons, bags, sacks, and wrapping papers	Bauer et al. (2022); Marsh and Bugusu (2007); Deshwal et al. (2019)
Metals (tin, steel, aluminum, and chromium)	May vary	Absolute	Absolute	Absolute	High mechanical stability	High-temperature stability	No	Metals surfaces are always coated with lacquers to prevent the migration of metal components into food products	Bottles, cans, tubs, caps	Bauer et al. (2022); Deshwal and Panjagari (2020)
Glass	Transparent Green Brown	Absolute			Brittle	High-temperature stability	No	Odorless and shows no chemical reaction with any food, i.e., chemically inert	Bottles, jars	Bauer et al. (2022); Marsh and Bugusu (2007)

LDPE low-density polyethylene, *HDPE* high-density polyethylene, *LLDPE* linear low-density polyethylene, *PET* polyethylene terephthalate, *PP* polypropylene, *MP* melting point

- low acidic products (white bread, whole wheat bread, chocolate nut bread, and date nut bread)
- non-acidic products (crumpets, banana nut bread, and carrot muffin) (Smith et al., 2004; Giannou et al., 2014).

Depending on the final moisture content of the product, bakery items can be classified as follows:

- low moisture products ($a_w < 0.6$, cookies and crackers)
- intermediate moisture products (0.6–0.85, chocolate coated doughnuts, Danish pastries, cream-filled cake, and soft cookies)
- high moisture products (a_w 0.85–0.99, bread, pita bread, fruit pies, carrot cake, cheesecake, pizza crust, and pizza) (Smith et al., 2004; Giannou et al., 2014).

Several factors influence bakery products' shelf life based on the type of products. Therefore, it is essential to understand the factors that cause spoilage and affect their shelf life. The term "food spoilage" denotes reactions triggered by different intrinsic and extrinsic factors. These reactions negatively affect the products' quality attributes and reduce shelf life. Figure 15.3 illustrates the four primary factors affecting the spoilage of different cereal and cereal-based products (Qian et al., 2021).

Products Properties

Cereal-based products (mostly bakery items) are perishable food stuffs that need appropriate intervention to avoid deterioration and extend their shelf life, particularly limiting the growth of spoilage and pathogenic microbes (Umaraw et al., 2020). The primary factors that influence the microbial growth in cereal-based products include the nutritional composition of the raw materials, water content, water activity (a_w), O_2 content, pH content, and additives used in the product

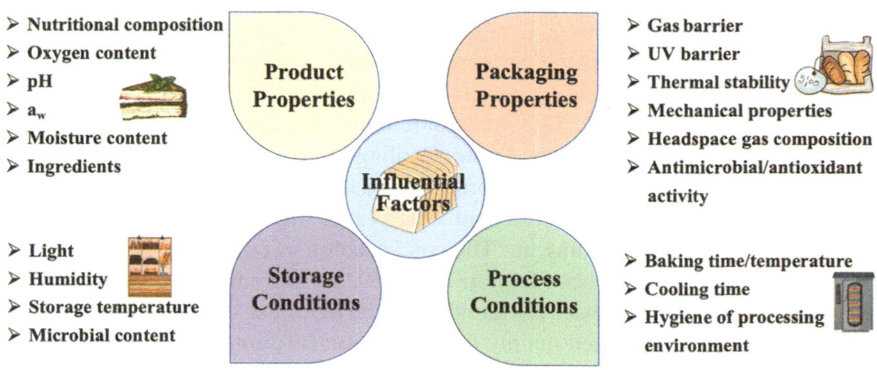

Fig. 15.3 Factors affecting the shelf life of bakery products. (Adopted from Reference (Qian et al., 2021) with permission from Elsevier 2022)

formulation. Microbes that grow in bakery products draw their energy by degrading the food components like carbohydrates, protein, and lipids. One crucial factor is the a_w that regulates the growth of microbes in cereal and cereal-based products. In general, bacteria, yeast, and mold grow in the bread and other bakery products with a high a_w in the range of 0.90–0.99. While fruit cakes and other products with intermediate a_w (0.6–0.85), facilitate the growth of osmophilic yeasts and molds (Smith et al., 2004). The presence of oxygen is critical, especially for the high-fat content products, as it promotes oxidative rancidity resulting in the development of free radicals and peroxides, and causes degradation and subequently loss of high-valued components, like vitamins and proteins.

Additionally, oxygen facilitates the growth of obligatory aerobes, yeast, and molds (Smith et al., 2004). Similarly, pH also regulates the growth of microbes in the foodstuffs, and a pH of 6.6–7.5 is optimal for microbial growth (Umaraw et al., 2020). Hence, bakery items like bread, having optimal pH and high $a_{w,}$ is highly susceptible to mold contamination (Axel et al., 2017). Moreover, the application of ingredients, such as sugar, salt, and other additives, inhibit microbial contamination and enhance the shelf life of the bakery products.

Processing Conditions

The different processing conditions, such as baking time and temperature, cooling period, and the baking process line's environment, influence the bakery products' shelf life (Qian et al., 2021; Dong & Karboune, 2021). Bread staling is a physical and chemical modification in the baked bread that reduces palatability and results in dry and leathery texture. Staling is a severe concern for bakery products and is influenced by the starching effect and cooling period. During baking, when heated in the presence of moisture, starch undergoes gelatinization or swelling and forms a crumb structure combined with gluten. When removed from the oven after baking, bread undergoes various changes, such as recrystallizing starch that influences the taste and aroma (Qian et al., 2021; Dong & Karboune, 2021; Taglieri et al., 2021). Therefore, baking time and temperature are vital in maintaining product quality. During the baking, mold, and spores are inactivated due to heat. However, insufficient baking time and temperature may result in the survival of microbes and spores that affect the product's quality and taste. Cooling time in the baking process is also an essential factor. After the baking process, bread removed from the oven undergoes rapid cooling. Carbon dioxide and water vapor diffuse through the crust and are replaced by the surrounding air. The condensation of escaped water vapor promotes the growth of microbes (Taglieri et al., 2021). Moreover, the flavor components present in the vapor phase also condense with the water vapor and are absorbed by the bread, then evaporated mainly from the crust, resulting in rapid degradation of bread aroma (Qian et al., 2021; Dong & Karboune, 2021). Therefore, setting the optimum bread cooling conditions is crucial for maintaining bread quality. Also, good sanitary practice and proper maintenance of hygiene by food handlers and processing ambient can significantly impact bakery products' quality and shelf life.

Storage Conditions

Storage parameters such as temperature, relative humidity, light, air, and initial microbial content in storage environments significantly impact bakery products' shelf life. Alpers, Kerpes (Alpers et al., 2021) reported that storage temperature significantly influences the kinetics of bread staling, and elevated humidity provokes excessive microbial growth. The relative humidity is so detrimental that even microbiological stable commodities such as biscuits deteriorate when exposed to high relative humidity (Vat, 2022). The bakery product contains a good portion of fats and is thus very sensitive to light, oxygen, moisture, and temperature. These factors trigger reactions such as rancidity and oxidation, causing discoloration of the commodities, producing undesirable off-flavors. In general, mold grows slowly in the bread. However, temperatures of 25–30 °C and 90% humidity or more promote toxic production. Chang, Song (Chang et al., 2020) demonstrated the impact of temperature on bread storage, where the total microbial count on bread stored at 4 °C for 60 h was 3.49 ± 0.37log CFU/g. At higher storage temperatures at 25 and 40 °C for 60 h, the total microbial count was 4.97 ± 0.59and5.54 ± 0.48log CFU/g respectively.

Overall, product properties (raw materials), processing and storage conditions, and properties of packaging material, as mentioned in Sect. 2, play a vital role in the maintenance of the quality and shelf life of cereal and cereal-based products. Based on these factors, appropriate packaging materials are selected.

5 Bakery Products and Specific Packaging Materials

Different bakery products have different requirements that need to be fulfilled by the packaging system, depending on their intrinsic properties and end-use. This section highlights some commonly consumed cereal and cereal-based products and their suitable packaging systems.

Bread

Bread is one of the most widely consumed fresh bakery products. Bread packaging is challenging as its quality and shelf life depend on moisture content. Moisture balance is a critical factor in bread packaging, as moisture prevents product drying while excess moisture can lead to crust softening and microbial spoilage. The most widely used material for bread packaging is glazed imitation parchment (GIP) paper. Both sides are impregnated with paraffin wax, resins, lacquers, or other additives to enhance the functional properties (Martins et al., 2021; Cauvain & Young, 2010). Another frequently used packaging material is LDPE, where the end is twisted and sealed with adhesive tape or a plastic clip. Similar to LDPE, several other packaging materials used for bread packaging include LLDPE, HDPE bags, polypropylene (PP) bags, and oriented polypropylene (OPP) (plain, pearlized, OPPcoex (co-extruded OPP), or Ac/OPP/Ac films) (Svanes et al., 2019). Perforated LDPE bags are used for the bread packaging to avoid the formation of leathery consistency on the bread surface because of the moisture migration from the crumb to the crust (Bauer et al., 2022; Cauvain & Young, 2010). Perforated PP films are suitable for fresh bread packed for immediate consumption, mostly used by the local manufacturer or at a local retailer (Pasqualone, 2019). In one of the studies, S.O. Agunbiade (Agunbiade et al., 2020) investigated the impact of packaging materials such as common cellophane bread bags (CCB), aluminum foil (ALP), transparent plastic bowls (TPB), and perforated plastic bowl (PPB) for bread packaging. The authors reported that CCB was the best packaging material in terms of sensory and higher integrity at ambient temperature. It was observed that for PPB packaging, a higher moisture loss through perforations resulted in least fungal growth rate however, the bread samples exhibited higher crumbliness and off odor. Morris (Morris, 2017) reported that PA (polyamide) acts as an aroma and taste barrier packaging material. Multi-layered films demonstrate high barrier properties against moisture and gases. Aluminum-coated LDPE is one of the most commonly used multi-layered packaging materials but has recently been substituted by acrylic-coated OPP, having a higher barrier property against oxygen and water vapor (Pasqualone, 2019).

Biscuits

Dry bakery products such as a biscuit (moisture content <6%, and $a_w = 0.30$) is fragile and highly hygroscopic. Also, biscuits have high-fat content and are susceptible to moisture and oxygen, causing oxidation and rancidity. Therefore, packaging material with high barrier properties against water and oxygen is preferred for biscuits. Similarly, light causes oxidation of fat resulting in discoloration and off-flavor development. Therefore, opaque materials are used for packaging of such products. Moreover, packaging materials should be free of residual solvents as some biscuits are prone to staining by inks, adhesives, and coatings applied in the packaging materials (ICPE, 2019).

Different flexible packaging materials in the form of wrappers and pre-formed or form-fill pouches are used in biscuit packaging. Traditionally, a regenerated cellulose fiber (RCF)(commercial name cellophane) coated with either LDPE or PVdC was used due to its high gas barrier, heat sealing properties, and gloss. However, due to its high cost and the introduction of new packaging material with superior packaging properties, cellophane is replaced with biaxially oriented polypropylene (OPP). In general, for less demanding commodities, OPP monofilm is used. In contrast, for higher quality products, duplex OPP or OPP combinations (pearlized or metalized) such as OPP/PE, OPP/PET, coextruded OPP (OPPcoex) film, or acrylic-coated (Ac) on both sides are used. Plain OPP films are coated with heat seal coating to expand seal-ability, whereas coextruded OPP demonstrates higher seal strength. Acrylic-coated OPP (AcOPP) is highly oxygen-impermeable and used for oxygen-sensitive products. Also, Ac and PVdC copolymer-coated OPP films also show higher barrier properties against flavor and aroma in contrast to uncoated OPP. Such high-barrier OPP packaging materials are used for biscuits and then put in a cardboard box as secondary packaging (Bauer et al., 2022; ICPE, 2019; Konstantas et al., 2019).

Various products like assorted biscuits, pastries, and cookies are packed in thermoformed plastic trays of polystyrene or PVC. These trays are either closed with lidding materials or a snap-on lid. It has also been reported that active packaging such as antimicrobial or antioxidant packaging can significantly enhance shelf life besides maintaining the quality of cereal and cereal-based products. Different flexible polymers such as PVDC coated nylon, polyester, LDPE, PP, EVA, and PS are frequently used with an active sachet for the packaging of bakery products (ICPE, 2019).

Cake

Cakes, pastries, and doughnuts are high moisture content products of various sizes, shapes, and forms. Therefore, packaging materials should be selected to discourage mold growth. Depending on the machinability and economics, various packaging materials, including PP, cast polypropylene (CPP), and PVC, are used (Bauer et al., 2022).

Whole Cereals and Cereal Products

Cereals and cereal products such as whole, broken, flaked, or milled grains (e.g., wheat, rye, rice, barley, millet, oats, and corn) are dry and have low fats. Therefore, these products are generally packed with paper bags, polyethylene bags as primary packaging, and cardboard boxed as secondary packages. When lined with LDPE, paper can be used for products that require high barrier property and can also prevent the migration of mineral oil to dry foods. Also, rigid laminates containing paper and plastic lids are usually used in snack product packaging (Bauer et al., 2022; Urbelis & Cooper, 2021). However, in breakfast cereals, the crispness is a critical factor besides preventing loss of aroma, vitamins, and formation of off flavors. Conventionally used packaging material for breakfast cereals includes fiberboard boxes with a super calendared, waxed, and glassine liner. Currently, the glassine liner has been replaced with HDPE. Currently, HDPE or HDPE coextruded with a thin layer of EVA (ethylene vinyl acetate) or PA (polyamide) and EVOH polymers are used for packaging breakfast cereals. EVA, ionomer, mPE (metallocene polyethylene), or blends permit low-temperature seals while offering easy consumer opening. In the case of products like flaked oat cereal, packaging material such as PVDC copolymer-coated with PP-LDPE retard lipid oxidation (Bauer et al., 2022; Kamal, 2021).

Pasta

Traditionally dried pasta is packaged in a paperboard carton having a plastic window. PE and OPP films are currently used for packaging most pasta. Also, the fresh pasta can be packaged in a rigid tray of PVC-LDPE sealed with PA-LDPE film when not intended for heating or pasteurization. For microwave heating, the rigid tray is usually made from crystalline polyethylene terephthalate (PET-C) or polystyrene-ethylene vinyl alcohol copolymer-LDPE (PS-EVOH-LDPE) laminate, and the film may be based on PVdC copolymer-coated PET, OPET-EVOH-LDPE, or PP (Bauer et al., 2022).

6 Advanced Packaging System to Enhance Shelf Life

As discussed in Sect. 5, selecting a suitable packaging material can tremendously impact cereal-based products' quality and shelf life. Products such as fresh pasta or fried snacks with high a_w and fat content are susceptible to microbial spoilage or oxidation. Advanced packaging such as MAP or active packaging (AP) and intelligent packaging (IP) or their combinations can enhance the shelf life of cereal products. An advanced packaging system can elevate the shelf life of specialized products such as ready-to-bake rolls and fine bakery products. Also, considering the growing needs of the food supply chain, it is crucial to develop novel packaging systems to prevent food loss. Therefore, use of modern packaging methods could address such shortcomings.

Advance packaging, such as MAP, uses enclosed mixtures of gas concentrations (different from the air) to retard spoilage and enhance the shelf life of food products (Czerwiński et al., 2021). MAP offers an advantage in reducing or eliminating chemical food preservatives, thus ensuring more natural and healthy food products. Commercially, MAP is used to pack the specialty bread in order to mitigate aerobic microbial spoilage. Carbon dioxide (CO_2) and nitrogen (N_2) are abundantly available at low cost, and their ability to retard microbial growth has made them commercially applicable.

MAP, along with the preservatives, creates a synergistic effect to prevent the microbial spoilage of bakery products. Rodriguez, Medina (Rodriguez et al., 2000) have evaluated the impact of different MAP gas compositions (air, 100% N_2, 20% CO_2/80% N_2, and 50% CO_2/50% N_2) with/without the incorporation of preservative (calcium propionate) on the shelf life of sliced wheat flour bread. MAP with 50% CO_2/50% N_2 and added preservative increased the bread shelf life by 195% when stored at 15–20 °C. Unfortunately, MAP application and its benefits have not been realised to their full potential in the bakery products due to the restriction posed by their porous structure. It is difficult to remove/replace the trapped oxygen within the structure and achieve desired gas composition in the package.

A food packaging material developed with the incorporation of active components which could release or absorb substance into or from the packaged product or the environment surrounding the food is known as active packaging (Qian et al., 2021). The active compounds are characterized based on their functions: antimicrobial components, antioxidant components, moisture absorbers, CO_2 emitters, and ethanol emitters. Moreover, nanocomposite and biodegradable-based packaging films with incorporated essential oils are finding commercial food packaging applications (Van Long et al., 2016). Heras-Mozos, Muriel-Galet (Heras-Mozos et al., 2019) coated the polyethylene films with the antifungal polymeric solution (0.25 garlic extract and 0.5% bread aroma). These films were further used to evaluate their impact on the storage quality of sliced white bread. Garlic extract was effective in preventing mold growth for 30 days. Hence, such bio-sourced active components could play a potential role in enhancing the product's shelf life.

However, the limited use of active components in food packaging is mainly due to their negative effect on sensory attributes of the product, risk of migration of nanoparticles to the food, and functional component sachet involves risk of direct ingestion and their uncontrolled release (Fig. 15.4).

Intelligent packaging materials can perform smart functions such as sensing, recording, detecting, communicating, tracing, and applying scientific logic (Yam et al., 2005). It can be accomplished by using an indicator that changes the color of the indicating mark or by producing a digital reading. Intelligent packaging can reflect product quality, temperature abuse during transportation and storage, gas

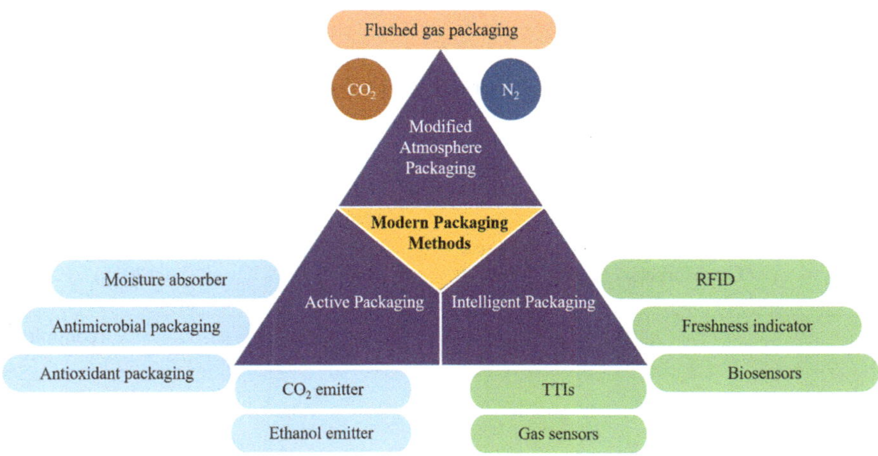

Fig. 15.4 Classification of different modern packaging methods. (RFID: radiofrequency identification; TTIs: time-temperature indicators)

Table 15.3 Application of different modern packaging methods for bakery products

Modern packaging methods	Packaging conditions	Target product	Outcomes	Reference
Modified atmosphere packaging	(a) 100% N_2 (b) 70% N_2/30% CO_2 (c) 50% N_2/50% CO_2 (d) 30% N_2/70% CO_2 (e) 100% CO_2	Sliced wheat bread	The bread slices packed with 100% CO_2 reduced the total viable count and yeast and mold count.	Degirmencioglu et al. (2011)
	(a) 100% CO_2 (b) 50% N_2/50% CO_2	Wheat bread	MAP was found to have the same staling rate as atmosphere air packaging.	Rasmussen and Hansen (2001)
	(a) 50% CO_2/50% N_2 (b) Sealing packaging (c) Wrapping packaging	Fermented rice cake	MAP extended the shelf life of the cake by 1–2 days. MAP restricted the weight and moisture loss compared to the other packaging methods.	Meng and Kim (2019)
	(a) 70% N_2,/30% CO_2 (b) 50% N_2/50% CO_2 (c) 90% N_2/10% CO_2	Milk cake	The milk cakes packed with gas composition (70% N_2, 30% CO_2) were found more suitable to increase the shelf life up to 28 days under refrigerated conditions.	Chawla et al. (2021)
Active packaging	Antimicrobial (Cinnamaldehyde)	Sliced wheat bread	Cinnamaldehyde effectively increased the shelf life and enhanced the physicochemical properties.	Balaguer et al. (2013)
	Antioxidant (oregano essential oil)	Preservative-free commercial bread	Oregano essential oil sachets were effective against yeasts and molds Unpleasant sensory attributes observed.	Passarinho et al. (2014)
	Ethanol emitter and O_2 absorbers	Chinese steamed bread	Unpleasant sensory changes	Sheng et al. (2015)

(continued)

Table 15.3 (continued)

Modern packaging methods	Packaging conditions	Target product	Outcomes	Reference
Combined modern packaging methods	MAP + active packaging + intelligent packaging Gas concentration: 10% CO_2 and 90% N_2, ethanol emitters, ethanol spray, and O_2 sensors	Commercial ciabatta bread	Ethanol emitters extended the shelf life to 16 days (packed with air). Ethanol emitters worked well in comparison to spray and control. Visible molds were seen after 30 days in MAP and ethanol emitter-packed food.	Papkovsky et al. (2013)
	MAP + active packaging Gas concentrations: 5, 10, and 21% O_2 balanced with N_2, added oxygen scavengers	Preservative-free white bread	A lower initial oxygen level (5%) enhanced the shelf life by 5–7 days. Using O_2 scavengers with one layer and three layers of LDPE was more effective under reduced O_2 concentration.	Upasen and Wattanachai (2018)

composition, and product movement monitoring (Yam et al., 2005). However, their commercial application in bakery products is uncommon (Table 15.3).

7 Future Trends

The sustainable use of resources as well as reducing food waste is crucial to fighting hunger. Hence, it becomes essential to save each slice of produced bread. Food packaging ensures containment and delivery to the last consumer with acceptable quality and product information. However, with increasing global supply, the need for modern food packaging methods is also escalated.

(a) MAP is a well-known method to extend the shelf life of food products. However, the porous structure of bakery products causes high residual O_2, which limits MAP's efficiency. Hence, MAP could be used along with the other modern packaging method to yield a synergistic effect.
(b) Explore and evaluate the potential of bio-sourced antimicrobial components. It will reduce the risk associated with using chemical-based active compounds and subsequently increase the acceptability of active packaging.

(c) Intelligent packaging of bakery products could provide solutions related to tracing, monitoring, and recalling in case of unacceptable product quality. However, it is important to consider the cost of the end product, which affects the consumer's decision.
(d) Use of other shelf life extension methods such as post packaging light emitting diode (LED) treatment could positively enhance the shelf life of bakery products.

It creates opportunities and challenges for scientists and manufacturers to find cost-effective, environmentally friendly, and non-hazardous food packaging solutions. A combination of active and intelligent packaging could be a possible approach for the shelf life extension of bakery products.

8 Conclusion

As discussed in this chapter, cereal and cereal-based products have a limited shelf life. These products undergo a series of physical, chemical, and microbiological changes throughout their shelf-life. Undesirable physical and chemical modifications of the bakery products cause a loss in their freshness, texture, and taste, while microbiological spoilage results in their undesirable appearance and cause food safety issues. Suitable packaging is one of the solutions to overcome these issues. Therefore, understanding packaging functions and properties are important. Appropriate use of packaging solutions can maintain quality and enhance shelf-life.

In general, packaging materials like glass, metal, paper, and plastics in different shapes (rigid, semi-rigid, and flexible) and sizes are explored for packaging cereal and cereal-based products. However, individual packaging system offers unique advantage and limitations. Also, considering the growing needs of the food supply chain, it becomes crucial to develop advanced packaging systems to prevent food loss and maintain quality. Current and future trends hint towards the use of advanced solutions such as MAP, active and intelligent packaging using natural ingredients singly or in combination with the basic packaging material to draw a balance between cost, efficiency, convenience, and acceptability at the industrial, distributor, and retail and consumer level. This necessitates further research that can be reproducible at the industrial scale to benefit all stakeholders in every sector of the food supply, distribution, and purchasing purview.

References

Agarwal, S., et al. (2020). Synthesis and characterization of tamarind kernel powder-based antimicrobial edible films loaded with geraniol. *Food Packaging and Shelf Life, 26*, 100562.

Agunbiade, S. O., Ojezele, O. J., & Barinemene, M. (2020). Storage environment/packaging materials impacting bread spoilage under ambient conditions: A comparative analysis. *Chemistry International, 6*(3), 131–136.

Ahmad, S., et al. (2022). Packaging design, fill rate and road freight decarbonisation: A literature review and a future research agenda. *Cleaner Logistics and Supply Chain, 4*, 100066.

Alpers, T., et al. (2021). Impact of storing condition on staling and microbial spoilage behavior of bread and their contribution to prevent food waste. *Foods, 10*(1), 76.

Axel, C., Zannini, E., & Arendt, E. K. (2017). Mold spoilage of bread and its biopreservation: A review of current strategies for bread shelf life extension. *Critical Reviews in Food Science and Nutrition, 57*(16), 3528–3542.

Balaguer, M. P., et al. (2013). Antifungal properties of gliadin films incorporating cinnamaldehyde and application in active food packaging of bread and cheese spread foodstuffs. *International Journal of Food Microbiology, 166*(3), 369–377.

Bauer, A.-S., et al. (2022). Cereal and confectionary packaging: background, application and shelf-life extension. *Foods, 11*(5), 697.

Belitz, H.-D., Grosch, W., & Schieberle, P. (2009). Cereals and cereal products. *Food Chemistry*, 670–745.

Bostik-Smart-Adhesive. (2022). *Why using a heat seal could benefit your packaging*. 11320 W. Watertown Plank Road, Wauwatosa, WI 53226, United States. Available online: https://industrial.bostik.com/why-using-a-heat-seal-could-benefit-your-packaging/. Accessed on 08 August 2022.

Cauvain, S. P., & Young, L. S. (2010). 13 – Chemical and physical deterioration of bakery products. In L. H. Skibsted, J. Risbo, & M. L. Andersen (Eds.), *Chemical deterioration and physical instability of food and beverages* (pp. 381–412). Woodhead Publishing.

Chang, H.-S., et al. (2020). Influence of sodium reduction and storage temperature on the growth of total microbes and Bacillus cereus in naturally contaminated hamburger patty and loaf bread. *Food Science and Biotechnology, 29*(10), 1433–1438.

Chawla, R., et al. (2021). Modified atmosphere packaging for milk cake: Assessment of ideal gas composition for extending shelf life. *British Food Journal, 123*, 2893–2906.

Czerwiński, K., et al. (2021). Towards impact of modified atmosphere packaging (MAP) on shelf-life of polymer-film-packed food Products: Challenges and sustainable developments. *Coatings, 11*(12), 1504.

Degirmencioglu, N., et al. (2011). Influence of modified atmosphere packaging and potassium sorbate on microbiological characteristics of sliced bread. *Journal of Food Science and Technology, 48*(2), 236–241.

Deschuyffeleer, N., et al. (2011). Identification and characterization of yeasts causing chalk mould defects on par-baked bread. *Food Microbiology, 28*(5), 1019–1027.

Deshwal, G. K., & Panjagari, N. R. (2020). Review on metal packaging: Materials, forms, food applications, safety and recyclability. *Journal of Food Science and Technology, 57*(7), 2377–2392.

Deshwal, G. K., Panjagari, N. R., & Alam, T. (2019). An overview of paper and paper based food packaging materials: Health safety and environmental concerns. *Journal of Food Science and Technology, 56*(10), 4391–4403.

Dong, Y., & Karboune, S. (2021). A review of bread qualities and current strategies for bread bioprotection: Flavor, sensory, rheological, and textural attributes. *Comprehensive Reviews in Food Science and Food Safety, 20*(2), 1937–1981.

Food-Packaging-Forum. (2013). *FOOD PACKAGING & HEALTH: Migration. Staffelstrasse 10, CH-8045 Zurich, Switzerland.* Issued date: April 22, 2013. Available online: https://www.foodpackagingforum.org/food-packaging-health/migration. Accessed on 08 Aug 2022.

Garcia-Arca, J., et al. (2021). Methodology for selecting packaging alternatives: An "action research" application in the industrial sector. *Central European Journal of Operations Research, 29*(4), 1427–1446.

Giannou, V., Lebesi, D., & Tzia, C.. (2014). Packaging and shelf-life prediction of bakery products. In *Bakery Products Science and Technology* (pp. 355–371).
Global-Plastic-Sheeting. (2013). *LDPE, HDPE AND LLDPE- What are the differences?; 1548 Jayken Way STE D, Chula Vista, CA 91911, United States*. Available online: https://www.globalplasticsheeting.com/hdpe-vs-lldpe-vs-ldpe. Accessed on 10 Aug 2022.
Gómez-Estaca, J., et al. (2014). Advances in antioxidant active food packaging. *Trends in Food Science & Technology, 35*(1), 42–51.
Heras-Mozos, R., et al. (2019). Development and optimization of antifungal packaging for sliced pan loaf based on garlic as active agent and bread aroma as aroma corrector. *International Journal of Food Microbiology, 290*, 42–48.
Hoque, M., et al. (2021). 3 – Biopolymer-based edible films and coatings for food applications. In K. Pal et al. (Eds.), *Food, medical, and environmental applications of polysaccharides* (pp. 81–107). Elsevier.
Hoque, M., et al. (2022a). Effect of cold plasma treatment on the packaging properties of biopolymer-based films: A review. *Applied Sciences, 12*(3), 1346.
Hoque, M., Sarkar, P., & Ahmed, J. (2022b). Preparation and characterization of tamarind kernel powder/ZnO nanoparticle-based food packaging films. *Industrial Crops and Products, 178*, 114670.
ICPE, I.C.f.P.i.t.E., Packaging of Bakery Products. (2019). *Nodal agency recognized by the Government of India to handle all issues related to Plastics and Environment in the country.* Available online available at: http://icpe.in/icpefoodnpackaging/pdfs/7_bakery.pdf. Accessed on 14 Aug 2022.
Ilhan, I., et al. (2021). Understanding the factors affecting the seal integrity in heat sealed flexible food packages: a review. *Packaging technology and science, 34*(6), 321–337.
Kamal, M. (2021). *The packaging of cereals, snack foods and confectionery.*
Konstantas, A., Stamford, L., & Azapagic, A. (2019). Evaluation of environmental sustainability of biscuits at the product and sectoral levels. *Journal of Cleaner Production, 230*, 1217–1228.
Marsh, K., & Bugusu, B. (2007). Food packaging—Roles, materials, and environmental issues. *Journal of Food Science, 72*(3), R39–R55.
Martins, I. E., et al. (2021). Effect of packaging materials and storage conditions on the microbial quality of pearl millet sourdough bread. *Journal of Food Science and Technology, 58*(1), 52–61.
Meng, L. W., & Kim, S. M. (2019). Effects of packaging methods and storage temperature on shelf life of fermented rice cake. *Journal of Food Processing and Preservation, 43*(8), e14047.
Morris, B. (2017). *Examples of flexible packaging film structures. The science and technology of flexible packaging* (pp. 697–709). Elsevier. ISBN 9780323242738.
Ncube, L. K., et al. (2020). Environmental impact of food packaging materials: A review of contemporary development from conventional plastics to polylactic acid based materials. *Materials, 13*(21), 4994.
Pålsson, H., & Hellström, D. (2016). Packaging logistics in supply chain practice–current state, trade-offs and improvement potential. *International Journal of Logistics Research and Applications, 19*(5), 351–368.
Papkovsky, D., Kerry, J., Hempel, A. W., & O'Sullivan, M. G. (2013). Use of smart packaging technologies for monitoring and extending the shelf-life quality of modified atmosphere packaged (MAP) bread: Application of intelligent oxygen sensors and active ethanol emitters. *European Food Research and Technology, 237*, 117–124.
Pasqualone, A. (2019). Bread packaging: Features and functions. In *Flour and breads and their fortification in health and disease prevention* (pp. 211–222). Elsevier.
Passarinho, A. T. P., et al. (2014). Sliced bread preservation through oregano essential oil-containing sachet. *Journal of Food Process Engineering, 37*(1), 53–62.
Qian, M., et al. (2021). A review of active packaging in bakery products: Applications and future trends. *Trends in Food Science & Technology, 114*, 459–471.
Rasmussen, P. H., & Hansen, A. (2001). Staling of wheat bread stored in modified atmosphere. *LWT-Food Science and Technology, 34*(7), 487–491.

Rodriguez, M., Medina, L., & Jordano, R. (2000). Effect of modified atmosphere packaging on the shelf life of sliced wheat flour bread. *Food/Nahrung, 44*(4), 247–252.

Roy, S., & Rhim, J.-W. (2021). Anthocyanin food colorant and its application in pH-responsive color change indicator films. *Critical Reviews in Food Science and Nutrition, 61*(14), 2297–2325.

Rüegg, N., Röcker, B., & Yildirim, S. (2022). Application of palladium-based oxygen scavenger to extend the mould free shelf life of bakery products. *Food Packaging and Shelf Life, 31*, 100771.

Salgado, P. R., et al. (2021). Recent developments in smart food packaging focused on biobased and biodegradable polymers. *Frontiers in Sustainable Food Systems, 5*, 125.

Sheng, Q., Guo, X. N., & Zhu, K. X. (2015). The effect of active packaging on microbial stability and quality of Chinese steamed bread. *Packaging Technology and Science, 28*(9), 775–787.

Smith, J. P., et al. (2004). Shelf life and safety concerns of bakery products—A review. *Critical reviews in food science and nutrition, 44*(1), 19–55.

Statisa, Bread & Cereal Products. (2022). *Issues on Jul* 2022. *Statista Inc., 3 World Trade Center, 175 Greenwich Street; 36th Floor, New York, NY 10007, United States.*

Svanes, E., Oestergaard, S., & Hanssen, O. J. (2019). Effects of packaging and food waste prevention by consumers on the environmental impact of production and consumption of bread in Norway. *Sustainability, 11*(1), 43.

Taglieri, I., et al. (2021). Overcoming bread quality decay concerns: Main issues for bread shelf life as a function of biological leavening agents and different extra ingredients used in formulation. A review. *Journal of the Science of Food and Agriculture, 101*(5), 1732–1743.

Umaraw, P., et al. (2020). Edible films/coating with tailored properties for active packaging of meat, fish and derived products. *Trends in Food Science & Technology, 98*, 10–24.

Upasen, S., & Wattanachai, P. (2018). Packaging to prolong shelf life of preservative-free white bread. *Heliyon, 4*(9), e00802.

Urbelis, J. H., & Cooper, J. R. (2021). Migration of food contact substances into dry foods: A review. *Food Additives & Contaminants: Part A, 38*(6), 1044–1073.

Van Long, N. N., Joly, C., & Dantigny, P. (2016). Active packaging with antifungal activities. *International Journal of Food Microbiology, 220*, 73–90.

Vat, J. (2022). *External factors influencing shelf life of bakery products.* Available at https://www.biscuitpeople.com/magazine/post/external-factors-influencing-shelf-life-of-bakery-products. Access on: 14 Auguest 2022).

Yam, K., Takhistov, P., & Miltz, J. (2005). R concise reviews/hypotheses in food science-intelligent packaging: Concepts and applications. *Journal of Food Science, 70*(1), R1–R10.

Index

M. A. Shah et al. (eds.), *Cereal-Based Food Products*,
https://doi.org/10.1007/978-3-031-40308-8

GPSR Compliance

The European Union's (EU) General Product Safety Regulation (GPSR) is a set of rules that requires consumer products to be safe and our obligations to ensure this.

If you have any concerns about our products, you can contact us on ProductSafety@springernature.com

In case Publisher is established outside the EU, the EU authorized representative is:

Springer Nature Customer Service Center GmbH
Europaplatz 3
69115 Heidelberg, Germany

The manufacturer's authorised representative in the EU is Springer Nature Customer Service Centre GmbH, Europaplatz 3, 69115 Heidelberg, Germany. If you have any concerns regarding our products, please contact ProductSafety@springernature.com

Printed and bound by CPI Group (UK) Ltd, Croydon, CR0 4YY
28/09/2026
02224056-0002